Chemical Water and Wastewater Treatment VI

Springer
Berlin
Heidelberg
New York
Barcelona
Hong Kong
London
Milan
Paris
Singapore
Tokyo

Hermann H. Hahn · Erhard Hoffmann
Hallvard Ødegaard (Eds.)

CHEMICAL WATER AND WASTEWATER TREATMENT VI

Proceedings of the
9th Gothenburg Symposium 2000
October 02 - 04, 2000
Istanbul, Turkey

Springer

Prof. Hermann H. Hahn, Ph. D.
Erhard Hoffmann
Universität Fridericiana zu Karlsruhe
Institut für Siedlungswasserwirtschaft
Am Fasanengarten
D - 76128 Karlsruhe

Prof. Hallvard Ødegaard
Norwegian University of Science and Technology
NTNU
Faculty of Civil Engineering
Dept. of Hydraulic and Environmental Engineering
S.P. Andersens vei 5
N - 7034 Trondheim

ISBN-13: 978-3-642-64126-8 Springer-Verlag Berlin Heidelberg New York

CIP data applied for
Die Deutsche Bibliothek - CIP-Einheitsaufnahme
Chemical water and wastewater treatment VI : proceedings of the 9th Gothenburg Symposium 2000, October 2 - 04, 2000, Istanbul, Turkey / Hermann H. Hahn ... (ed,) - Berlin ; Heidelberg ; New York , Barcelona ; Hong Kong ; London ; Milan ; Paris ; Singapore ; Tokyo : Springer, 2000

ISBN-13: 978-3-642-64126-8 e-ISBN-13: 978-3-642-59791-6

DOI: 10.1007/978-3-642-59791-6

Springer-Verlag Berlin Heidelberg New York
a member of BertelsmannSpringer Science+Business Media GmbH

Softcover reprint of the hardcover 1st edition 2000

Typesetting: Camera-ready by editors
Cover layout: de`blik, Berlin
Printed on acid-free paper SPIN: 10765717 02/3020 - 5 4 3 2 1 0

Preface

It is with great pleasure that the editors of the Ninth International Gothenburg Symposium present these proceedings. It is a testimony to the viability of the idea of these symposia: to bring together all concerned for the protection of our environment and for the search for a sustainable solution for development both in industrialized areas and in countries on the road of development. From the start, the goal of the symposia has been to bring together those who come up with ideas in the field of chemical treatment of water and wastewater and those who put the ideas into practice. Before these symposia were organized, there was no platform for the exchange of ideas and experiences among decision makers, operators and researchers. We, the organizers and editors, are proud to provide this opportunity in this important area.

The contents of these proceedings reflect, on the one hand, the problems recognized by all concerned and, on the other hand, the response of specialists to develop and provide solutions. When the table of contents of this book is compared with that of previous symposia, one detects a change in topic emphasis away from a singular observation of one sectional problem and toward the consideration of all directly and indirectly connected areas. Instead of a purely natural sciences based, technical solution, the authors consider solutions that also involve sociological insights and the participation of the public at large. We, the editors, are particularly pleased and grateful that the host country has contributed experiences on the highly complex area of integrated industrial production and environmental protection. We are equally appreciative of the contributions from all over the world regarding novel ideas such as automatic control and management of all types of installations, such as the integration of biological and chemical processes and such as the impressively growing reuse of materials found in water and wastewater and eliminated or separated in the course of treatment.

The team of editors is deeply saddened by the sudden death of the invaluable Adam Leinz, whose care and attention have been an integral part of all the proceedings since the very beginning. We would like to dedicate this preface to his memory. We have been fortunate, however, to find new support in the form of Jörg Kegebein, who will try to maintain Adam's high standards. We are also fortunate to still have Karin Knisely on board. Revisions made in response to her comments removed some of the ambiguity in the communications, making them more readable. More thanks are extended, as always, to the authors, to the printers and publishers, and again to Kemira Kemi for providing financial support.

Karlsruhe, Germany
Trondheim, Norway
July 2000

H.H. Hahn, E. Hoffmann
H. Ødegaard

Members of the Organising Committee

Prof. H.H. Hahn, University Fridericiana, Germany

Prof. T. Hedberg, Chalmers University of Technology, Sweden

Mr. K. Stendahl, Kemira Kemwater, Sweden

Prof. H. Ødegaard, The Norwegian University of Sciences and Technology, Norway

Members of the Scientific Committe

Prof. T. Asano, USA

Prof. P. Balmér, Sweden

Prof. M. Boller, Switzerland

Dr. N. Booker, Australia

Mr. R. Brenner, Germany

Ms. V. Brutu da Costa, Portugal

Prof. J.C. van Dijk, The Netherlands

Dr. P. Dolejs, Czech Republic

Prof. S.S. Ferreira Filho, Brazil

Dr. N. Graham, England

Prof. P. Harremoës, Denmark

Mr. I. Karlsson, Sweden

Prof. R. Laukkanen, Finland

Prof. R. Mujeriego, Spain

Prof. H. Ødegaard, Norway

Prof. D. Ohron, Turkey

Prof. Ottaviani, Italy

Mr. R. Pujol, France

Mr. F. Rogalla, Brazil

Prof. Y. Watanabe, Japan

Contents

Coagulation and Floc Separation Processes

Floc Characteristics

Part III: Mixing, Flocculation and Floc-Separation Reactors

Oxidation Processes

Control of Chemical Treatment Processes

Drinking Water Treatment

Municipal Wastewater Treatment

Industrial Wastewater Treatment

Sludge Treatment

Resources Reuse

Coagulation and Floc Separation Processes

Polyaluminum Coagulants for Drinking Water Treatment: Chemistry and Selection

J.K. Edzwald, D.J. Pernitsky and W.L. Parmenter

Department of Civil and Environmental Engineering, University of Massachusetts, Amherst, MA 01003-5205 USA
edzwald@snail.ecs.umass.edu

Abstract

Selection of polyaluminum (PAXs) coagulants is examined for different raw water types and solid-liquid separation processes. PAXs were tested with various basicities and sulfated versus non-sulfated. Four widely different water supplies were studied in terms of alkalinity, turbidity, and concentration and nature of natural organic matter (NOM). The performance of the PAXs was evaluated for turbidity and NOM removal. Solubility and Al speciation of the PAXs differ from alum. High basicity PAXs, for example, are more soluble than alum below pH 7 and contain a highly charged Al_{13} polymer for coagulation. High basicity PAXs produce less filter bed headloss than alum and other PAXs in direct filtration applications. An important finding is direct filtration of low turbidity supplies is not feasible, even when the TOC is low (2.5-3.5 mg/L), if the NOM is composed of some aquatic humic matter. DAF is very effective in treating low turbidity supplies of low and high alkalinity, and supplies high in natural color. All of the coagulants gave good DAF performance, but generally the medium and high basicity PAXs were particularly effective. Generally, best sedimentation performance is achieved using medium and high basicity PAXs with sulfate. A silica containing PAX can also improve settling. NOM controls coagulant dosages for high TOC supplies containing aquatic humic matter. For the latter type supply, stoichiometric relationships were found between the optimum coagulant dosage and raw water TOC or UV (254 nm). Good removals of NOM were found, and generally a broader dosing range of good treatment was found when using high basicity PAXs.

Introduction

Polyaluminum coagulants (PACls or PAXs) have been used for several years. Advantages over alum and ferric salts include reduced acidity, preformed positively charged monomers and polymers, formation of rapid and denser floc, and reduced sludge. In the USA and other countries, the use of coagulants has changed from the traditional role of turbidity removal to one requiring multiple coagulation objectives - maximizing particle, pathogen, and turbidity removal, maximizing organic carbon removal (*Enhanced Coagulation*), and minimizing sludge production. Differences in raw water chemistries among water supplies complicate selection of the best coagulant. Furthermore, water plants employ a variety of solid-liquid separation processes so that evaluation of coagulants must consider the differences among these processes. Our research investigated these issues. The primary goal of the paper is to present findings on selecting PAXs to treat different raw water types. A second goal considers tailoring the PAX coagulation and chemistry to the solid-liquid separation method.

Raw Water Quality

The chemistry of the water supply greatly influences coagulant selection and performance. The important raw water quality variables affecting coagulation are coagulant demanding substances, the nature of the natural organic matter (NOM), and bulk water chemistry properties of pH, alkalinity, and temperature. When a coagulant is added to a raw water, positively charged coagulant species react with particles and colloidal and dissolved NOM. These substances exert a coagulant demand, and therefore affect the required coagulant dose. Coagulants that destabilize particles via charge neutralization will have a dose dependence related to the turbidity, specifically to the particle surface area concentration. For coagulants that function through sweep floc coagulation, turbidity has less of an effect on the coagulant dose.

The composition of the NOM has a significant impact on whether NOM controls coagulant doses and on NOM removal by coagulation. Edzwald and co-workers [1-4] developed the concept of specific UV absorbance (SUVA) as an operational indicator of the nature of NOM and of the potential effectiveness of coagulation in removing NOM. SUVA is defined as the normalized UV absorbance of a water sample in m^{-1} divided by its DOC in mg/L. From guidelines of Edzwald, it is the nature of the NOM (characterized by SUVA) and the concentration of the NOM (measured by TOC, DOC, or UV) that dictate the role of NOM in coagulation. For supplies with high SUVA, NOM controls coagulation so that there is a strong dependence between the coagulant dosage and the NOM concentration; high NOM removal is expected. For supplies with a low SUVA, there is little or no dependence between coagulant dosage and NOM; poor NOM removal is expected.

Experimental

Model Water Supplies and Solid-Liquid Separation

Four surface water supplies from various regions of the USA were selected for this research. These supplies were chosen for their widely different water quality characteristics as summarized in Table 1. The high quality reservoir supply represents low turbidity, low TOC, low alkalinity waters. Solid-liquid separation following coagulation by direct filtration and dissolved air flotation (DAF) treatment were studied. The high natural color, high TOC, low alkalinity supply was studied with particle separation by conventional sedimentation versus DAF treatment. It is common to treat turbid river waters with sedimentation, so this was studied for the low alkalinity, turbid river supply. Finally, the low turbidity, high alkalinity supply was studied primarily with conventional sedimentation with some evaluation of DAF treatment.

Table 1. Model Raw Waters

Class	Turbidity ntu	TOC or DOC mg/L	UV m^{-1}	SUVA m^{-1}/mg/L	Alkalinity $CaCO_3$ mg/L
High Quality Reservoir	L 0.3 – 2.0	L 2.5 – 3.5	L-M 7 – 12	M 2.5 – 3.5	L < 30
High Color Reservoir	L 0.6 – 1.7	M-H 6.0 – 8.0	M-H 23 – 34	H 3.8 – 4.8	L < 30
Turbid River	L-M-H 3.0 – 60.0	L-M 1.3 – 4.0	L 3 – 5	L 1.9 – 2.8	L < 30
Low Turb High Alk Supply	L 0.6 – 2.7	L 2.9 – 3.5	L 4 - 5	L 1.4 – 2.0	H > 100

L, Low; *M*, Medium; *H*, High

Coagulants

A variety of PAXs were studied in terms of basicity, sulfated, non-sulfated, and silica; two polyaluminum sulfates were also evaluated. A summary is presented in Table 2 of the PAXs and their general properties. Alum was used in all experiments as a reference coagulant. Not all PAXs were tested for each water supply, but always different basicities and sulfated versus non-sulfated PAXs were assessed for each supply.

Table 2. Coagulants and Properties

Coagulant	Al O %	s.g.	Basicity %	Sulfate %
Alum	8.3	1.33	-	23.5
LBNS (Low Basicity Non-Sulfated)	10	1.24	12 – 30	0
MBNS (Medium Basicity Non-Sulfated)	10	1.20	39 – 44	0
MBS (Medium Basicity Sulfated)	10	1.20	39 – 45	2.4 - 2.6
HBNS (High Basicity Non-Sulfated)	10	1.24	60 – 75	0
HBS (High Basicity Sulfated)	10	1.24	65 – 70	1.3 - 1.8
PAX-XL19 High Basicity (Al Chlorohydrate)	24	1.36	83	0
PAX-XL60 (Si-PAX)	13.4	1.31	40	0
UPAS-1220 (Poly Al Sulfate)	7.6	1.37	30	NA
UPAS-1620 (Poly Al Sulfate)	7.6	1.36	43	NA

NA, Not Available

Chemistry

PACls are produced by the controlled titration of $AlCl_3$ solutions with base. They are characterized by the ligand ratio r or their relative basicity.

$$r = \frac{[OH^-]}{[Al]} \qquad \textbf{(1)}$$

$$basiscity = \frac{(r)}{(3)} * 100\% \qquad \textbf{(2)}$$

Commercial PACls are available with basicities ranging from about 15 to 85%. The basicity affects the aluminum speciation and alkalinity consumption of the water upon addition of the coagulant. As the ligand number increases from r = 0 to 2 (basicities of 0 to 70%), the total Al in the form of Al_{13} ($Al_{13}O_4(OH)_{24}^{+7}$) polymer increases [5]. Other polymers and monomeric Al may be present, but Al_{13} polymer is the dominant form of Al for high basicity PACls. For r > 2.1, the Al_{13} polymer begins to decrease as colloidal precipitate forms. Therefore, by setting the PACl basicity, coagulants with different amounts of Al monomers, small polymers, and Al_{13} polymer can be produced. Al_{13} polymers with their high positive charge should be quite effective in bringing about coagulation by charge neutralization. Sulfate is added to PACls in small amounts (typically, < 5%) to enhance precipitation of $Al(OH)_3$(am) and settling of floc. Silica can be added to improve particle aggregation and settling [6].

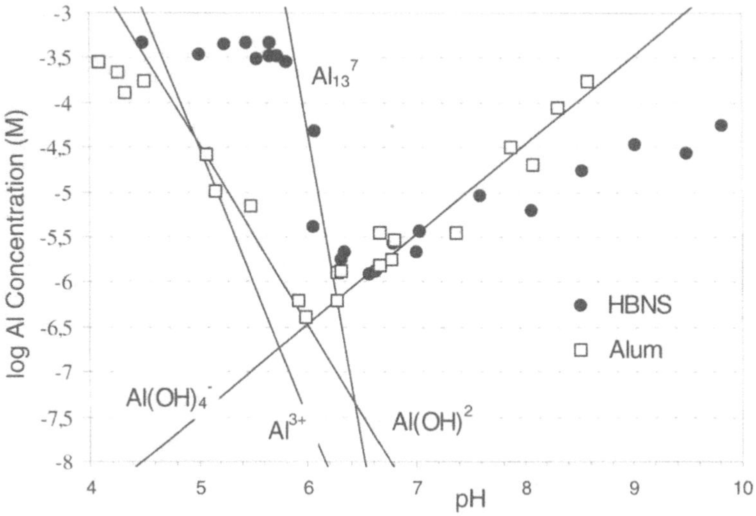

Fig. 1. Solubility of alum and high basicity non-sulfated PACl (25°C). Points are experimental; lines are calculated for equilibrium with $Al(OH)_{3(am)}$; $pK_{so} = 33.1$

Figure 1 summarizes the solubility of alum and the HBNS PACl for warm water temperature coagulation. The figure shows the experimental data and the theoretical Al speciation. The figure shows that with alum, Al is least soluble at pH 6 to 6.3; consequently, coagulation at this pH converts most of the added Al to solid floc particles and minimizes residual soluble Al. It is advantageous to coagulate with alum in this pH region for sweep floc coagulation, to minimize residual Al, and to utilize a greater fraction of dissolved positively charged Al compared to higher pH coagulation. As the water temperature decreases, the pH of minimum solubility for alum shifts to higher pH in the region of 6.5 to 7 [4, 7]. The HBNS data (Fig. 1) show a much different solubility response. At pH below 7, it is very soluble and exists primarily as the highly charged Al_{13} polymer. Coagulation below pH 7 means that this highly charged Al_{13} polymer would be available to destabilize particles by charge neutralization and to react with negatively charged NOM colloids and macromolecules. In the region of about pH 7.5 to 7, the HBNS PACl hydrolyzes decreasing the charge density of the polymer and the soluble Al concentration. Above pH 7, it is not in equilibrium with the same solid phase that alum produces. The HBNS PACl is less affected than alum by decreasing temperature with respect to its primary mode of coagulation. The other PACls have a mixture of monomers and polymers, and their solubility curves fall between the alum and HBNS curves.

High Quality Reservoir

Direct filtration and DAF were examined as solid-liquid separation processes. Direct filtration was investigated with small-scale pilot filters at a flow rate of 40 mL/min with a 0.13 m bed of sand. Coagulant was added ahead of an in-line static mixer. No flocculation mixing was used. Pilot runs were of short duration due to the shallow laboratory filter bed depth. DAF treatment was evaluated with an Aztec Flotation Jar Test apparatus. This bench-scale unit allows for well-controlled rapid mixing, flocculation, and flotation to be carried out in the same jar without transfer of the test water. No granular media filtration followed DAF, and must be accounted for in the interpretation of the DAF data.

Figure 2 compares direct filtration performance for alum versus the HBNS PACl. Alum was used with pH control to optimize its chemistry; no pH adjustment was used for the PACl. Turbidity performance was good for both coagulants. The headloss data are of particular interest. They show less headloss with the HBNS PACl compared to alum. In general, we found that the high basicity PACls (HBNS and HBS) produced less filter bed headloss compared to alum and to the LBNS and MBS PACls. Better turbidity and NOM removal performance was found for the high basicity PACls.

A significant finding of the research is that while historically this low turbidity type supply has often been treated by direct filtration, this type supply is no longer a good candidate for direct filtration when you consider removal of NOM. While the TOC is low, the supply has a medium SUVA value indicating that the NOM influences optimum coagulant dosages for removal of NOM as well as turbidity. Experiments were performed over several seasons, and the results showed that optimum coagulant dosages were 1.5 mg/L as Al or greater. These coagulant dosages are too high for direct filtration treatment since they will cause an excessive headloss from the solids produced by coagulation. The high basicity PACls reduced the headloss compared to alum, but the headloss was still too high.

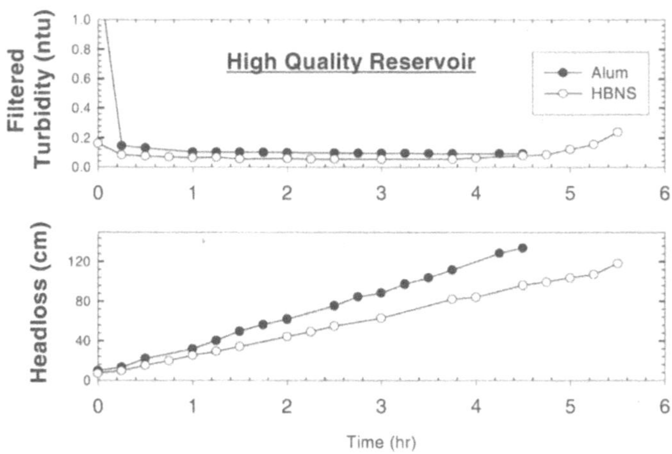

Fig. 2. Direct filtration comparison of alum versus HBNS PACl (constant dose of 1.5 mg/L as Al)

DAF, however, was quite effective in treating this type of supply producing floated water turbidities < 0.5 ntu with good removals of NOM (~ 50% DOC) for both warm and cold water conditions. All the coagulants produced good DAF performance, but the HBNS was particularly effective over a broad dosing range. Because DAF produces low turbidity water that is then applied to granular filters in full-scale plants, the filter design can be integrated with the DAF to yield cost effective plant designs.

High Natural Color Supply

The same Aztec Flotation apparatus was used for DAF while a Kemira Flocculator 90 system was used to study sedimentation. Best coagulation pH conditions are at 6 to 7 to treat a low alkalinity, high TOC, high color supply. In treating this type of supply, medium and high basicity PACls have an advantage over alum because less base is needed to control coagulation pH.

Coagulation of a supply high in aquatic humic matter generally produces low density floc that may settle poorly under cold water conditions. Figure 3 compares settled water turbidities versus DAF following coagulation and flocculation (20 min for sedimentation, 10 min for DAF) for alum and several PACls. For low sedimentation overflow rates (long settling times), all the coagulants produced good settled water turbidities at optimum dosing. However, examination of the settled water turbidity data at high overflow rates showed that the MBS and HBS PACls produced floc that settled more rapidly compared to the other coagulants. DAF data were collected for only one overflow rate (1 m/hr, 3-10 times higher than settling), and showed good turbidity performance for all the coagulants.

Several other experiments were performed for warm and cold water conditions, and evaluated NOM removals as well as turbidities. In summary, it was found that the high basicity PACls had a broader coagulant dosing range for good treatment. The medium and high basicity PACls with sulfate performed best for sedimentation since they produced denser more rapid settling floc. DAF was more effective than sedimentation in treating this type supply; lower turbidities and often lower NOM levels were found for DAF. Better performance was found for DAF using the MBS and the high basicity PACls.

New US EPA regulations [8] require *Enhanced Coagulation* to remove TOC and thus control formation of disinfection by-products. TOC requirements depend on raw water TOC and alkalinity levels. For this model water of low alkalinity and medium to high TOC, the requirements are 45-50 % TOC removal. SUVA values for this model supply were 3.8 to 4.8 (Table 1) indicating that the NOM is composed mostly of aquatic humic matter. This leads to the following important findings that apply to this and similar type waters. The high humic content (SUVA > 4) and high TOC means that the NOM controls coagulant dosages for effective treatment of these type water supplies. A stoichiometric relationship was observed between the optimum coagulant dosage and raw water TOC or UV. Specific relationships are presented in Table 3 at pH 6.3-6.5, and are generally 0.7-1 mg Al per mg TOC or 15-20 mg Al per cm^{-1} of UV.

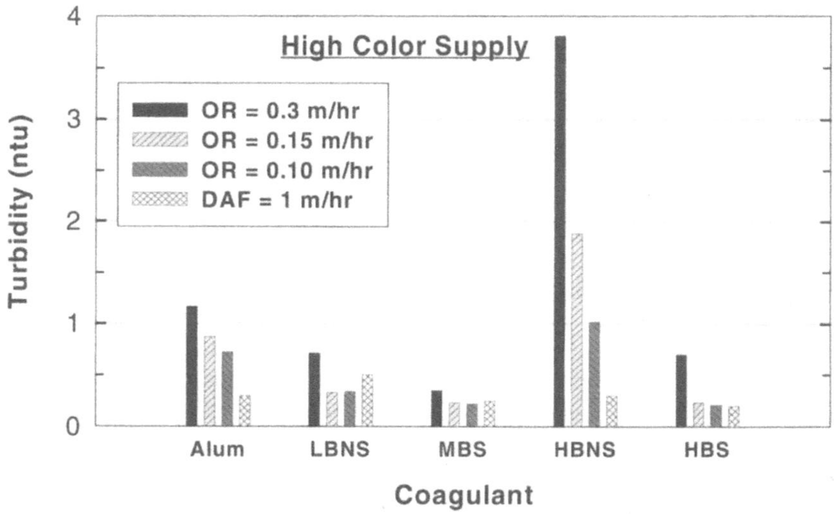

Fig. 3. Clarification performance for various sedimentation overflow rates versus DAF (constant coagulant dose of 6 mg/L as Al, temperature of 4° C)

These relationships should yield good estimates of coagulant dosing ranges for use in jar tests, pilot studies, and full-scale plant trials. They also are in general agreement with literature relationships [9,10]. With the high SUVA, NOM removals by coagulation should be good. This was confirmed by the data in Table 3 showing that TOC removals were > 50% exceeding the EPA requirements.

Experimental data for coagulation of this type supply at higher pH (7-8) showed poorer removals of NOM, compared to coagulation at pH 6.3-6.5, and higher coagulant dosages were needed. Lower settled water turbidities were found with the medium and high basicity PACls with sulfate.

Table 3. Summary of Normalized Optimum Coagulant Doses and Removals of NOM for the High Natural Color Supply

Coagulant	pH	Optimum Dose* per UV ($mgAl/cm^{-1}$)	Optimum Dose* per TOC (mgAl/mg)	Removal UV %	Removal DOC %	Removal TOC %
Alum	6.3 - 6.5	15 - 19	0.7 - 0.9	80 - 90	65 - 70	65 - 70
LBNS	6.3 - 6.5	15 - 26	0.8 - 1.1	80 - 85	65 - 75	65 - 70
MBS	6.3 - 6.5	15 - 19	0.7 - 0.9	65 - 85	50 - 75	40 - 75
HBNS	6.3 - 6.5	15 - 19	0.7 - 0.9	75 - 85	65 - 85	60 - 75
HBS	6.3 - 6.5	15 - 18	0.7 - 0.8	70 - 85	50 - 75	45 - 75

*Lower doses are for DAF

Turbid River

This supply, as indicated in Table 1, has low NOM concentrations and generally a low SUVA. Experiments conducted when the raw water DOC and UV were less than 2 mg/L and 4 m^{-1} found removals of 20% for DOC and 40-60% for UV. NOM does not control coagulation for this supply, and removal of organic matter is not an issue because of the low NOM concentrations. The TOC after treatment was generally < 1.5 mg/L. Thus, for this supply the main coagulation issue is removal of turbidity.

Experiments focused on the ability of PAXs to coagulate and settle turbidity from this turbid river supply under warm and cold water temperatures. This supply is of low alkalinity so that alum addition required pH control; consequently alum coagulation was done at pH 6.3-6.7, depending on water temperature. No pH adjustment was required with the PAXs; pH was 6.3-7.1, depending on the PAX basicity.

Figure 4 illustrates settling performance for the various coagulants for the more difficult case of cold water coagulation and settling. Data are shown for three sedimentation overflow rates. The data show that all coagulants performed well with turbidities < 1 ntu when low overflow rates (long settling times) were used. At high overflow rates, however, alum and the non-sulfated PAXs gave poor performance. Excellent performance was found for the HBS and the medium basicity polyaluminum sulfate (UPAS-1620) PAXs. These two PAXs are apparently able to produce a more dense, rapid settling floc.

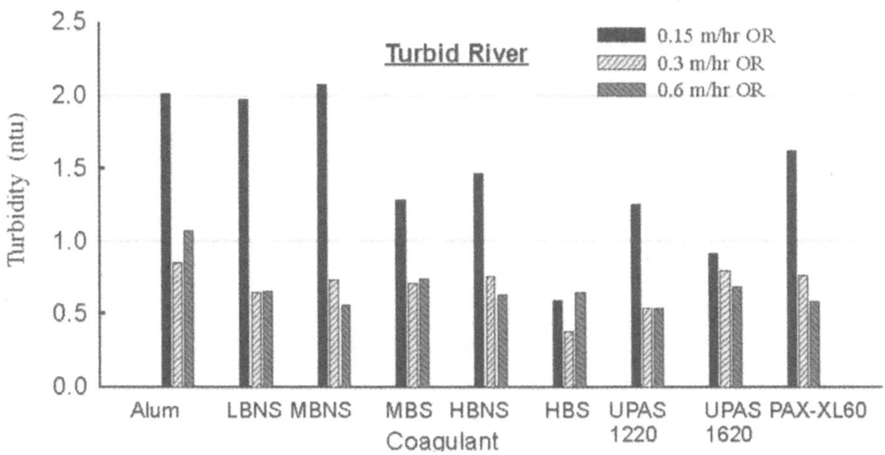

Fig. 4. Sedimentation performance for various overflow rates (constant dose of 0.8 mg/L as Al, temperature of 4°C)

Low Turbidity, High Alkalinity Supply

This supply has high hardness and alkalinity (Table 1). Its chemistry affects the nature of the NOM found in the supply as well as coagulation chemistry. First, the NOM has a very low SUVA indicating a low fraction of aquatic humic matter. The NOM present does not control coagulation, and is not removed well by coagulation. We found UV and TOC removals of 20-40% and 10-20%, respectively. Second, the high pH (7.8-8.4) and alkalinity means coagulation is carried out at unfavorable pH conditions unless acid is added. Our experiments focused on coagulation without pH adjustment to evaluate PAX performance against alum.

Figure 5 compares sedimentation to DAF for various coagulants. The data show poor settling performance for alum and the non-sulfated PAXs. Good settling performance was found for the medium and high basicity PAXs and for the silica containing PAX. DAF performance was less affected by choice of coagulant except that alum did not perform well. Excellent DAF performance was found for all of the PAXs.

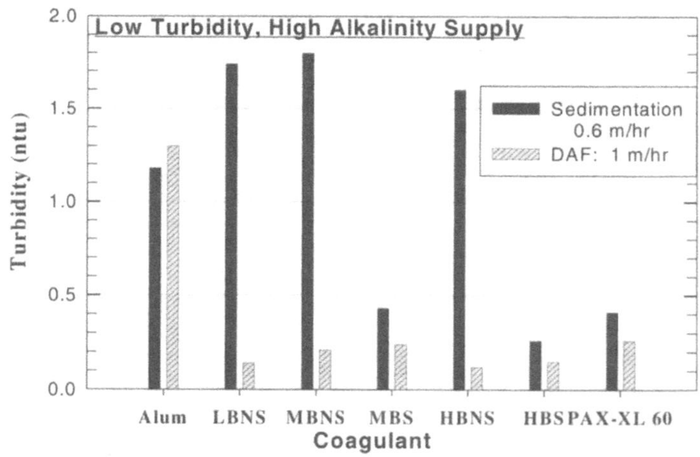

Fig. 5. Comparison of sedimentation to DAF for various coagulants for the low turbidity, high alkalinity supply (constant dose of 1.5 mg/L as Al)

Conclusions

Selection of polyaluminum (PAXs) coagulants was examined for different raw water types and solid-liquid separation processes. PAXs were tested with various basicities and sulfated versus non-sulfated. Solubility and Al speciation of the PAXs differ from alum. High basicity PAXs, for example, are more soluble than alum below pH 7 and contain a highly charged Al_{13} polymer for coagulation. High basicity PAXs produce less filter bed headloss than alum and other PAXs in direct filtration applications. An important finding is direct filtration of low turbidity supplies is not feasible, even when the TOC is low (2.5-3.5 mg/L), if the NOM is composed of some aquatic humic matter. DAF is very effective in treating low turbidity supplies of low and high alkalinity, and supplies high in natural color. All of the coagulants gave good DAF performance, but generally the medium and high basicity PAXs were particularly effective. Generally, best sedimentation performance is achieved using medium and high basicity PAXs with sulfate. A silica containing PAX can also improve settling. NOM controls coagulant dosages for high TOC supplies containing aquatic humic matter. For the latter type supply, stoichiometric relationships were found between the optimum coagulant dosage and raw water TOC or UV (254 nm). Good removals of NOM were found, and generally a broader dosing range of good treatment was found when using high basicity PAXs.

Acknowledgments

The authors are pleased to acknowledge the financial support of the Holland Company (Adams, MA, USA), Kemwater North America (Savannah, GA, USA), and Kemira Kemi (Helsingborg, Sweden). We are grateful to Tom and Dan Holland (Holland Company), Britt Muotaka (Kemwater NA), and Lars Gillberg (Kemira Kemi) for their assistance. We thank Kristen Berger, Mary Donovan, and especially Mark Janay for their contributions to this research.

References

1. Edzwald, J.K., Becker, W.C., Wattier, K.L.: Surrogate Parameters for Monitoring Organic Matter and THM Precursors. J. Amer. Wat. Works Assoc. 77 (4) (1985) 122
2. Edzwald, J.K., Van Benschoten, J.E.: Aluminum Coagulation of Natural Organic Matter. In: Chemical Water and Wastewater Treatment, H.H. Hahn and R. Klute (Eds). Springer Verlag, New York 1990, pp. 341-359
3. Edzwald, J.K.: Coagulation in Drinking Water: Particles, Organics, and Coagulants.Wat. Sci. Tech. 27 (11) (1993) 21
4. Edzwald, J.K., Tobiason, J.E.: Enhanced Versus Optimized Multiple Objective Coagulation. In: Chemical Water and Wastewater Treatment V, H.H. Hahn, E. Hoffmann, and H. Ødegaard, (Eds). Springer Verlag, New York 1998, pp. 113-124
5. Bottero, J.Y., Cases, J.M., Fiessinger, F., Poirier, J.E.: Studies of Hydrolyzed Aluminum Chloride Solutions. 1. Nature of Alumin Species and Composition of Aqueous Solutions. J.of Physical Chem. 84 (1980) 2933
6. Tang, H.X., Luan, Z.K., Wang, D.S., Gao, B.Y.: Composite Inorganic Polymeric Flocculants. In: Chemical Water and Wastewater Treatment V. H.H. Hahn, E. Hoffmann, and H. Ødegaard, (Eds). Springer Verlag, New York 1998, pp. 25-34
7. Van Benschoten, J.E., Edzwald, J.K.: Chemical Aspects of Coagulation Using Aluminum Salts 1. Hydrolytic Reactions of Alum and Polyaluminum Chloride. Wat. Res. 24 (1990) 1519
8. USEPA (1998) National Primary Drinking Regulations: Disinfectants and Disinfection Byproducts; Final Rule. 40 CFR Parts 9, 141, and 142, Federal Register, **63** (241), 69390-69476 (December 16, 1998)
9. Edzwald, J.K., Bunker, D.Q., Dahlquist, J., Gillberg, L., Hedberg, T.: Dissolved Air Flotation: Pretreatment and Comparisons to Sedimentation. In: Chemical Water and

Wastewater Treatment III, R. Klute and H. H. Hahn (Eds), Springer Verlag, New York 1994, pp. 3-18

10. Eikebrokk, B., Fettig, J.: Treatment of Coloured Surface Water by Coagulation/ Direct Filtration: Effect of Water Quality, Type of Coagulants and Filter Aids In: Chemical Water and Wastewater Treatment, H. Hahn and R. Klute (Eds), Springer Verlag, New York 1990, pp. 361-376

Properties and Evaluation of Polyferric-Silicate-Sulfate (PFSS) Coagulant as a Coagulant for Water Treatment

B. Gao*, Q.Yue*, H. Zhao* and Y. Song**

*Dep. of Environmental Engineering, Shandong University, Jinan 250100, P. R. China, bygao@sdu.edu.cn
**Dep. of Chemical Engineering, Shandong University, Jinan 250100, P. R. China, songyonghui@hotmail.com

Abstract

Polyferric-Silicate-Sulfate (PFSS), as a new type of coagulant, was prepared by using sodium silicate, sulfuric acid and ferric sulfate as materials. Three topics were investigated: (1) the zeta potential of hydrolyzate of PFSS under different pH values, (2) the effects of Fe/Si molar ratio and dosage of PFSS on turbidity removal, and (3) the relationship between the stability of PFSS and Fe/Si molar ratio in PFSS solutions. There is a relationship between the optimum coagulation pH range and Fe/Si molar ratio, and the coagulation mechanism of PFSS was discussed.
The experimental results show that Fe/Si molar ratio has an effect on the zeta potential of hydrolyzate, the stability of PFSS, the coagulation performance and the optimum coagulation pH range of PFSS. If the Fe/Si ratio is too high or too low, it will not be beneficial to the stability of PFSS solutions. A Fe/Si ratio of about 1.0 gives the most stable PFSS solution. It was also found that PFSS gave the best turbidity removal effect when its Fe/Si molar ratio was 1.5.
The results suggest that the composite of ferric sulfate and polysilicic acid can enhance aggregating efficiency due to higher molecular weight, but is weaker in terms of charge effectiveness in the coagulation process due to charge neutralization occurring when hydrolyzed ferric species of different charge and polysilicic acid are combined. Therefore, it is important to consider the aggregating efficiency and charge effectiveness when developing a PFSS coagulant.

Introduction

Activated silica, which is a special form of polysilicic acid (PSA), has been employed as an effective and non-hazardous flocculant aid for many years to treat water[1]. Usually, activated silica is used separately with aluminum sulfate or ferrous sulfate in water treatment. However, because of its negative charge and pre-production only on-site, activated silica can not often be used as a common coagulant alone and can not become a commercial product. Some investigators have tried to delay the gelation time of activated silica and have succeeded more or less, depending on the preparation procedure and the silica-to-metal molar ratio [2,3].

Activated silica is actually a sort of polysilicate. Polysilicate, one kind of inorganic anionic polymer, can be combined with the inorganic cationic polymers of Al and Fe to improve its aggregating function or to retard its hydrolysis reaction. The composites of polysilicate with ferric sulfate, or polyferric-silicate-sulfate, which are prepared by introducing ferric sulfate into polysilicic acid (PSA), may be called PFSS.

The strongly acidic range is generally chosen to prepare stable ferric-silicate polymer composite coagulants. In this paper, PFSS was prepared in our laboratory and its properties as a coagulant were studied. A jar test was carried out to evaluate the coagulation performance of PFSS in water treatment. It is hoped that the preliminary results of this paper will be useful for the development of a new, stable, and highly effective ferric-silicate polymer composite coagulant in the future.

Experimental Methods

The following instruments and reagents were used: JS94E Micro-Electrophoretic Mobility Detector (Jiecheng Company, Ltd., Shanghai, China), UV-754 Spectrophotometer (Shanghai Analytical Instrument General Factory, Shanghai, China), YZD-1A Turbidimeter (Ning-Ca Agricultural & Environmental Instrument, Ltd., Yingchuan, China), DC-506 Laboratory Stirrer (Huashui New Technological Development Co., Shanghai, China), pHS-2 pH meter (Shanghai Leici Instrument Factory, Shanghai, China), sodium silicate (industrial products, modulus was 3.21), kaolin (chemical purity grade), other chemicals were analytical reagent grade.

Preparation of PFSS

5.11g of sodium silicate (containing 28% SiO_2) was diluted to 2.66% as SiO_2, and then its pH value was adjusted to 4.0 with sulfuric acid. After aging for 2 h, 4.439g of ferric sulfate was added and the mixture was heated to dissolve the ferric sulfate. After aging for 2 h, PFSS coagulants (liquid) containing 2.0% SiO_2 was prepared. The amount of ferric sulfate was then changed and this procedure was repeated to obtain PFSS coagulants with different Fe/Si molar ratios. The physical properties that characterize the PFSS are summarized below:

Appearance:	brown and clear liquid	Fe O content:	1.5~2.7 wt %
SiO content:	2.0 wt %	Molar ratio Fe/Si:	0.25~1.75
Density(25°C):	1.05~1.15		

Jar Test Experiments

Coagulation tests were performed by using the test water that was prepared with kaolin and distilled water. After adding 1g of kaolin into 1 liter of distilled water, 0.032 g of Na_2CO_3 was added to adjust the alkalinity of the test water to 0.8 to 1.0 meq/l. A certain amount of PFSS was added to 500 ml of the test water. After that, the test water was rapidly stirred at 120 rpm for 2 min, slowly stirred at 60 rpm for 7 min, and then 50 ml of the test water was withdrawn to measure its zeta potential. After the flocs had settled quiescently for 10 min, a sample of the supernatant was withdrawn from a position of about 2 cm below the water surface and its turbidity was measured. Distilled water was used as the blank and the turbidimeter was calibrated with a standard hydrazine solution. The pH value of test water was adjusted with dilute HCl solution or NaOH solution.

Measurement of Zeta Potential

The zeta potential of colloidal particles was measured by JS94E Micro-Electrophoretic Mobility Detector. In the detector, a 0.5 cm cuvette made of glass was used and the electrodes were placed in the cuvette. The voltage between the electrodes could be adjusted according to experimenter's need (usually 10 mv). A probe was used to measure environmental temperatures automatically and continuously, and the information was returned to computer to adjust automatically the parameters to compute zeta potential. In the computing process, a multimedia technique was used: the microparticles (with a radius range from 0.5 μm to 10 μm) which had been magnified 1500 times already were automatically and continuously "photographed", and four pseudocolor "photos" in two dimensions were provided for the computation and analysis.

Results and Discussion

Size Distribution of Kaolin Particles

The molar ratio of silicon to aluminum in the kaolin reagent was 1.43, measured by scanning electron microscopy (SEM). Its size distribution, as measured by size distribution detector, is shown in Table 1. The data show that the radii of most (about 60%) of the kaolin particles were in the range from 4 μm to 9 μm, and the radii of 75.70% of them were smaller than 10 μm. This is similar to the condition of natural water (most of the soil particles which can bring about turbidity are smaller than 10 μm in radius [4]).

Table 1. Size distribution of kaolin particles

Radius (μm)	1~2	2~3	3~4	4~5	5~6	6~7	7~8	8~9	9~10	>10
Size distribution (%)	1.50	3.90	6.05	9.86	12.25	14.25	15.64	9.80	2.45	24.30

Zeta Potential of PFSS under different pH Values

The PFSS samples were diluted (about 100-fold) to make the ferric concentration in the range from $1.67 \cdot 10^{-3}$ to $3.34 \cdot 10^{-3}$ mol/l. The concentration in diluted samples should not be too high or too low, since that would lead to too many colloidal particles or excessively small ones in the diluted samples. This, in turn, would make difficult to determine the zeta potential. The pH values of diluted samples were adjusted to 2.5-8.5 range with dilute HCl solution or NaOH solution before the measurement. Because some of the dissolved hydrolysis products from PFSS are adsorbed onto the surface of the sediment of PFSS hydrolysis products, zeta potential of the sediment could provide information similar to the change in zeta potential of all PFSS hydrolysis products. Figure 1 shows the zeta potential of PFSS coagulant at different pH values, and also compares PFSS to an $Fe_2(SO_4)_3$ solution (containing $3.34 \cdot 10^{-3}$ mol/l of ferric) and to a PSA solution (containing $3.33 \cdot 10^{-4}$ mol/l of SiO_2), when the Fe/Si molar ratio of the PFSS was 0.5 and 1.0, respectively. It can be seen that the zeta potential of PFSS coagulant was greater than that of the PSA solution and less than that of the $Fe_2(SO_4)_3$ solution at the same pH value. Furthermore, as the Fe/Si molar ratio increased, the zeta potential approached that of the $Fe_2(SO_4)_3$ solution, and as it decreased, it approached that of the PSA solution. These results demonstrate that charge neutralization occurred when the different charged chemicals were combined. This means that the composite of ferric sulfate and PSA can enhance aggregating efficiency due to higher molecular, but is weaker in terms of charge effectiveness in coagulation processes due to charge neutralization occurring when hydrolyzed ferric species of different charge and polysilicic acid are combined.

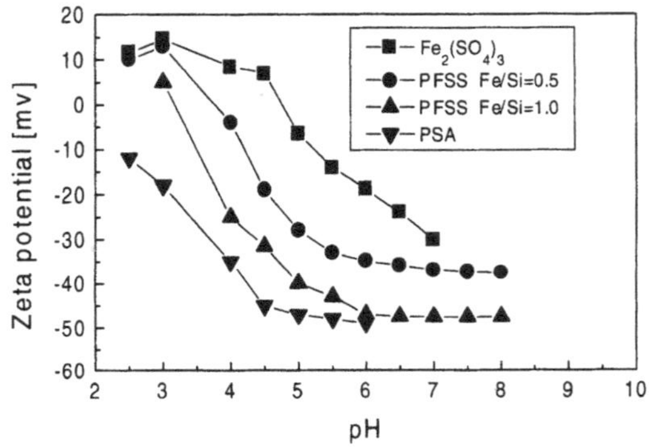

Fig.1. Zeta potential of coagulants as a function of pH value

Effect of Fe/SiO_2 Molar Ratio on Coagulation Performance of PFSS

PFSS (0.25 ml) of PFSS was added to 1 liter of test water. The pH value of the test water was then adjusted to about 8.0 by dilute HCl or NaOH solution. Figure 2 shows that the coagulation performance of PFSS was improved as the Fe/Si molar ratio increased, and was optimal when the ratio was about 1.5. When PFSS was used as coagulant in the above procedure, the residual turbidity of the treated test water was 98.1 mg/L, suggesting that PFSS with a suitable Fe/Si molar ratio can have a much better coagulation performance than $Fe_2(SO_4)_3$.

Effect of PFSS Dosage on Coagulation Performance

Figure 3 shows the effect of Fe/Si molar ratios in PFSS of 0.5,1.0 and 1.5, respectively, on the residual turbidity of the test water. The results again showed that the greater the Fe/Si molar ratio, the better coagulation performance of PFSS. A dose of 0.19 ml/l of the PFSS with a Fe/Si molar ratio of 1.5 gave the best turbidity removal effect. Higher doses could not restabilize the kaolin suspension. When PFSS with lower Fe/Si molar ratios were used, successively higher doses were required in order to achieve the same effect.

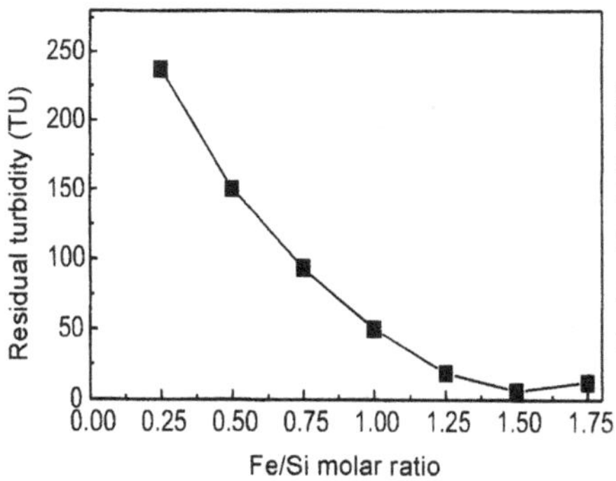

Fig. 2. Effect of Fe/Si molar ratio on the coagulation performance of PFSS

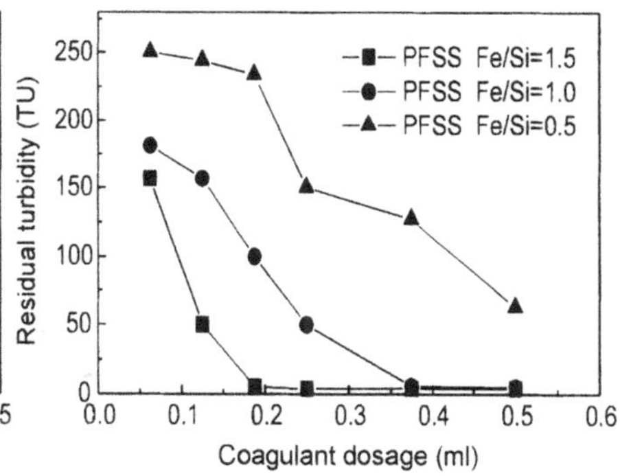

Fig. 3. Effect of PFSS dosage and Fe/Si molar ratio on coagulation performance

The Stability of PFSS Solutions and the Effect of Gelation on the Coagulation of PFSS

PSA, which was often used as a flocculant aid in water treatment, gels in a short time during storage. For example, it was found that a PSA solution containing 20 g/l of SiO_2 at a pH of about 3.0 gelled after it was aged for 6-7days at room temperature (about 30°C).When ferric sulfate was added to the PSA solution to form PFSS solution, however, it can be stored for a much longer time, owing to the interaction between PSA and ferric ion [5].

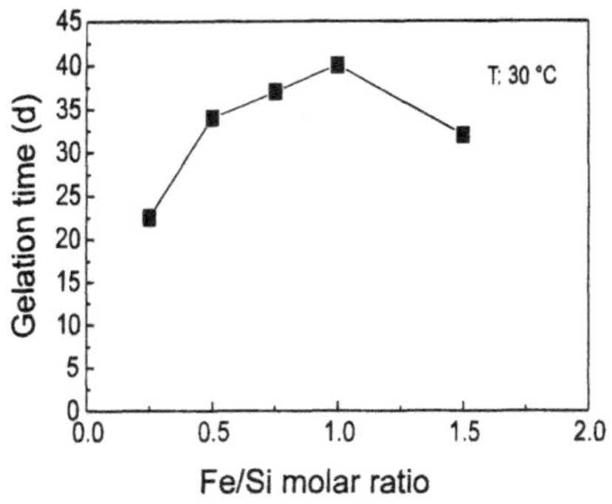

Fig.4. Effect of Fe/Si molar ratio on the stability of PFSS solution

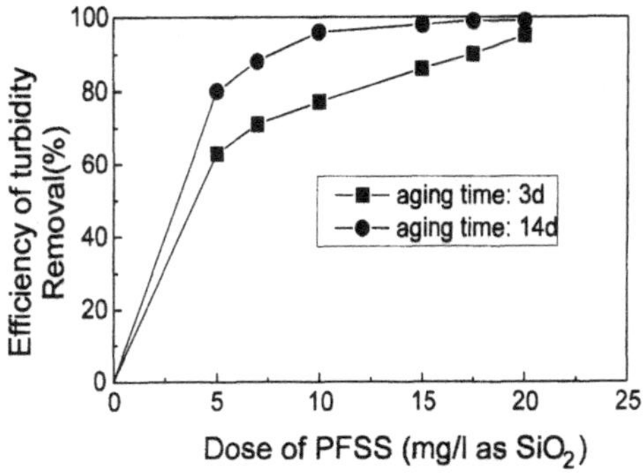

Fig. 5. Effect of aging time on the coagulating effect of PFSS with Fe / Si = 0.33 at pH 6.5

Figure 4 shows the relationship between the Fe/Si molar ratio and the stability of the PFSS solution. It can be seen that an Fe/Si molar ratio of about 1.0 gives the most stable PFSS solution, while higher or lower ratios decrease stability. A PFSS solution with Fe/Si = 1.0 containing 20 g/l SiO_2 can be stored for 40 days at room temperature (about 30°C) without gelation. PFSS solution can be stored for a longer time at lower temperature.

Figure 5 shows the effect of aging time on the coagulating effect of a PFSS solution with Fe/Si = 0.33 at a pH of 6.5. It can be seen that before the gelation, the longer the aging time, the better the coagulation effect of PFSS. This may be because the degree of polymerization of PFSS increases with the aging time. For higher polymerized polymer, stronger bridge-formation can occur in the flocculation step.

In order to investigate how gelation affects the coagulating performance of PFSS, the coagulating effect of the gelled PFSS solution was compared with that of a PFSS solution, which had aged for 20 days but did not gel. The dose of coagulant PFSS solution was1.0 mg/l as Si, and the turbidity and pH of the test water were 300 TU and 7.8, respectively. The results are summarized in Figure 6. It can be seen that gelled PFSS removes less turbidity than PFSS without gelation. During the experiment, it was also found that the flocs formed by gelled PFSS were smaller than those formed by PFSS without gelation when treating the test water. This suggests that the gelled PFSS possesses weaker adsorption and bridge-formation properties.

Effect of pH on the Coagulation Performance of PFSS

Figure 7 shows how the Fe/Si molar ratio and pH affected turbidity removal when the PFSS dose was fixed at 1 mg/l (as SiO_2). PFSS with a low Fe/Si molar ratio gave an excellent coagulation performance in the low pH range. As the Fe/Si molar ratio increased, the pH range for optimal coagulation shifted slightly toward higher pH region. The results suggested that the Fe/Si molar ratio is a factor in producing a pH range suitable for coagulation.

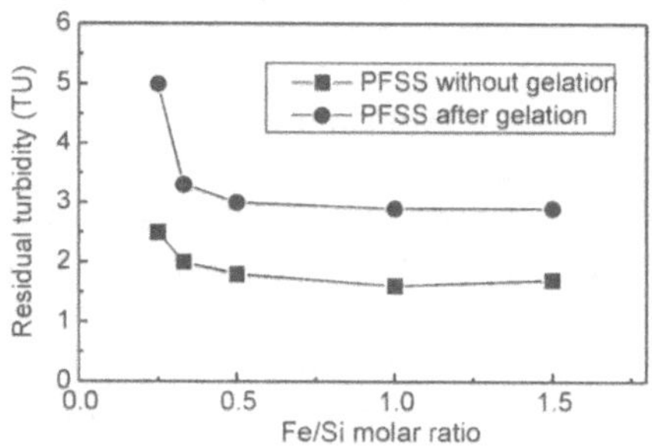

Fig. 6. Comparative coagulation performance of PFSS before and after gelation at pH 7.8

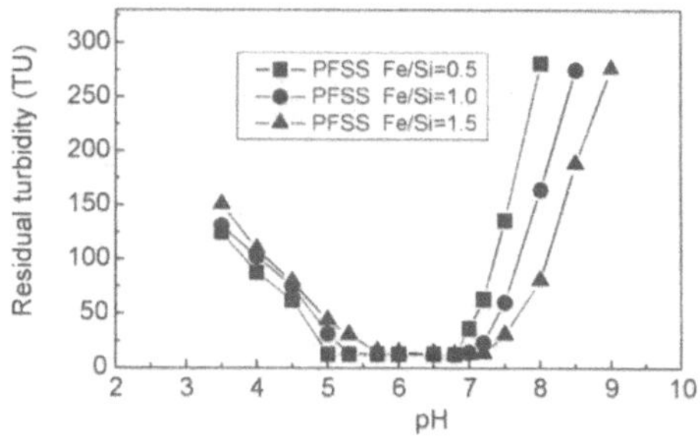

Fig. 7. Change of residual turbidity of kaolin flocs with pH

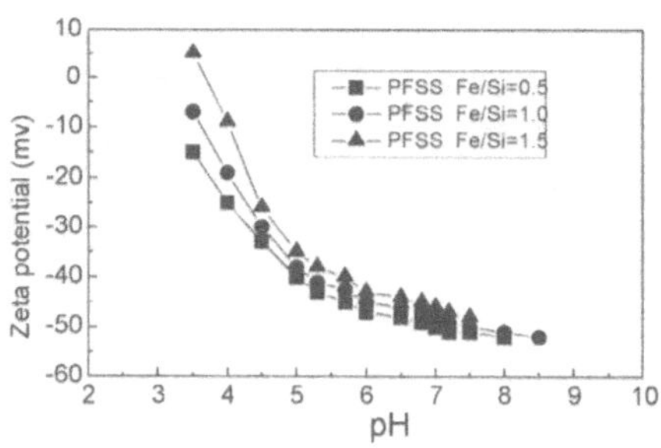

Fig. 8. Change of zeta potential of kaolin flocs with pH

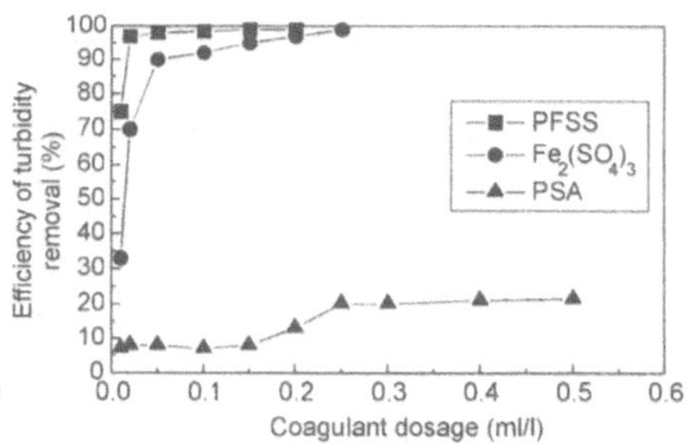

Fig. 9. Comparison of turbidity removal efficiency of PFSS, $Fe_2(SO_4)_3$ and PSA

In Figure 8, the curves relating zeta potential to pH demonstrate that PFSS coagulant is an excellent turbidity remover from kaolin suspensions, even when the flocs formed during coagulation/flocculation are charged by relatively more negative electrons. This means that PFSS may be more effective in the aspect of bridge-formation.

Comparison of Coagulation Performance of PFSS, Ferric Sulfate and PSA

The coagulation performance of a PFSS solution with Fe/Si = 1.0 containing 20 g/l SiO_2 and 18.65 g/l Fe^{3+} was compared to that of a ferric sulfate solution and a PSA solution using coal washing wastewater taken from the Yangzhuang Coal-Mine, Mineral Bureau of Feicheng, China. The turbidity and pH of the wastewater were 580 TU and 7.76, respectively. The concentration of the ferric sulfate solution was 18.65 g/l as Fe^{3+}, and the concentration of PSA was 20g/l as SiO_2. The experimental results are shown in Figure 9. It can be seen that PFSS achieved greater removal of turbidity than the other coagulants at a lower dosage. At higher dosages, however, the differences between the PFSS and ferric sulfate were slight, and these two coagulants performed much better than the PSA.

Discussion and conclusions

The following factors should be considered when evaluating the PFSS as a coagulant in water treatment : (1) the general characteristics of PSA as a polymer; (2) the general characteristics of ferric salt; (3) the interaction between silicon hydroxyl of PSA, ferric ions and ferric hydrolysates. After PFSS is added as a coagulant to the water, two different processes may proceed simultaneously. In one process: (1) a change in degree of hydrolysis and the transformation of species of ferric salt will be brought about by the dilution and the increase in pH; (2) the ferric hydrolyzates will be incorporated into PSA to form polysilicates containing ferric ions; (3) the increase in pH will lead to further polymerization of PSA until gelation. In the second process, (1) all kinds of ferric hydrolyzates will be adsorbed onto the suspended particles in the mixing process, destabilizing the particles; (2) after ferric hydrolyzates are adsorbed onto the suspended particles, large molecules or the gelled PSA will bridge and stick the particles together, forming large flocs which precipitate, leaving clean water. These two processes can proceed simultaneously and be finished rapidly. In the optimum pH range for turbidity removal, the aggregated suspended substance having relatively more negative electrons indicate that the coagulation mechanism of PFSS is obviously different from that of conventional ferric salt coagulants. The PFSS coagulant shows typical adsorption-bridge formation and stick-sweeping action characteristics.

The results of this study show that PFSS, as a new type of inorganic polymer coagulant, gave an excellent coagulation performance. The Fe/Si molar ratio in PFSS significantly affected coagulation performance, electrophoretic nature of hydrolyzates, stability ,and the optimum pH range of PFSS for turbidity removal in water treatment. Other advantages of PFSS are the simple preparation process and its non-toxicity.

Future investigations might address the chemical speciation of PFSS, the interaction between PSA and the hydrolysis products of ferric ions, special mechanisms of coagulation and flocculation, and optimization of production technology.

References

1. Baylis, J. R.: Silicates as aid to coagulation. J. AWWA 29 (1939)1355-1396.
2. Takao Hasegawa: Method and Flocculant for Water Treatment, US Patent, 4,923,629, 1990 05-08
3. Takao Hasegawa, Takuya Onitsuka, Minoru Suzuki: Water Nagoya'89, ASPAC IWSA. 31S-05-A8, (1989) 152-161.
4. Baojiu Xu : Principles of applied and waste water treatment at present time, Advanced Education Press, Beijing (1990)
5. Iler, R. K.: The Chemistry of Silica. Wiley, New York, 1979

Cationic Organic Polymers for Flocculation of Municipal Wastewater - Experiments and Scenario Study

A.R. Mels* and A.F. van Nieuwenhuijzen**

*Sub-department of Environmental Technology, Wageningen University and Research, PO-Box 8219, 6700 EV Wagenin, The Netherlands
Adriaan.Mels@algemeen.mt.wag-ur.nl
**Department of Sanitary Engineering, Delft University of Technology, PO-Box 5048, 2600 GA Delft, The Netherlands
A.F.vanNieuwenhuijzen@CiTG.TUDelft.nl

Abstract

Organic polymers can be favourable flocculants in comparison with inorganic (metal-)flocculants because chemical sludge production is absent and the increase in salt concentration of the effluent is low. Jar test experiments showed that a higher weight ($8 \cdot 10^6$ g/mol; (+) 24 %) cationic polyacrylamide gave good flocculation results, with turbidity- and $COD_{particulate}$ removal efficiencies of 65-90 % and a TSS removal of > 90 % at relatively low dosing concentrations. Rapid coagulation mixing conditions were optimal with a mixing time of 30 s ($G = 800\ s^{-1}$). A decrease in low flocculation mixing time from 180 s to 30 s ($G = 50\ s^{-1}$) resulted in a decrease in particle removal of 25 %. A scenario study was conducted to estimate the costs involved and to quantify the environmental impacts. The scenario study showed that the total costs of a complete wastewater treatment system based on coagulation-flocculation with organic polymers are in the same range (-5 %) as those of a system with $FeCl_3$. Total sludge production is 15 % lower. Salinisation is of the order of +0.2 mg Cl^- per litre effluent for dosing polymer and +29 mg Cl^- per litre effluent for dosing $FeCl_3$.

Introduction

In the first phase of a research project on sewage treatment methods based on physical-chemical pre-treatment, a large number of possible treatment scenarios was identified and evaluated [1]. The general conclusions of this study were that wastewater treatment systems based on enhanced particle removal in the pre-treatment can be applied more energy efficiently (savings up to 80%), can be designed much smaller and can financially compete with commonly applied systems. Similar conclusions were also reported by Ødegaard and Karlsson [2].

There are, however, also some problems with physical-chemical pre-treatment when (inorganic) metal salts like $FeCl_3$ are applied for coagulation/flocculation:

- an increase in inorganic (chemical) sludge production;
- a quite high consumption of ferric- or aluminium salts;
- an increase in the salinity of the effluent.

Experimental research and a scenario study are used to consider whether it might be feasible to replace metal salts with cationic organic polymers. Cationic polymers are highly promising with respect to solving part of the identified problems, since chemical sludge production is absent and salinisation is low.

Cationic organic polymers for flocculation of municipal wastewater – Experiments

Methodology

In order to determine the feasibility of cationic organic polymers for flocculation of municipal wastewater, jar tests under different process conditions were conducted.

Characterisation of organic polymers

For coagulation/flocculation a linear low weight cationic polyamine ($0.25 \cdot 10^6$ g/mol; charge +100 weight %) and three linear polyelectrolytes based on polyacrylamides ($4 \cdot 10^6$, $6 \cdot 10^6$ or $8 \cdot 10^6$ g/mol; charge +24 weight %) were tested (all CYTEC).

Experimental set-up

The jar tests were conducted in a set up with 6 stirred beakers of 2.5 l capacity. The beakers were filled with 2.0 l of wastewater. Cationic polyelectrolyte was dosed simultaneously to all beakers under rapid coagulation mixing conditions (mixing intensity $G = 800\ s^{-1}$; 120 s for polyamine polymers; 20 s for polyacrylamide polymers). After coagulation mixing, slow flocculation mixing was applied ($G * t = 50\ s^{-1} * 180\ s$). After 900 s of settling the supernatant was analysed for turbidity, pH and TSS.

Analysis

Turbidity was measured with a HACH turbidimeter. COD, TSS, pH and temperature were measured according to Standard Methods [3].

Table 1. Wastewater characteristics of the Bennekom WWTP (June 1998 - March 1999)

Parameter	Average concentration	Fluctuation
Turbidity	130 NTU	75 – 220 NTU
COD_{total}*	525 mg O_2/l	220 – 670 mg O_2/l
Total Suspended Solids (TSS)	160 mg d.s./l	100–300 mg d.s./l
pH	7.9	7.4 - 8.5

* 59 % as particulate COD (> 0.45 μm)

Wastewater characteristics

For the experiments, wastewater from the municipal wastewater treatment plant (WWTP) of the village of Bennekom (in The Netherlands) was used. Table 1 gives an overview of relevant parameters.

Results of the experiments

Comparison of cationic organic polymers

Figure 1 shows the turbidity removal after flocculation and settling with the tested cationic polymers.

Figure 1 shows that the polyacrylamide polymers of 6 and $8 \cdot 10^6$ g/mol gave a good turbidity removal at relatively low dosing rates. The average TSS-concentration of the supernatant was 15 ± 3 mg dry solids per litre.

The experimental work that is described in the following paragraphs focuses on this type of organic flocculant.

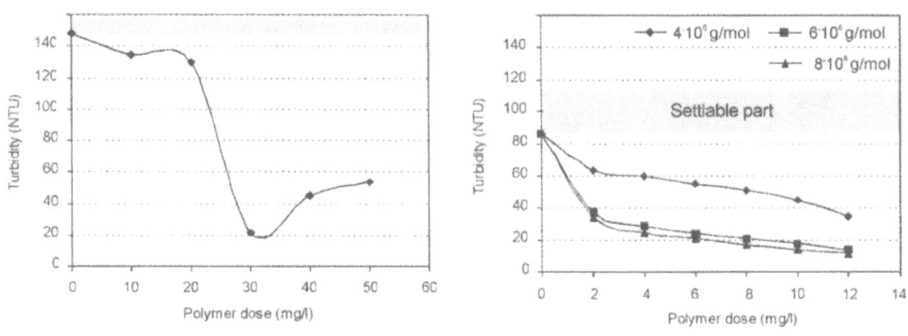

Fig. 1. Turbidity removal after flocculation and settling with different cationic polymers. Left: low molecular weight polyamine ($0.25 \cdot 10^6$ g/mol); right: polyacrylamide- based polymers with increasing molecular weight (4, 6 and $8 \cdot 10^6$ g/mol)

Particle removal efficiency of a high weight ($8 \cdot 10^6$ g/mol) organic polymer

For comparing the results of jar test experiments with different influent turbidities we used a methodology that was developed within the context of a parallel investigation into dissolved air flotation (DAF) with organic flocculants in the same experimental period [4]. In this study, we found (i) a linear relationship between particulate COD > 0.45 μm and turbidity (see fig. 2a) and (ii) hardly any removal of dissolved COD after flocculation with organic polymers (i.e. less than 10 %). Based on these results we correlated the turbidity removal rate of the jar test experiments to the polymerdose to influent-NTU ratio (mg polymer/100 NTU).

Figure 2b gives the results of 5 experiments with different influent composition. The figure shows that there is good agreement among the different experiments. The flocculation with the $8 \cdot 10^6$ g/mol polymer results in a particle removal of 65 - 90 % at dosing rates of 1 – 10 mg polymer per 100 NTU. This is equivalent to a total COD removal of 38 - 53 % for the tested wastewater (with a particulate COD fraction of 0.59). For settling without flocculant addition, a particle removal of approximately 30 % was obtained, which is equivalent to a total COD removal of ca. 17 %.

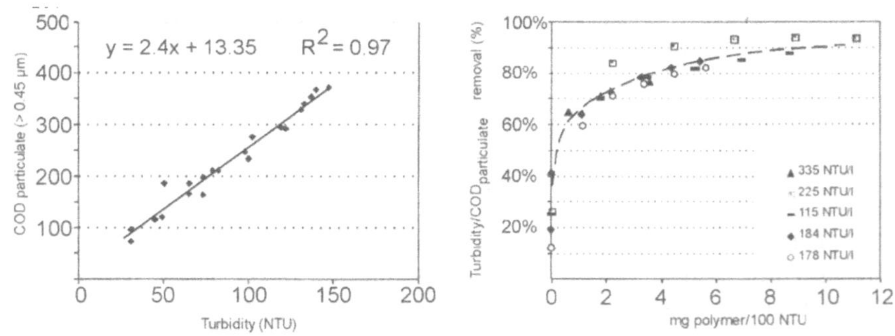

Fig. 2. Left (2a): Relationship between particulate COD (> 0.45 μm) and turbidity for the tested wastewater; Right (2b): Comparison of five experiments based on the polymerdose-$NTU_{influent}$-ratio (in these experiments a $8 \cdot 10^6$ g/mol polyacrylamide was used)

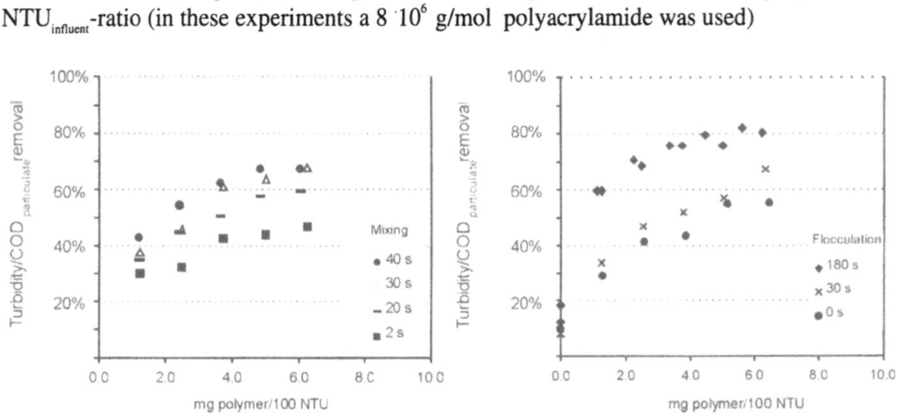

Fig. 3. Left (3a): Effect of coagulation mixing time on particle removal (G = 800 s^{-1}; flocculation mixing = 30 s·50 s^{-1}); Right (3b): Effect of flocculation mixing time (G = 50 s^{-1}; coagulation mixing 20 s * 800 s^{-1})

Particle removal under different process conditions

Within the scope of the jar test experiments, the effect of different process conditions on particle removal was investigated. Figure 3 shows how decreased coagulation mixing time (Figure 3a) and decreased flocculation mixing time (Figure 3b) result in reduced particle removal. The left picture shows that rapid coagulation mixing time > 30 s was optimal (note: the flocculation mixing time in these experiments was 30 s). The right picture indicates that the particle removal decreases by circa 25 % if the flocculation time is decreased from 180 s to 30 s and 0 s (at $G = 50\ s^{-1}$). The settling time could be reduced from 900 s to 300 s without a decrease in turbidity removal (result not shown).

The Replacement of Metal Salts with Cationic Organic Polymers – Scenario Study

A scenario study was carried out in order to study how replacing $FeCl_3$ with cationic organic polymers affects the costs and environmental impacts of a complete wastewater treatment system. With the information derived from the experiments, two scenarios based on coagulation-flocculation (with cationic polymer and with Fe^{3+}) were set up in the evaluation programme DEMAS. DEMAS is a spreadsheet-based evaluation tool, which was developed within the framework of former scenario studies [1, 5]. Total costs and environmental impacts were calculated for both scenarios and an additional reference scenario.

Selected treatment scenarios

Table 2 shows an overview of the scenarios and the design criteria, which were used in the calculations.

The surface load of the pre-settlers of all scenarios was set at 3 m/h (although the results of the experiments indicate that the load may be higher). The mixing conditions of both scenario I and II were set for 30 s rapid mixing and 180 s flocculation, chosen because of the experimental results.

For post-treatment, activated sludge systems with biological BOD-, N- and P-removal were designed ($N_{total} < 10$ mg N/l, $P_{total} < 1$ mg P/l). In scenario II however, phosphate is assumed to be merely removed by precipitation as $FePO_4$. The sludge load of the activated sludge systems was calculated according to the HSA-methodology [6,7] using the pre-settler effluent values of Table 3. When this methodology is used, the sludge load of the systems with pre-treatment can be designed 25 % higher. The higher sludge load is possible because pre-treatment leads to an increased performance of the activated sludge system. Pre-treatment results in a higher sludge age (better nitrification) because fewer inorganics are entering the system, which results in a higher VSS/TSS ratio [2].

Table 2. Selected treatment scenarios, including design criteria

	Dosing	Surface load pre-settler m/h	Sludge load activated sludge kg BOD/ kg d.s. per day	Surface load post-settler m/h
Reference		3	0.06 *	0.7
Scenario I	4 mg/l cationic polymer	3	0.08 *	0.7
Scenario II	15 mg Fe^{3+}/l + 0.5 mg/l anionic polymer	3	0.08 *	0.7

* Calculated according to the HSA-methodology with a sludge concentration of 4 kg per m^3 and a minimum process temperature of 10 °C.

Influent and effluent composition of the pre-settler

Table 3 shows the assumed influent and effluent concentrations for COD, BOD, N, P and TSS. A distinction was made between particulate and dissolved matter. 70 % of COD was assumed to consist of particulate material [8]. The effluent composition was calculated based on particle removal rates of 40 % for the reference scenario and 70 % for scenario I and II (result of the experimental study). We assumed that a BOD/N ratio of 2.5 would still allow sufficient denitrification in the activated sludge part.

Table 3. Composition of influent and effluent of the pre-settler

	Influent	Reference ($R_{particles}$ 40 %)	Scenario I ($R_{particles}$ 70 %)	Scenario II ($R_{particles}$ 70 %)
CODtotal (mg O_2/l)	**600**	**432**	**306**	**306**
CODparticulate	420	252	126	126
CODdissolved	180	180	180	180
BOD5 (mg O_2/l)	**220**	**172**	**135**	**135**
BODparticulate	121	73	36	36
BODdissolved	99	99	99	99
Ntotal (mg N/l) *	**63**	**58**	**54**	**54**
Nparticulate	14	8	4	4
Ndissolved	50	50	50	50
Ptotal (mg P/l) *	**11.0**	**10.0**	**9.2**	**2.0**
Pparticulate	2.7	1.6	0.8	0.8
Pdissolved	8.3	8.3	8.3	1.2
TSS (mg d.s./l)	200	100	20	20
BOD/N	3.5	3.0	2.5	2.5

* Including N and P from reject water (N:+9 mg/l, P:+2 mg/l)

Plant size and system boundaries

The scenarios were designed for a 100,000 population equivalents (p.e.) plant and had an average flow of 19,500 m^3 per day. During dry weather conditions the maximum flow was 1,000 m^3/h, while during storm weather conditions the maximum flow increased to 3,000 m^3/h (SWF = 3 x DWF).

The system boundaries, which are shown in Figure 4, include all on-site water and sludge treatment processes. Water treatment was combined with sludge treatment.

The sludge handling procedure consisted of mechanical thickening of the combined primary and secondary sludge to 6 % dry solids. The thickening was followed by sludge digestion. In the digester (retention time 20 days) 50 % of the organic fraction of the primary sludge and 30 % of the secondary sludge were degraded. The energy value of the biogas was 22,000 kJ/m^3, of which 32 % was converted to electricity in the TE-installation. The remaining energy (i.e. heat) was used for sludge heating.

After digestion the final sludge was dewatered by centrifugation to 25 % dry solids. The energy consumption for dewatering was assumed to be 4 kWh/m^3. To improve dewatering 7.5 kg polymer per ton dry solids was dosed. The dewatered sludge was transported for central incineration and stored as chemical waste.

Cost calculations

To estimate the costs, the market value method was used. The market value reflects all aggregated costs over a total depreciation period of thirty years with an interest rate of 5.5 % and an inflation rate of 2.5 %. The market value was divided into costs for construction and electro-mechanical facilities, energy, chemicals, sludge handling and personnel (staff/maintenance).

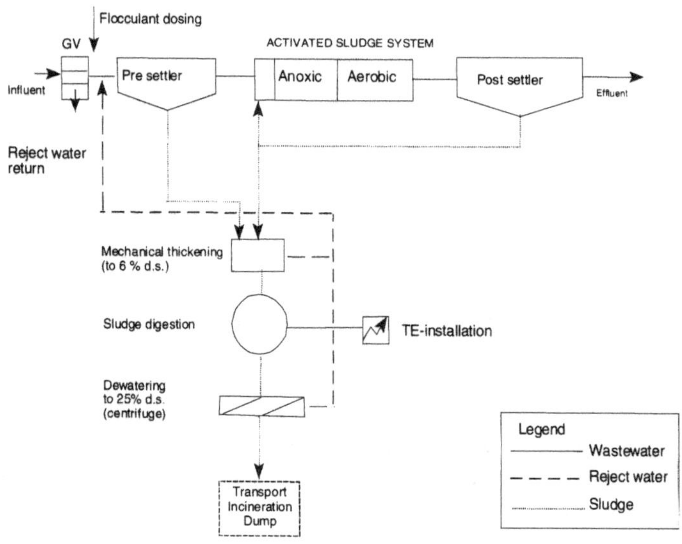

Fig. 4. System boundaries of the scenarios

Table 4. Assumptions for the cost calculations (Euro)

Item	**Costs** *Construction**	*Electro-mechanical**
Fine screen	Euro 136,400 (total)	227,300 (total)
Flocculation	455 per m^3	341 per m^3
Pre- and post-settlers	273 per m^2	68 per m^2
Activated sludge system (per m^3)	136 per m^3	91 per m^3
Sludge thickening (per m^3)	909 per m^2	227 per m^2
Digester	182 per m^3	91 per m^3
TE-installation	46 per m^3 biogas/day	227 per m^3 biogas/day
Energy	0.073	per kWh
$FeCl_3$ (40 %)	455	per ton active Fe^{3+}
Polyelectrolyte	3,410.-	per ton
Final sludge disposal	365.-	per ton dry solids
Staff	36,400.-	per staff member per year
Maintenance	0.5 % of civil and 1.5 % of e/m- investments	

* For construction and electro-mechanical investments an 'overhead' factor of 0.7 was assumed.

Results of the scenario study

Table 5 shows the results of the calculations.

Costs

A crucial factor in (the consideration of) using organic polymers in the wastewater treatment process is the cost. Figure 5 shows the relative costs compared to the total costs of the reference scenario. The calculations distinguish among total costs of (civil and electro-mechanical) investments, chemical use, sludge disposal and personnel. The results show that including a particle flocculation and removal step leads to ca. 15 % reduction of the investment costs. An enhanced particle removal step makes it possible to design a smaller activated sludge system, which results in significant cost reduction. The investment costs include the additional required costs for sludge handling equipment (thickening, digestion, dewatering) due to the increased volume of (primary) sludge.

Table 5. Calculated market values and environmental impacts for each scenario

Criteria		Reference	Scenario I	Scenario II
Total market value Euro 10^6		36.6	35.2	37.0
- civil + e/m investment		18.2	15.2	15.6
- chemical use		1.1	3.1	2.6
- sludge disposal		11.6	12.3	14.0
- personnel		5.8	4.6	4.7
Energy balance	MWh/y	- 1,033	- 529	- 544
Sludge production	ton/y	1,565	1,657	1,897
- organic		1,103	1,195	1,195
- inorganic		462	462	462
- chemical		--	--	240
Chemicals	ton/y			
- $FeCl_3$		--	--	310
- Polyelectrolyte*		11.7	40.9	14.2
Effluent quality	mg/l			
Cl^-		+ 0	+ 0.2	+ 29
Staff	no.	3	3	3
Footprint	m^2	9,680	8,200	8,250

* Including polyelectrolytes which are used for dewatering.

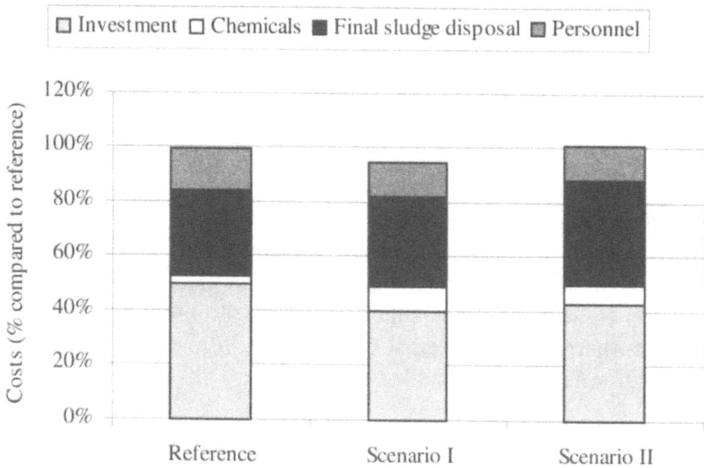

Fig. 5. Relative costs of the three scenarios (total costs in reference = 100 %)

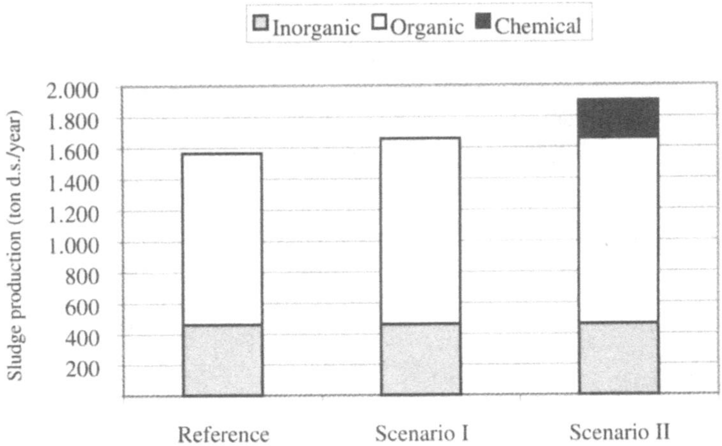

Fig. 6. Sludge production of the three scenarios

The calculations show that scenario I with polymer dosing has higher chemical costs than scenario II. These extra costs are, however, compensated by the lower costs for final sludge disposal at the incinerator. The total costs for scenario I appear to be even slightly (ca. 5 %) lower than the costs for both other scenarios

Environmental impacts

Compared to the reference scenario, scenarios I and II can be designed 15 % smaller and use ca. 50 % less energy (this includes energy recovery from biogas generation). Figure 6 shows that the addition of $FeCl_3$ in scenario II leads to an increase of the total inorganic sludge fraction (incl. chemical sludge) of 52 %. Compared to scenario I with polymer addition the amount of final sludge after digestion and dewatering is 15 % higher.

The value for salinisation of the effluent of scenarios I and II is based on the chlorine release from the used flocculants. Based on the charge composition and the molecular weight it can be calculated that cationic polyacrylamide-based polymers with a charge of +24 weight % release 0.05 mg Cl^- per mg of polymer. With $FeCl_3$ addition 1.9 mg Cl^- is added per mg Fe^{3+}.

Conclusions

The experimental study shows that cationic organic polymers are suitable flocculants for pre-treatment of municipal wastewater. Polyacrylamide polymers of 6 and $8 \cdot 10^6$ g/mol result in good flocculation results, with turbidity- and $COD_{particulate}$ removal efficiencies of 65-90 % and a TSS-removal of > 90 % at relatively low dosing concentrations. Coagulation mixing conditions were optimal at a mixing time of 30 s (G of 800 s^{-1}). A decrease in flocculation mixing time from 180 s to 30 s (G = 50 s^{-1}) resulted in a 25 % decrease in particle removal.

Although the costs for polymers are approximately 7.5 times higher than Fe^{3+} per mg active product, the scenario study showed that the total costs of a scenario with polymer addition and a scenario with $FeCl_3$ addition are in the same range (-5 %). Total sludge production was 15 % lower. Salinisation of the effluent was of the order of +0.2 mg Cl^- for dosing polymer and +29 mg Cl^- for dosing $FeCl_3$.

References

1. Van Nieuwenhuijzen, A.F., Mels, A.R., Van der Graaf, J.H.J.M., Klapwijk, B., De Koning, J., Rulkens, W.H.: Identification and Evaluation of Wastewater Scenarios based on Physical-Chemical Pre-treatment, In: Proceedings of the 8th Gothenburg Symposium Chemical Water and Wastewater Treatment V (Eds. H.H. Hahn, E. Hoffmann, H. Ødegaard), Springer-Verlag, Berlin (1998)
2. Ødegaard, H., Karlsson, I.: Chemical Wastewater Treatment – Value for Money, In: Proceedings of the 6th Gothenburg Symposium Chemical Water and Wastewater Treatment III (Eds. R. Klute, H.H. Hahn), Springer-Verlag, Berlin (1994)
3. APHA: Standard Methods for the Examination of Water and Wastewater (1998)
4. Mels, A.R., Van der Meer, A.K., Rulkens, W.H., Van Nieuwenhuijzen, A.F., Klapwijk, A.: Flotation with polyelectrolytes as a first step of a more sustainable wastewater treatment system, IWA-conference 'Paris 2000', 3-7 July 2000
5. Mels, A.R., Van Nieuwenhuijzen, A.F., Van der Graaf, J.H.J.M., De Koning, J., Klapwijk, A., Rulkens, W.H.: Sustainability criteria as a tool in the development of new sewage treatment methods. Water Science & Technology 39 (5) (1999) 243-250
6. ATV-A131 Arbeitsblatt A131: Bemessung von einstufigen Belebungsanlagen ab 5000 Einwohnerwerten, Regelwerk der ATV, St. Augustin, Febr. (1991) (in German)
7. STOWA: Evaluation of the HSA-model for application in The Netherlands, Report no. 95-17 (1995) (in Dutch)
8. STOWA: Characterisation of raw and pre-treated wastewater, Report no. 99-13 (1999) (in Dutch)

Chemistry, Function, and Fate of Acrylamide-Based Polymers

S.K. Dentel, B.M. Gucciardi, N. J. Griskowitz, L. L. Chang, D. L. Raudenbush and B. Arican

Department of Civil and Environmental Engineering, University of Delaware, Newark, DE 19716 USA
dentel@udel.edu

Abstract

Acrylamide-based polymers are used widely in water and wastewater treatment to enhance flocculation processes. It is often assumed that a polymer's molecular weight and its type and amount of electrical charge are the crucial parameters governing its performance. Yet it is clear that the actual chemical structure of a polymer plays a key role in both process performance and in the subsequent fate of the substance. This paper relates specific aspects of polymer chemistry, such as ionizable sites, degree of cross-linkage, and hydrolyzable pendant groups, to effectiveness in flocculation. The degree of biodegradability of cationic polyacrylamides - and their ultimate environmental fate - are also related to chemical structure, and this should be taken into account in order to minimize undesirable performance and effects when using these polymers.

Background

In any common text providing an overview of the use of organic polyelectrolytes for coagulation or flocculation, the two properties invariably cited as key polymer properties are *molecular weight* and amount (or, perhaps, only the type) of *ionic charge* on the molecule. There are good reasons for this generalization.

Firstly, both of these properties can be quantified. The molecular weight (or, more properly, molecular mass) has long been related to rheological, osmotic, and light scattering properties by straightforward equations (e.g., [1]). Dentel et al. [2] cite a simple viscosimetric method for determining the average molecular mass of acrylamide-based polymers. The ionic charge concentration (often denoted as charge density) of a polymer can be determined by a procedure known as a colloid titration, first described by Terayama [3] and presented recently in a procedures manual [2].

Secondly, these methods are readily associated with the two well-known conceptual models of flocculation. A polymer's charge density should be indicative of the amount of polymer charge available to effect particle destabilization via the "patch model," and if its molecular weight is sufficient, the "bridging model" should apply [4,5] (see ref. [6] for a reconsideration of these models). However, reliance solely on these two parameters does not provide an adequate basis for understanding many important aspects of flocculation. In this paper, several other significant properties of flocculant polymers are described relevant to their performance in water and wastewater treatment processes. A specific focus is placed on polymer conditioning of wastewater sludges (biosolids).

Chemical Properties of Polymers

Monomer type

It is widely understood that most polymers used in water and wastewater treatment are acrylamide-based. Figure 1 shows the basic acrylamide monomer (Structure 1) which, polymerized, yields polyacrylamide. The hydrolysis of this structure gives acrylic acid monomers which comprise polyacrylic acid. This polymer is primarily nonionic at pH <4 but deprotonation to the anionic form (polyacrylate, Figure 1) occurs at increasing pH [7]. Figure 2 shows this by use of a charge titration analysis at three different pH levels [8]. Acrylamide and acrylate monomers may be combined to provide polymers of varying anionicity. The ratio is generally expressed as a mole percent of the charged monomer; for example, a 10 mole % PAA/PAM polymer is a low charged anionic polymer. These polymers are employed as coagulant aids or filter aids in water treatment or as conditioners of water treatment sludges (residuals). They may be synthesized by a number of different methods [6,9].

Structures 2 – 5 in Figure 1 show various acrylamide-based cationic polymers. Structure 2 depicts the polymer often termed a "Mannich" polymer because the nonionic polyacrylamide is post-reacted with formaldehyde and dimethylamine to give the aminomethylated polyacrylamide via the Mannich reaction. This tertiary amine is only cationic at low or moderate pH due to its deprotonation at increasing pH, unless it is quaternized to give the second indicated structure. The quaternization reaction may utilize methylchloride, dimethyl sulfate, or ethyl or benzyl analogs of these quaternizing agents [9,10].

Structures 3-5 exemplify monomers that can be copolymerized with acrylamide to create a variety of copolymers. Though related, these exhibit somewhat differing properties based on their monomer characteristics and on the manner in which the copolymerization reaction takes place. Structure 6, the DADMAC monomer, is commonly used as a homopolymer (polyDADMAC) for coagulation processes in water treatment due to its high cationic charge content (Figure 2). However, it can also be copolymerized with acrylamide to create a cationic flocculant [9].

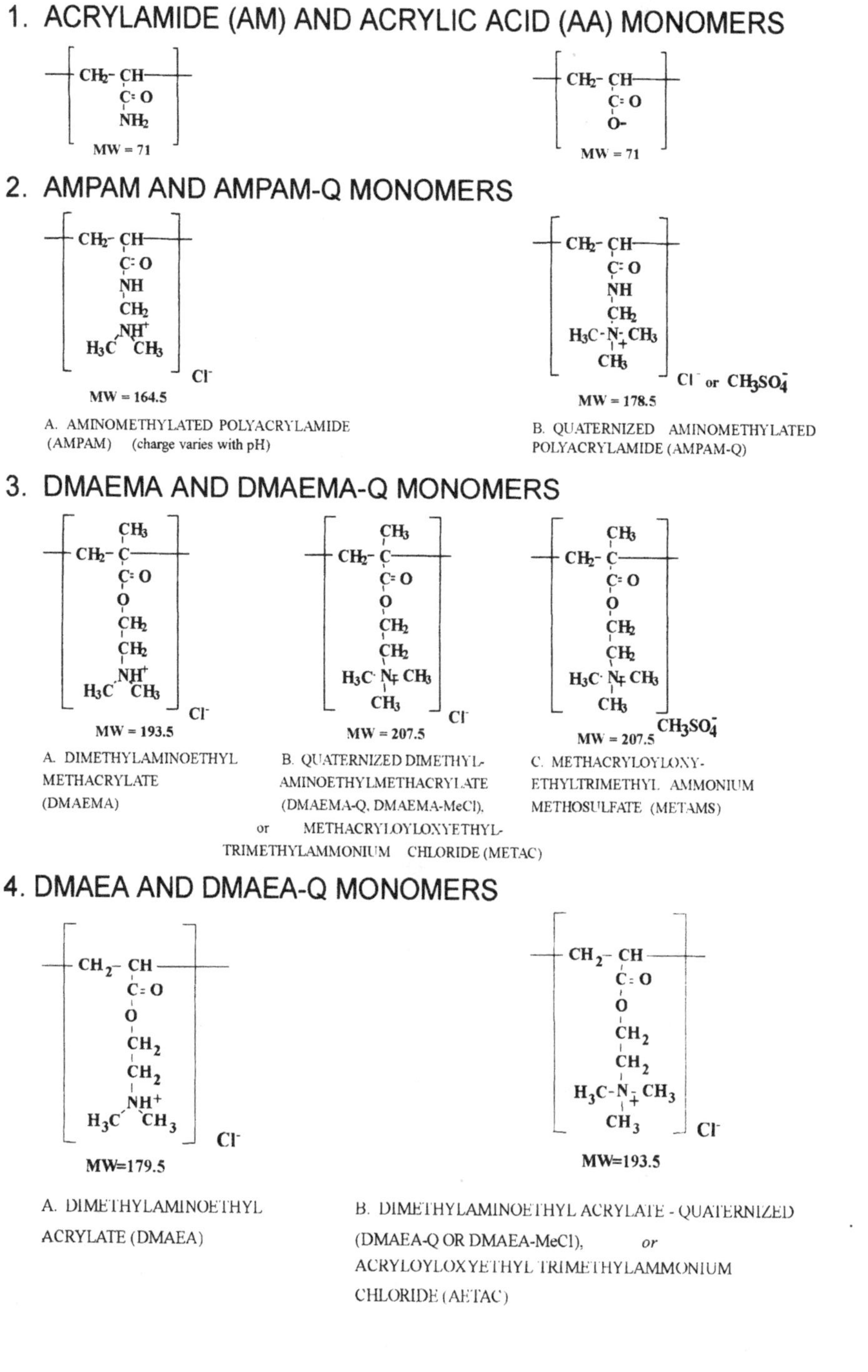

Fig. 1. Chemical structures of monomers used in flocculant polymers.

5. MAPTAC MONOMER

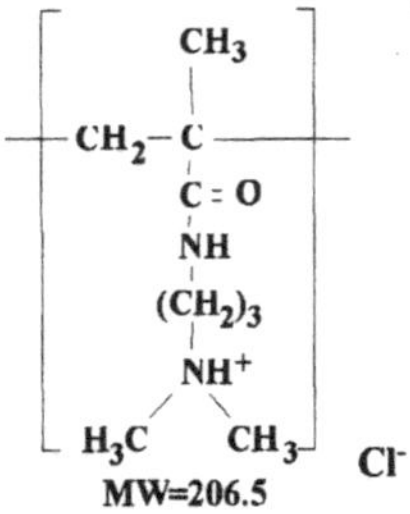

A. DIMETHYLAMINOPROPYL METHACRYLAMIDE

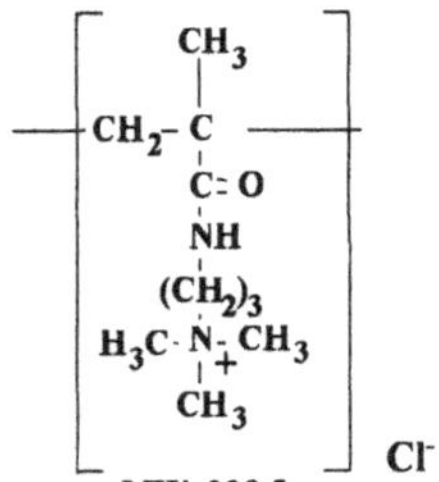

B. METHACRYLAMIDOPROPYL TRIMETHYL AMMONIUM CHLORIDE (MAPTAC)

6. DADMAC MONOMER

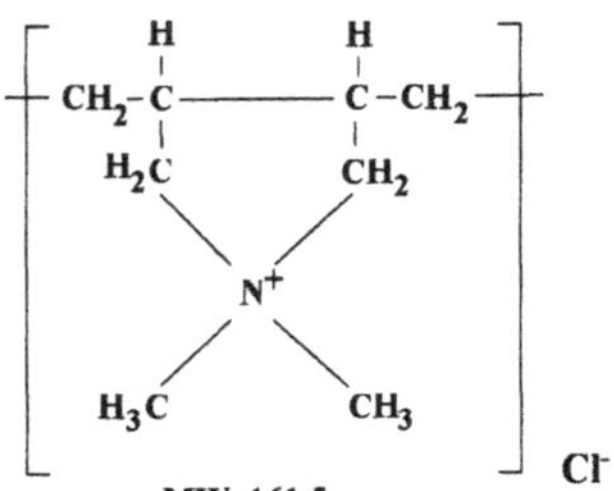

Fig. 1. (continued) Chemical structures of monomers used in flocculant polymers

Polymer Structure and Function in Flocculation Processes

If the charge density and molecular weight were the only important parameters, then there would not be so many polymer types in use. It is therefore useful to consider what other performance characteristics might be affected by the monomer type used and by other aspects of polymer product formulation. In this paper, some of these characteristics are related to experimental results. In some cases, only conjectural information is available (e.g. from issued patents) due to the proprietary value of the information.

Consider that two polymers with the same mole % monomer composition, and the same average molecular weight, may differ in numerous other respects:

- Monomer type: as indicated in the previous section, a variety of monomers are used in formulating different polymers that may have the same overall charge;
- Monomer distribution: for example, a 50/50 mole % composition could consist of many polymer molecules, each having 50% of each monomer, but could also consist of two homopolymers combined at equal concentrations;
- Monomer location: a 50/50 mole % product could have alternating polymers of two types, or most of the like monomers clustered together on the polymer chain;

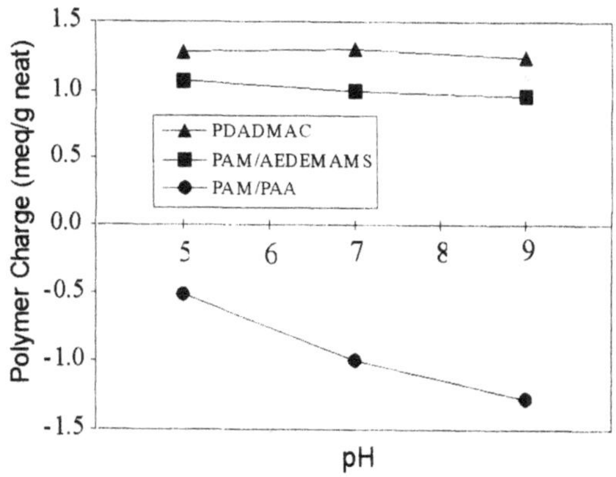

Fig. 2. Examples of polymer charge as a function of pH as determined by colloid titration. PDADMAC is pH-invariant; PAM/AEDEMAMS, a copolymer of acrylamide (90%) and of acryloyloxyethylmethyldiethylammonium methyl chloride (10%), loses a small amount of charge at high pH. The PAM/PAA's anionic charge is highly dependent on pH.

- Molecular weight distribution: the molecular weights of all molecules of one polymer preparation may be very uniform, and another may have a wide variation, but both have the same average value;
- Degree of cross-linkage as opposed to a completely linear configuration;
- Other aspects of product formulation, such as additives affecting pourability and dispersibility of dry products, and surfactants enabling micro-emulsion product forms vs. emulsion [11].

The most important of the above factors will be considered in the remainder of this paper.

Monomer Type and Protonation/Deprotonation of Charged Sites

Even though two polymers may have the same amount of ionic charge at one pH, the charge levels of same monomer types are pH-dependent. As indicated previously, acrylic acid anionic monomers, and all of the non-quaternized cationic monomers, exhibit acid-base behavior. If the polymer charge is an important aspect of its functioning, it is important to know the pH range over which the charge is maintained. The transition between the charged and the uncharged species would occur over a 2-unit pH span for a monomeric acid/base pair, but for a polymer, this may be affected by the proximity of ionizable monomer groups on the chain, since charge repulsion inhibits the charging of neighboring sites. The solution ionic strength also affects the transition through electrical double-layer effects.

Figure 3 shows the typical ionic charge for three cationic acrylamide-based polymers. This indicates the distinction between a Mannich polymer, which deprotonates at high pH, and a comparable polymer that has been 85% quaternized. A copolymer that is nearly all quaternized, but has a 55 mole % charge content, is also shown.

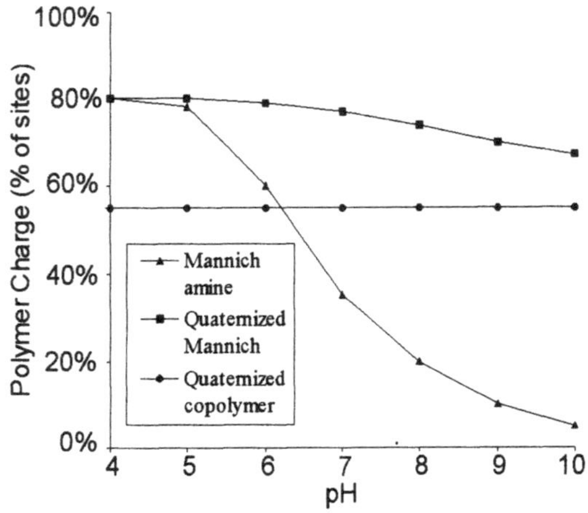

Fig. 3. Charge trends typifying various types of cationic polymers. Mannich polymer charge is highly pH dependent unless quaternized, and then slightly so. The quaternized copolymer maintains constant charge.

These differing behaviors as a function of pH may be important if different sludges are to be conditioned at the same site or if significant pH fluctuations occur over time. For wastewater sludges, some amount of positive charge is required to accomplish adsorption whether it is for charge neutralization, bridging, or both.

Monomer Type and Product Properties

The chemical characteristics of monomers comprising a polymer affect more than the overall ionicity. For example, the ease with which a polymer may be dispersed in water will be affected by both the hydrophobicity of the pendant groups (the smaller functional chains extending from the main polymer "backbone") and the type of counter-ions associated with charged sites. Some monomers are more compatible with others in synthesizing polymers that have random monomer distributions along the backbone [9]. This, in turn, is likely to affect the overall polymer conformation and its proclivity for patching or bridging.

Hydrophobic segments of polymers may also play a role in either facilitating attachment to relatively hydrophobic colloidal surfaces, or in fostering interactions with hydrophobic molecules. For example, Zemaitaitiene et al. [12] report the removal of disperse (uncharged) dyes from water by appropriate doses of PDADMAC. A combination of hydrophobic and charge interactions seems likely when polymers form complexes with long chain organic acids and anionic surfactants [12,13], such that it may be possible to tailor polymers for specific interactions of this type.

Cross-Linkages and Polymer Function

In addition to the use of selected monomers, the synthesis of flocculant polymers can also be affected by the inclusion of small amounts of a branching agent. The branching agent may be characterized as a polyfunctional analog of the component monomers, such that incorporation in the polymer chain provides a branching site. Agents such as methylene-*bis*-acrylamide are commonly used (e.g., [14-16]). A chain transfer agent may also be included if the length of the overall structure is to be limited; this may be advantageous since extensive cross-linking can decrease solubility.

Figure 4 [16,17] indicates effects of cross-linkage of a 50 mol % PAM/PDADMAC copolymer, as indicated by filtration tests at the optimum dosage for each polymer formulation. The increased dewaterability is obvious. It is also suggested that there is an optimum amount of cross-linkage, and that it is particularly beneficial when larger proportions of WAS are present in a combined sludge.

The results in Figure 4 do not indicate dose or shear-related effects of cross-linkage. Figure 5 presents results using an AM/AETAC copolymer, Percol (Zetag) 778, in conventional linear form and as Percol 778FS, which has been highly cross-linked to create what is termed a "structured" polymer [14]. The CST measurements followed different degrees of mixing to simulate dewatering under varying amounts of shear [18]. These results show that the structured polymer requires a higher dose for optimum flocculation (due to its nonlinear configuration and concealment of some cationic sites), but that it imparts a significant improvement in floc strength. These types of cross-linked or structured polymers have proven to be quite successful in high-shear dewatering devices. Some suppliers are known to indicate the cross-linked forms with terms such as XL, FS, FL, and FLX [6].

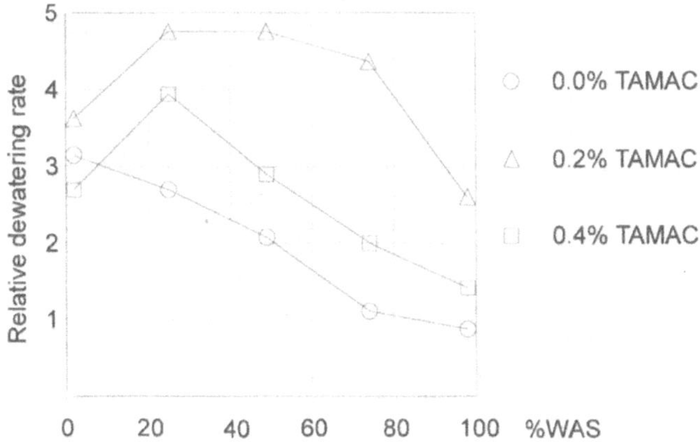

Fig. 4. Effect of adding cross-linking agent (triallylmethyl ammonium chloride, TAMAC) to a cationic copolymer. Results of Buchner funnel tests using combinations of primary/waste activated sludge (WAS) [16,17].

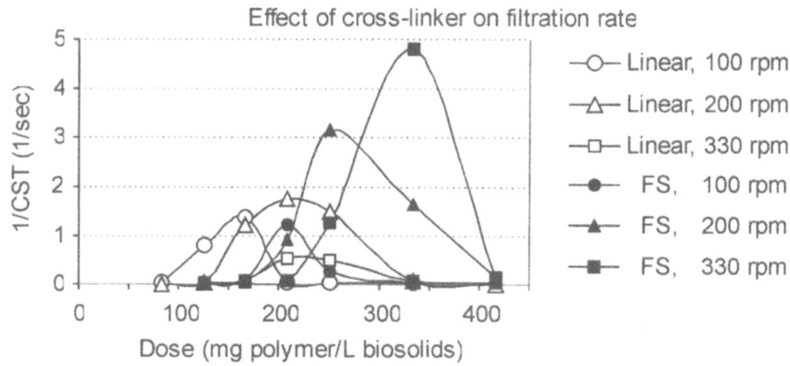

Fig. 5. Dewaterability of anaerobically digested biosolids from Wilmington, Delaware USA, various polymer dosages and extent of shear. Percol 778 (linear) and 778FS (structured) used as flocculants. Original sample at 2.78% total solids and 750 sec CST.

Polymer Structure and Post-Usage Effects

The chemical structures of flocculant polymers are important beyond their direct application for improvement of clarification, thickening, and dewatering processes. Since these additives are intended to act by adsorption onto particulate matter, their fate is ultimately to be disposed of with the sludge or biosolids. At typical doses, it is not unusual for them to comprise 1% of the sludge by dry weight fraction. In the U.S., this represents over 3 x 10^7 kg of polymer annually [19]. Thus it is pertinent to ask whether these chemicals have any direct or indirect environmental impacts, which requires that their chemical structures and interactions be examined.

It has recently been demonstrated that some portions of cationic co-polymers are biodegradable under both anaerobic and aerobic conditions [18]. For example, Figure 6 shows the results of anaerobically digesting a seeded polymer solution: significant gas production occurred. Interestingly, the gas production was less than would be observed if the polymer were completely degraded, suggesting a partial degradation. Figure 7 shows the deduced hydrolysis and cleavage of a cationic pendant group from an AM/AETAC copolymer. This reaction is likely to occur not only by enzymatic attack, but also at high pH [20], such as when biosolids are lime stabilized. The figure also shows deamination of the polyacrylamide, which can be enzymatically mediated [21]. Evidence also suggests that the cleaved pendant group (dimethylaminoethanol) can be broken down into the odorous trimethylamine [18]. Thus the following implications may be drawn:

- Use of polymers in which cationic groups are bound by ester linkages may exacerbate alkylamine odors and lead to loss of adsorptive capacity.
- Any polyacrylamide-based polymer can act as a source of mobile and biologically available nitrogen.
- Any cationic acrylamide-based polymer may eventually be transformed into PAM or PAA, which are not readily biodegraded.

It may be concluded that many polymer properties influence polymer performance in a treatment facility, but also determine its side-effects and environmental fate. There are many aspects of this that are poorly understood; this paper has explored several in which recent evidence has shed some light on a polymer's structure-behavior relationships.

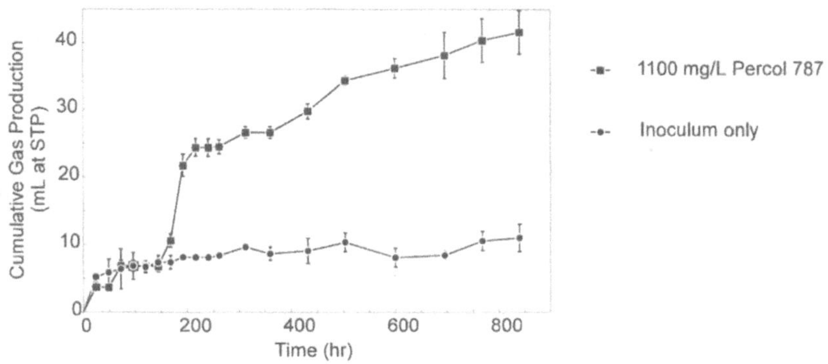

Fig. 6. Anaerobic digestion results for Percol 787, an AM/AETAC copolymer. Batch digestions utilizing the serum bottle method [19]

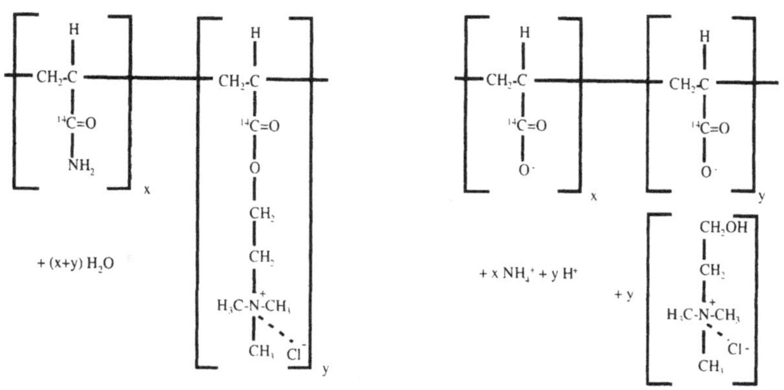

Fig. 7. Probably reactions partially degrading copolymers of AM and AETAC

References

1. Hiemenz, P.C., Rajagopalan, R.: Principles of Colloid and Surface Chemistry. Dekker, New York (1997)
2. Dentel, S.K., Abu-Orf, M.M., Griskowitz, N.J.: Guidance Manual for Polymer Selection in Wastewater Treatment Plants. Water Envt. Research Foundation, (1993) Alexandria, Va.
3. Terayama, H.: Method of colloid titration (a new titration between polymer ions). J. Polymer Sci. 7 (1952) 243-253

4. Mabire, F., Audebert, R., Quivoron, C.: Flocculation properties of some water-soluble copolymers toward silica suspensions: a semiquantitative interpretation of the role of molecular weight and cationicity through a "patchwork" model. Jour. Coll. Int. Sci. 97 (1984) 120-136
5. Lindquist, C.M., Stratton, R.A.: The role of polyelectrolyte charge density and molecular weight on the adsorption and flocculation of colloidal silica with polyethylenimine. J. Colloid Interface Sci. 55 (1976) 45-59
6. Dentel, S.K.: Conditioning. In: From Sludge to Biosolids, Spinosa, L, Vesilind, PA (eds) London: IWA (2000)
7. Barvenik, F.W.: Polyacrylamide characteristics related to soil applications. Soil Science 158 (1994) 235-243
8. Gucciardi, B.M.: The development, description, and application of practical methods for characterizing organic polyelectrolytes used in water treatment. Master's thesis, University of Delaware(1990)
9. Mortimer, D.A.: Synthetic polyelectrolytes – a review. Polymer International 25 (1991) 29-41.
10. Murray, P.G., Whipple, W.L.: Flourescent water soluble polymers. U.S. Patent 5,986,030 assigned to Nalco Chemical Company (1999)
11. Davies, W.B., Healy, J.E., Miller, G.K.L., Kokakiewicz, J.J., Ryles, R.G.: Spray drying of polymer-containing dispersions, water-in-oil emulsions and water-in-oil microemulsions, and dry polymer products formed thereby. U.S. patent 6,011,089 assigned to Cytec Technology Corp (2000)
12. Zemaitaitiene, R., Barkauskaite, J., Mazoniene, E., Zliobaite, E.: Cationic polymers and their complexes in wastewater treatment. In: Chemical water and wastewater treatment V (Hahn, H.H., Hoffmann, E., and Ødegaard, H., eds). Springer, Berlin Heidelberg New York (1998)
13. DiVincenzo, J.P., Dentel, S.K.: Sorption-desorption of 1,2,4-trichlorobenzene on soil: anionic surfactant and cationic polyelectrolyte effects. Journal Environmental Quality 25 (1996) 1193-1202.
14. Flesher, P., Farrer, D., Field, J.R.: Flocculation processes. U.S. Patent 4,943,378 assigned to Allied Colloids Ltd (1990)
15. Neff, R.E., Pellon, J.J., Ryles, R.G.: High performance cationic polymer flocculating agents. U.S. Patent 5,945,494 assigned to Cytec Technology Corp. (1999)
16. Morgan, J.E., Yorke, M.A., Boothe, J.E.: How cationic polymer structure relates to dewatering efficiency of waste-activated sludges. In: Ions in polymers (A. Eisenberg, ed.) Advances in Chemistry Series No. 187, American Chemical Society.
17. Rey, P.A.,Varsanik, R.G.: Application and function of synthetic polymeric flocculants in wastewater treatment. In: Water soluble polymers (J.E. Glass, ed.) Advances in Chemistry Series No. 213, American Chemical Society. (1986)
18. Baskerville, R.C., Bruce, A.M., Day, M.C.: Laboratory techniques for predicting and evaluating the performance of a filterbelt press. Filtration & Separation 15 (1978) 447-454
19. Dentel, S.K., Chang, L.-L., Raudenbush, D. R., Junnier, R.W., Abu-Orf, M.M.: Influence of polymer chemistry on biosolids and the environment. Water Envt. Research Foundation (2000)Alexandria, Va.
20. Aksberg, R., Wagberg, L.: Hydrolysis of cationic polyacrylamides. Journal of Applied Polymer Science 38 (1989) 297-304
21. Grula, M.M., Huang, M., Sewell, G.: Interactions of certain polyacrylamides with soil bacteria. Soil Science 158 (1994) 291-300

Floc Characteristics

Investigating the Chemistry of Aluminium-Based Coagulants from Ga K-Edge Absorption Spectroscopy

L.J. Michot*, E. Montargès-Pelletier*, B.S. Lartiges*, V. Kazpard*,
J.B. d'Espinose de la Caillerie**, V. Briois***

*Laboratoire Environnement et Minéralurgie INPL-ENSG-CNRS UMR 7569, BP 40, 54501 Vandoeuvre Cedex, France.
laurent.michot@ensg,inpl-nancy.fr.
**Laboratoire de Physique Quantique CNRS ESA 7069, ESPCI, 10 rue Vauquelin, 75231 Paris Cedex 05, France
***Laboratoire pour l'Utilisation du Rayonnement Electromagnétique Bât 209D Centre Universitaire Paris-Sud B.P. 34 - 91898 Orsay Cedex, France

Abstract

Using gallium salts, it is possible to combine ^{71}Ga NMR and Ga K-edge x-ray absorption spectroscopy for determining the speciation of the solution along the hydrolytic pathway. Using such experimental methods, we can propose a logical sequence of oligomers leading to the formation of Keggin-type polycations. The data obtained make it possible to test the coagulating efficiency of gallium salts with increasing hydrolysis ratios while knowing precisely the nature of the active species in solution. In view of the analogies observed between the aqueous behavior of aluminum and gallium, such a fundamental study could provide useful insight leading to a better control of aluminum salts based coagulation processes in drinking water treatment operations.

Introduction

Aluminium salts are the most widely used coagulants in drinking water treatment. Upon addition to the raw water, the aluminium salt undergoes dissociation, hydrolysis, and polymerisation, to yield the actual coagulant species. A clear understanding of the coagulation-flocculation mechanism can only be achieved if the active species are precisely identified. In the case of aluminium salts, ^{27}Al NMR and small angle x-ray scattering investigations of aluminium hydrolysis [1,2] have revealed that, in a given pH/concentration range, base-hydrolysed solutions of aluminium chloride and aluminium nitrate contain a Keggin-type polycation formed with 12 aluminium atoms in octahedral co-ordination surrounding a central aluminium tetrahedron. For higher hydrolysis ratios, these polycations aggregate as fractal inorganic polymers of increasing fractal dimension [3], before the formation of aluminium hydroxide.

Although there is a general agreement about this reaction scheme, many details remain unclear. The nature of precursors leading to the formation of the polycation is still a matter of controversy, the binding of Al_{13} units within coagulant species is not known, and the actual transformation mechanisms from Al_{13} fractal aggregates to the hydroxide remains unexplained.

Solution NMR is very sensitive to changes in the co-ordination of the investigated nucleus and has provided univocal evidence for the existence of the polycationic structure [1-3]. Unfortunately, due to its nuclear properties, aluminium exhibits rather broad NMR resonances. The octahedral signal can then not be directly assigned to various oligomeric species. On the other hand, as shown by recent studies on the hydrolytic chemistry of chromium [4], iron [5], lanthanum [6], titanium [7], extended x-ray absorption fine structure (EXAFS) spectroscopy is extremely well suited to analyse the nature of species in solution and their evolution from the first hydrolysis stages, but is much less sensitive to changes in the co-ordination of the atom. Aluminium is a light atom, however, so it is very difficult to analyse by EXAFS. To circumvent this difficulty, it is possible to use the proven analogies between aluminium and gallium chemistry. Indeed, ^{71}Ga NMR and potentiometric studies [8,9] showed that aluminium and gallium hydrolytic chemistry were parallel and that a Ga_{13} polycation structurally analogous to the Al_{13} entity was formed.

The aim of the present paper is therefore to illustrate how the combination of ^{71}Ga NMR and Ga K-edge EXAFS measurements can be used to better understand the hydrolytic chemistry of aluminium-based coagulants.

Materials and Methods

Sample preparation

Gallium salts used in this study were high purity anhydrous GaCl3 (99.99%+) and hydrated $Ga(NO_3)_3$ (99.99 %+ H_2O 21wt %), purchased from Aldrich Chemicals. Stock solutions of 0.6 M gallium salts were obtained by dilution, using ultrapure water (Millipore).

As the synthesis conditions have a strong influence on the nature and relative concentrations of the species formed [10,11], hydrolysis reactions were carried out under carefully controlled conditions. A small polymethacrylate reactor of standard geometry [12] was built. The solutions were stirred using a rectangular helix turning at 500 rpm. 50 ml of a NaOH (Normapur Prolabo) solution at various concentrations was added to 50 ml of the stirred gallium solution at a rate of 50 ml/hour using a syringe pump (Harvard apparatus type 22). After base addition, the solution was stirred for an additional 45 minutes before collection for further analysis. The pH was recorded after each run to draw a rough titration curve.

NMR experiments

Nuclear magnetic resonance experiments were performed on a Bruker ASX500 spectrometer at 11.7 T. In order to obtain quantitative measurements, weighted amounts of solution were placed in a 5 mm diameter glass tube of 1 cm length occupying only the middle part of the radio frequency coil of a Bruker static probe with a high quality factor. ^{71}Ga one-pulse experiments were performed with a $\pi/4$ pulse duration of 1.7 μs and a recycle time of 1 s. Between 2000 and 8000 acquisitions were performed depending on the intensity of the signal. A line broadening of 100 Hz was applied and the spectra were referenced to a 1 M aqueous solution of $Ga(NO_3)_3$. The resulting NMR spectra were fitted using pure lorentzian curves.

EXAFS measurements

EXAFS measurements were carried out at the Laboratoire pour l'Utilisation du Rayonnement Electromagnétique (LURE, Orsay, France) on the D44 station of the DCI storage ring (1.85 GeV and 300 mA). X-ray absorption spectra were recorded in transmission at room temperature and ambient pressure around the Ga K edge (10367 eV) from 10250 to 11240 eV with 2 eV steps and 2 seconds collecting time. All solutions were placed in a special cell [13] with an adjustable path length.

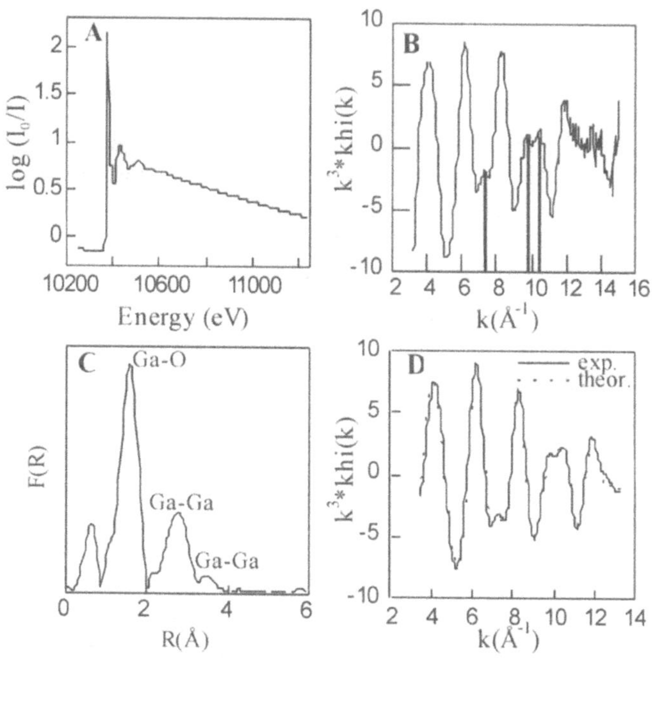

Fig. 1. EXAFS data analysis for $GaCl_3$, r = OH/Ga = 1.1. A: Raw absorption spectrum. B: Extracted EXAFS oscillations. C: Radial distribution function. D: Comparison between filtered experimental spectrum and theoretical spectrum

As described elsewhere [14], absorption spectra (Figure 1A) were reduced using standard procedures [15] with software written by Michalowicz [16] to yield the EXAFS signals (Figure 1B). A Kaiser window (3.5-14.7$Å^{-1}$) was used for deriving Fourier transforms from EXAFS spectra. The radial distribution functions (RDF, Figure 1C) obtained were not corrected for phase shifts, which lead to peaks shifted down by ≈ 0.3 Å compared with crystallographic distances.

For modeling experimental spectra, theoretical phase shifts and amplitude backscattering functions were calculated from the α-GaOOH structural model using FEFF 6 code [17]. Numbers and distances of nearest neighbors were derived from least-squares fitting. Fitting was first carried out on individual contributions (oxygen neighbors shell and gallium neighbor shells) and parameters were finally refined on the complete spectrum (Figure 1D). According to classically admitted values [15], the accuracy on distances is taken as ± 0.01Å and the accuracy on Debye-Waller factors and number of neighbors as ± 10%.

Coagulation Experiments

Silica suspensions were prepared from a concentrated silica sol, Ludox HS 40, supplied by SPCI (France), consisting of spherical particles with a mean diameter of 14 nm. The stock silica suspension was diluted to 300 ppm with deionized water (MilliQ), and a concentration of 336 ppm $NaHCO_3$ was introduced to provide ionic strength buffer similar to that of typical river water. pH was adjusted to 6.0 using 0.5 M HCl.

The flocculation experiments were conducted at room temperature in a standard 1L glass beaker fitted with 4 Plexiglas baffles. Stirring was carried out using a rectangular paddle located at one-third of the beaker height from the bottom. The required amount of coagulant was added with a microsyringe at a point just below the free surface of the suspension. The mixing of coagulant solution and silica suspension was carried out for 20 minutes at 100 rpm. The resulting flocculated suspension was poured into an Imhoff cone. The turbidity of the supernatant and floc volume were measured after 24h settling.

Results and discussion

NMR results

The NMR spectra of $GaCl_3$ solutions hydrolyzed at selected r values are presented in Figure 2A. For r ≥ 1.3, a signal around 173 ppm appears. This was previously assigned [8] to a gallium atom in a distorted tetrahedral environment corresponding to the central atom of a Ga_{13} polycation.

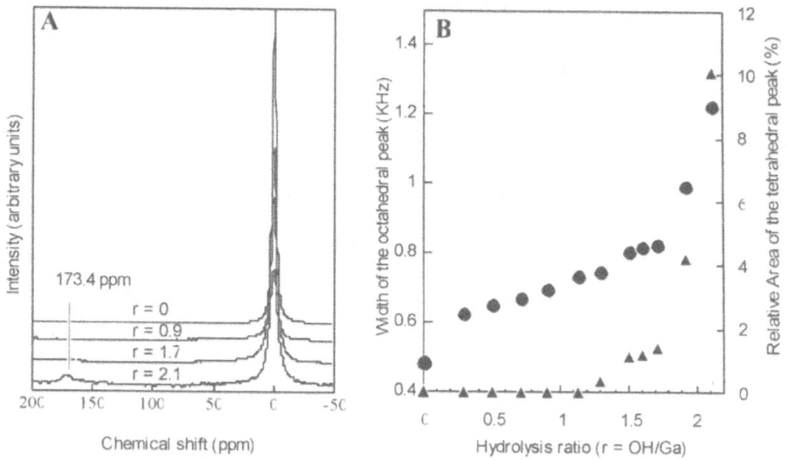

Fig. 2. ^{71}Ga NMR results. A: Raw spectra. B: Evolution of the width of the octahedral peak (●) and of the relative area of the tetrahedral signal (▲) with r = OH/Ga

For r = 2.0 and 2.2, the Ga^{VI}/Ga^{IV} ratios (Figure 2B) are close to 12, which suggests that the Ga_{13} polycation is the main species in solution. For r values ≤ 1.3 the width of the octahedral signal increases, which reveals the formation of hydrolyzed species, whose nature cannot be determined by NMR.

EXAFS results

With increasing hydrolysis ratio, the EXAFS signals show new resonances appearing around k = 7.2, 9.7 and 10.5 $Å^{-1}$ [18]. These changes in the EXAFS spectra are revealed in the radial distribution functions (RDF) (uncorrected for phase shifts). The first peak corresponds to oxygen neighbors and changes very little with increasing hydrolysis ratio. A set of peaks at greater distances, which can be assigned mainly to gallium-gallium contributions, grows as hydrolysis progresses.

The simulation of experimental spectra yields the following information:

- For r ≤ 1.8, the first coordination shell consists of 6 oxygens at 1.95 Å in agreement with the dominant octahedral coordination and literature data [19].
- For r = 2.0 or 2.2, better modeling results are obtained by splitting the 6 oxygens into 4 neighbors at 1.92 ± 0.01 Å and 2 neighbors at 2.03 ± 0.01 Å. This two-shell modeling fits the reported structure of Al_{13} sulfates crystals [20].
- For the second coordination shell, the values obtained suggest the presence of three atomic shells of gallium neighbors: a first shell Ga-Ga_1 at 3.01-3.05 Å, a second one Ga-Ga_2 at 3.47-3.53 Å and a third one Ga-Ga_3 at 3.90-3.95 Å. The first distance corresponds to edge sharing gallium octahedra [18]. The second one around 3.5 Å could be assigned either to single corner sharing between a gallium tetrahedron and a gallium octahedron in the Ga_{13} structure, or to double

corner sharing gallium octahedra [18]. The 3.9 Å distance can be assigned to single corner sharing gallium octahedra and to multiple scattering effects.

Figure 3 shows the evolution of the number of neighbors corresponding to these three distances as a function of the hydrolysis ratio.

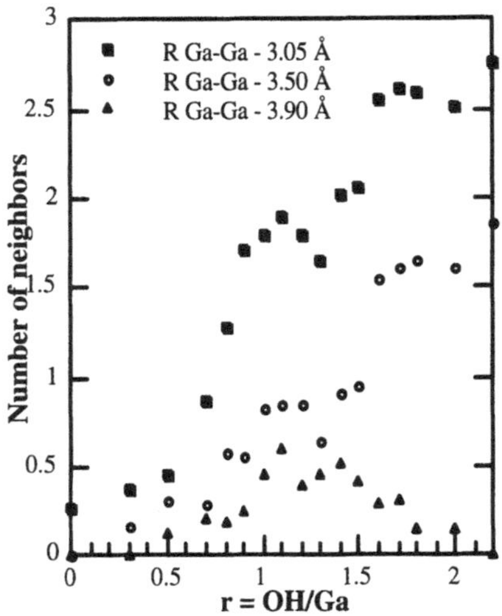

Fig. 3. Evolution of the number of gallium neighbors in the different shells of hydrolyzed gallium chloride as a function of the hydrolysis ratio r = OH/Ga

The number of neighbors at 3.05 Å (edge sharing octahedra) exhibits two strong increases for $0.6 \leq r \leq 0.9$ and for $1.3 \leq r \leq 1.7$. The number of neighbors at 3.50 Å (double corner sharing octahedra or single corner sharing between Ga^{VI} and Ga^{IV}) increases more or less linearly up to r = 1.5 and then increases steeply. The number of neighbors at 3.9 Å increases strongly for $0.9 \leq r \leq 1.1$, remains approximately constant up to r = 1.4 and then decreases regularly down to 0 at r = 2.2. The total number of gallium neighbors quickly becomes greater than 2, which suggests that the oligomeric species formed become rapidly at least trimeric even in the first hydrolysis stages. The evolution for $r \geq 1.4$ can certainly be assigned to the formation of increasing amounts of the Ga_{13} polycation as revealed by ^{71}Ga NMR experiments that showed that this species appears for $r \approx 1.3$ and increases thereafter. It must be pointed out that for r = 2.2 the Ga EXAFS spectra can be simulated by taking into account only this species. This suggests that for a total gallium concentration of 0.3 M, at a hydrolysis ratio r = OH/Ga = 2.2, nearly 100% of the gallium atoms in the chloride solution are present as the Ga_{13} polycation.

According to the previous analysis, the nature of the oligomeric precursors present in the solution before the formation of Ga_{13} polycations can be approached from a careful study of the r domain before 1.3.

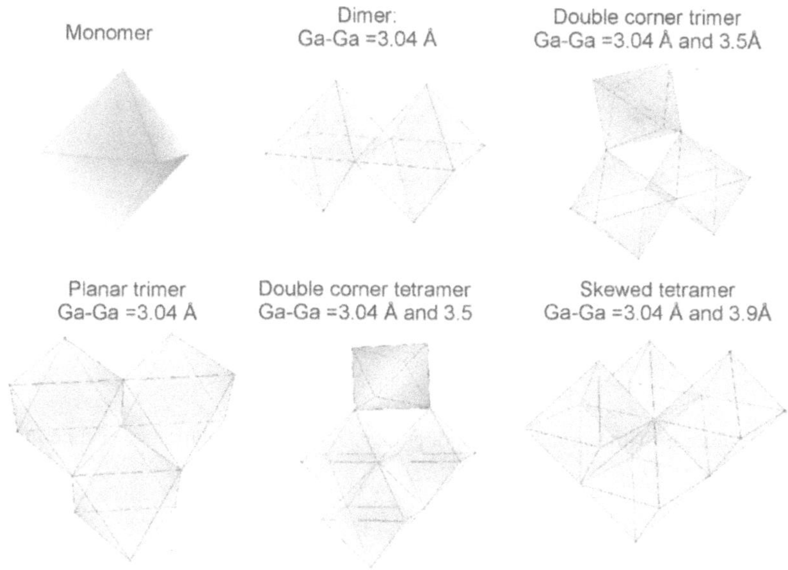

Fig. 4. Structure of the oligomers considered

The oligomeric species we considered are presented in Figure 4. These species were selected on the basis of distances and number of neighbors as well as on literature data [18]. Using the Feff 7.0 software [21], the EXAFS spectra corresponding to each oligomer can be calculated. Simple combination of these calculated signals can then be used to reproduce the experimental EXAFS spectra for each r value < 1.3. A consistent picture of the evolution of the composition of hydrolyzed $GaCl_3$ solutions can then be obtained (Figure 5).

For r = 0, the solution already contains 20% of edge sharing dimers; trimers start being present in the solution in the very early stages of the hydrolytic pathway (r ≥ 0.3). Two types of trimers could be formed: a double corner one, explaining the observed Ga-Ga distance around 3.5 Å and a planar one explaining the strong increase in the number of neighbors at 3.05 Å. For r around 0.9, a double corner tetramer appears, which explains the concomitant increase in both the contributions at 3.04 and 3.5 Å. For r = 1.2, the solution contains only tetramers and trimers. Further hydrolysis leads to the formation of a Ga_{13} polycation by association between those species. The Keggin structure would then form by condensation of trimers and tetramers, which disagrees with all the previously published models that assume the existence of a monomer, either tetrahedral or octahedral.

With higher hydrolysis ratios, increasing amounts of Ga_{13} are formed, resulting in an increase of the contributions at 3.04 Å and 3.50 Å and a concomitant decrease of the other species. For r = 2.0 nearly all gallium atoms would then be engaged in Ga_{13} polycations.

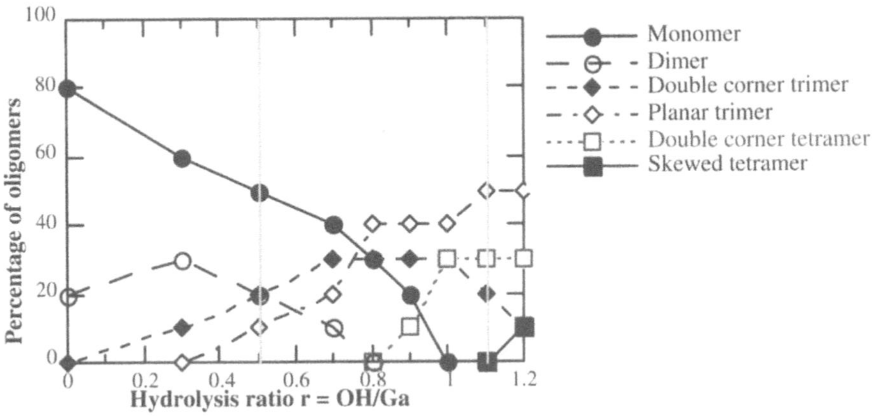

Fig. 5. Evolution of the proportion of oligomers in a $GaCl_3$ solution as a function of the hydrolysis ratio r=OH/Ga

Coagulation Experiments

Using the data from Figure 5, we were able to test the coagulating efficiency of hydrolysed gallium salt solutions while knowing the exact speciation of chemical species. For jar tests we chose three different hydrolysis ratios: r = 0.5, where the solution contains mainly trimers and monomers; r = 1.1, where it contains trimers and tetramers; and r = 2.0, where it contains nearly only Ga_{13} polycations.

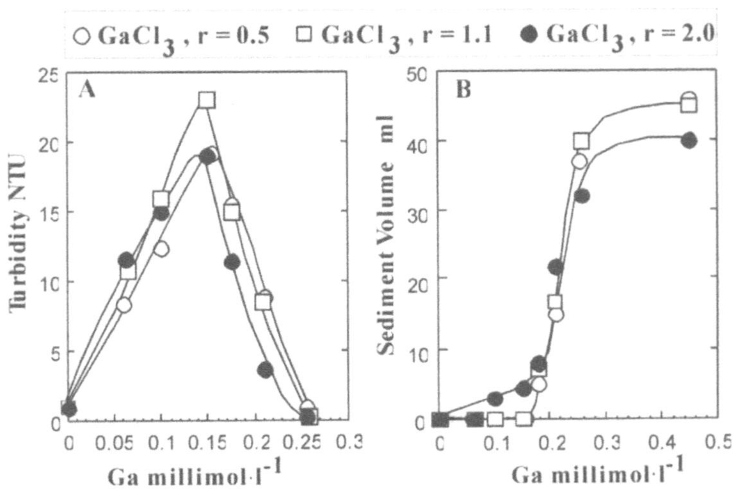

Fig. 6. Jar tests using $GaCl_3$ solutions at various hydrolysis ratios as coagulant

The turbidity evolution (Figure 6A) reveals that polycations were more efficient coagulants than smaller oligomers. In addition, the concomitant evolution of sediment volume shows that the flocs formed with Ga_{13} have a different behaviour. They are denser as they lead to a smaller floc volume at the optimum concentration (Figure 6B). Investigation of the Al_{13}/ludox HS system [22] has revealed complex destabilisation mechanisms involving the formation of negatively charged aluminosilicate sites and subsequent bridging of the particles with the coagulant species. Such a complex mechanism could also be invoked in the hydrolysed Ga/silica colloid system. It is then likely that the proportion of gallium species attached to the silica surface and within bridges depends on the hydrolysis ratio.

Conclusion and Perspectives

This study shows the potential of combining NMR and EXAFS for determining the chemical nature of species in hydrolysed $GaCl_3$ solutions. A parallel investigation on $Ga(NO_3)_3$ has revealed a similar evolution [18]. In view of the analogies observed between the aqueous behavior of aluminum and gallium, this approach can be used to deal with other systems relevant to water clarification. This could lead to clarify the hydrolytic pathway of $Ga_2(SO_4)_3$ as well as the hydrolysis of gallium salts in the presence of organic ligands. As shown from preliminary jar test results, knowing the nature of coagulating species should improve the understanding of destabilisation mechanisms involving hydrolysing metal salts.

References

1. Akitt, J.W., Farthing A.: New ^{27}Al nuclear magnetic resonance studies of the hydrolysis of the aluminium(III) cation. J. Magn. Reson. 32 (1978) 345
2. Bottero, J.Y., Cases, J.M., Fiessinger, F., Poirier, J.E.: Studies of hydrolysed aluminium chloride solutions. 1. Nature of aluminum species and composition of aqueous solutions. J. Phys. Chem. 84 (1980) 2933
3. Bottero, J.Y., Axelos, M., Tchoubar, D., Cases, J.M., Fripiat, J.J., Fiessinger, F.: Mechanisms of formation of aluminum trihydroxide from Keggin Al_{13} polymers. J. Colloid Interf. Sci. 117 (1987) 47
4. Roussel, H.: Ordre cationique dans des hydroxydes doubles lamellaires et études de la formation de la phase [Zn-Cr-Cl]. Thèse, Université de Clermont Ferrand. (1999)
5. Rose, J., Flank, A.M., Masion, A., Bottero, J.Y., Elmerich, P.: Structure and mechanisms of formation of $FeOOH(NO_3)$ oligomers in the early stages of hydrolysis. Langmuir 13 (1997) 1827
6. Ali, F., Chadwick, A.V., Smith, M.E.: EXAFS analysis of the structural evolution of gel-forms La_2O_3. J. Mater. Chem. 7 (1997) 285
7. Chemseddine, A., Moritz, T.: Nanostructuring titania: control over nanocrystal structure, size, shape, and organization. Eur. J. Inorg. Chem. 2 (1999) 235
8. Bradley, S.M., Kydd, R.A. and Yamdagni, R.: Detection of a new polymeric species formed through the hydrolysis of gallium(III) salt solutions. J. Chem. Soc. Dalton Trans. (1990) 413

9. Bradley, S.M., Kydd, R.A., Yamdagni, R.: Comparison of the hydrolysis of gallium(III) and aluminum(III) solutions by nuclear magnetic resonance spectroscopy. J. Chem. Soc. Dalton Trans. (1990) 2653
10. Kloprogge, J.T., Seykens, D., Jansen, J.B.H., Geus, J.W.: ^{27}Al nuclear magnetic resonance study on the optimization of the development of the Al_{13} polymer. J. Non-Cryst. Solids 142 (1992) 94
11. Morgado, Jr, E., Lam, Y.L., Nazar, L.F.: Formation of peptizable boehmites by hydrolysis of aluminum nitrate in aqueous solution. J. Colloid Interf. Sci. 188 (1997) 257
12. Holland, F.A.and Chapman, F.S.: Liquid mixing and processing in stirred tanks. Rheinhold Publishing, New-York (1966)
13. Villain, F., Briois, V., Castro, I., Helary, C., Verdaguer, M.: Multipurpose x-ray absorption cell. Anal. Chem. 65 (1993) 2545
14. Montargès, E., Michot, L.J., Ildefonse, P.: Ga_{13}-PEO-modified smectites: relationship between physico-chemical properties and local pillar structure. Microporous and Mesoporous Materials 28 (1999) 83
15. Teo B. K.: EXAFS: Basic principles and data analysis. In: Inorganic Chemistry Concepts, Vol.9, Springer-Verlag (1986)
16. Michalowicz A.: Méthodes et programmes d'analyse des spectres d'absorption des rayons X (EXAFS). Applications à l'étude de l'ordre local et du désordre cristallin dans les matériaux inorganiques. PhD Thesis, Université Paris Val de Marne (1990)
17. Mustre, J., Rehr, J.J., Zabinsky S.I., Albers, R.C.: Ab initio curve-wave x-ray absorption fine structure. Phys. Rev. B, 44 (1991) 4146
18. Michot, L.J., Montargès-Pelletier, E., Lartiges, B.S., D'Espinose de la Caillerie, J.B., Briois, V.: Formation mechanism of the Ga_{13} polycation: a combined NMR and EXAFS study. J. Am. Chem. Soc. (In press). (2000)
19. Lindqvist-Reis, P., Muñoz-Páez, A., Diaz-Moreno, S., Pattanaik, S., Persson, I., Sandström, M.: The structure of the hydrated gallium(III), indium(III), and chromium(III) ions in aqueous solution. A large angle x-ray scattering and EXAFS study. Inorg. Chem. 37 (1998) 6675
20. Johansson, G.: On the crystal structure of some basic aluminum salts. Acta Chem. Scand., Short communications 14 (1960) 771
21. Zabinsky, S.I., Rehr, J.J., Ankudinov, A., Albers, R.C., Eller, M.J.: Multiple-scattering calculations of x-ray-absorption spectra. Phys. Rev. B 52 (1995) 2995
22. Lartiges, B.S., Bottero, J.Y., Derrendinger, L.S., Humbert, B., Tekely, P., Suty, H.: Flocculation of colloidal silica with hydrolyzed aluminum: an ^{27}Al solid state NMR investigation. Langmuir 13 (1997) 147

Monitoring Flocs Produced by Water Treatment Coagulants

J. Gregory*, L. Rossi** and L. Bonechi**

*Department of Civil and Environmental Engineering, University College London, Gower Street, London WC1E 6BT, UK
j.gregory@ucl.ac.uk
**Dipartimento di Inginegneria Civile, Universita degli Studi di Firenze, 50139 Firenze, Italy

Abstract

Pre-polymerized inorganic coagulants, such as polyaluminum chloride (PAC), are claimed to have several advantages over conventional aluminum and iron salts. However, their mode of action is not fully understood and progress in this area has been made largely by empirical approaches. In this study we have followed the dynamics of flocculation of model clay/humic acid suspensions following the addition of different coagulants, including aluminum sulfate, ferric sulfate and two commercial PAC products. Experiments were conducted using controlled stirring rates in a modified jar test procedure, in which the formation, break-up and re-formation of flocs could be monitored continuously by a flow-through optical technique. The results show significant differences in the behavior of the coagulants, notably in the rate of floc formation and floc strength. In all cases, only very limited re-growth of flocs occurred after breakage. Further experiments along these lines should help to elucidate the mode of action of different forms of coagulant and to aid in their selection for specific applications.

Introduction

There is growing interest in the use of alternative coagulants to the traditional hydrolyzing metal salts used in water treatment. Among these, polymerized forms of aluminum and iron are now quite widely used [1]. Polyaluminum chloride (PAC) can be prepared by controlled neutralization and hydrolysis of aluminum chloride solutions and is commercially available from several sources. There are distinct benefits reported for the use of PAC over conventional coagulants such as aluminum sulfate. These include superior performance at low temperatures and reduced production of sludge.

However, it is not yet clear how the mechanisms of coagulation differ among different products. It is known that in many forms of PAC a predominant species is the so-called 'Al_{13}' form, $Al_{13}O_4(OH)_{24}^{7+}$, which has been identified by ^{27}Al NMR studies, among other techniques [2]. This species carries a high positive charge and is strongly adsorbed by negatively charged colloids. Most discussion of its mode of action has focused on charge neutralization as a destabilizing mechanism. However, the performance of traditional coagulants such as aluminum sulfate ('alum') is strongly related to the formation of a hydroxide precipitate, giving 'sweep flocculation' [3]. The role of precipitation in the action of the polymerized coagulants is not yet entirely clear.

As part of a comprehensive study of polymeric inorganic coagulants, we have carried out dynamic testing of some commercial PAC products and compared their behavior with that of aluminum sulfate under similar conditions. As reported previously [4] a simple flow-through optical monitoring technique gives detailed information on the growth and break-up of flocs in a modified jar test procedure. A useful comparison between the performance of different coagulants can be made, which gives some insight into their mode of action. Most previous work on polymerized inorganic coagulants has used a standard jar test approach, which gives only limited information. However, Matsui et al. [5] used a similar dynamic approach to the one used here, to compare the performance of alum and PAC.

Experimental

Materials

Suspension: Kaolin clay (Imerys, St Austell, Cornwall, UK) was used as a model suspension. 200 g of kaolin was dispersed in 500 mL of deionized water in a high-speed blender. To obtain full dispersion it was necessary to raise the pH of the suspension to about 7.5, which was achieved by adding 5 mL of N/10 NaOH. After blending at 4000 rpm for 20 minutes the clay suspension was diluted to 1 L with deionized water and allowed to stand overnight in a measuring cylinder. The top 500 mL was decanted and its solids content was determined gravimetrically and found to be 135 g/L. This was diluted to give a final solids content of 50 g/L. The particles were mostly below about 5 μm in size, with a mean size of about 2 μm, determined by an Elzone 280 PC particle counter.

The stock suspension was diluted in London tap water to give a clay concentration of 100 mg/L and a turbidity of about 160 NTU, as determined by a Hach Model 18900 turbidimeter. London tap water has high total hardness (ca. 280 mg/L as $CaCO_3$) and alkalinity (ca. 240 mg/L as $CaCO_3$) and a pH of around 7.4. This is a useful medium for studies with hydrolyzing coagulants, since there is a high buffer capacity and the pH is close to optimum for hydrolyzing coagulants. However, the high calcium content (around 2 mM Ca^{2+}) causes destabilization of the kaolin particles, which slowly coagulate. This would give problems in interpreting the results with hydrolyzing coagulants. To avoid this difficulty, a small amount of commercial humic acid (Aldrich) was added to the stock kaolin suspension. Humic acid adsorbs onto the clay particles and gives enhanced stability against divalent metal ions such as Ca^{2+}. Humic acid solution was included in the stock 50 g/L kaolin concentration to give a concentration of 1 g/L. Thus the diluted samples for all of the coagulation experiments contained 100 mg/L clay and 2 mg/L humic acid in London tap water.

Coagulants: The following coagulants were used:

- Aluminum and ferric sulfate (Fisons, UK), both prepared as M/10 solutions in deionized water.
- Two commercial polyaluminum chloride (PAC) products from Kemira Kemi AB, Helsingborg, Sweden, which had different degrees of neutralization, r (= OH/Al) and are designated here as:
  - PAC 1 (r = 2.1), supplied as an aqueous solution with 5.0 wt % Al
  - PAC 2 (r = 2.3), supplied as a 11.5 wt % Al solution.

These solutions were used directly, without prior dilution. (Dilution can have a significant deleterious effect on the performance of these products).

Methods

Apparatus: The test procedure was similar to that described previously using an on-line flocculation monitor (PDA 2000, Rank Brothers Ltd, UK) in a modified jar test procedure [4]. The test suspension was contained in a 500 mL square (6 × 6 cm) plastic bottle, with holes drilled in the top to take sample tubing. The suspension was stirred with two cross-shaped glass paddles on a glass rod. Each cross had four blades of about 2 × 2 cm. Stirring was by a motor with continuously adjustable speed (0-200 rpm). During a test, sample was circulated through transparent plastic tubing (3 mm i.d.) by means of a peristaltic pump. The tubing was clamped in the PDA 2000 instrument so that the flowing sample was illuminated by a narrow light beam (850 nm wavelength). The PDA measures the average transmitted light intensity (dc value) and the rms value of the fluctuating component. The ratio (rms/dc) provides a sensitive measure of particle aggregation. In this work, this ratio will be called the Flocculation Index (FI).

Procedure: 500 mL of test suspension (100 mg/L kaolin and 2 mg/L humic acid in London tap water) was placed in the flocculation vessel and stirred at the required speed. Sample was pumped at about 25 mL/min through the tubing and the average (dc) and fluctuating (rms) components of the transmitted light intensity were monitored by the PDA instrument. Readings were taken every two seconds and the results stored in a computer for subsequent spreadsheet analysis (Excel). After allowing time for steady readings to be established, coagulant at the required dosage was pipetted into the suspension. In the experiments reported here, coagulant was added at a certain stirring speed, which was then maintained for several minutes (i.e. without a separate rapid mix phase). After flocs had formed, the stirring speed was increased, so that some floc breakage occurred. In some cases, the stirring speed was then reduced in order to observe the re-formation of flocs. Details of the stirring regimes will be given with the appropriate results.

Results and Discussion

Effect of coagulant type and dosage

Flocculation results for various dosages of 'alum' are shown in Figure 1. Stirring conditions were the same in all cases. Coagulant was added while stirring at 100 rpm and this stirring rate was maintained for 180 s. After this time the stirring speed was increased to 200 rpm. These conditions were chosen on the basis of investigations of stirring rate effects (see later). At 100 rpm no significant decrease of floc size occurred over a period of three minutes. 100 rpm is also sufficiently fast to avoid the need for a separate 'rapid mixing' phase.

The curves in Figure 1 show characteristic features. With increasing alum dosage the rate of rise in the FI value becomes steeper and the plateau value increases. At around 2 mg/L (as Al) there is little further increase in these parameters, which are known to correlate well with residual turbidities in traditional jar test procedures [6] and with floc size [5] When the stirring rate is increased to 200 rpm there is an immediate and fairly rapid decrease in FI, as a result of floc break-up under the increased shear conditions.

Curves of similar form were obtained for the two PAC products, under the same stirring conditions, but with larger values of FI, especially in the case of PAC 2 [7]. It has been shown that during the floc growth phase, the rise in FI with time can be fitted to a sigmoid curve, from which certain parameters can be derived [6].

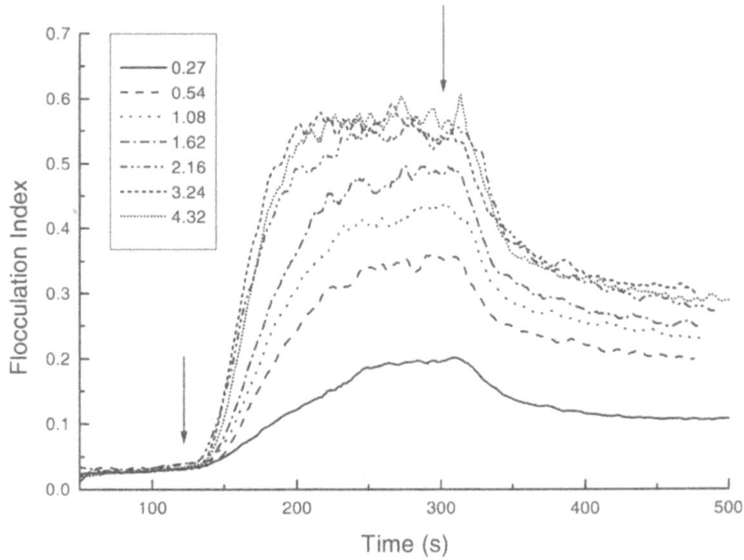

Fig. 1. Flocculation by 'alum' at a range of dosages (mg/L Al, as shown in the legend). The stirring rate was initially 100 rpm and then (at 300 s) it was increased to 200 rpm.

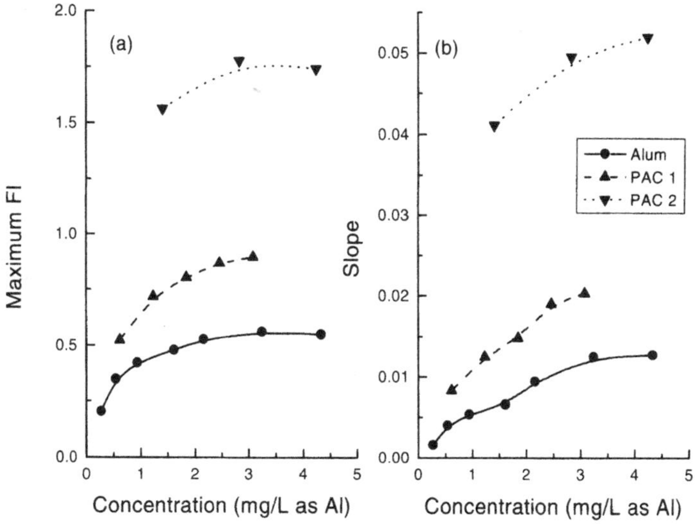

Fig. 2. Parameters derived by empirical fitting of FI results to a sigmoid curve, plotted as a function of coagulant dose. (a) Plateau value of Flocculation Index. (b) Rate of increase of FI. (See text)

The most important of these are the maximum (or plateau) FI value and the greatest rate of increase of this value (i.e., the maximum slope). These parameters are plotted against coagulant dosage for alum, PAC 1 and PAC 2 (all expressed as mg/L Al) in Figure 2. The plots show that the plateau value ('Maximum FI') and the greatest rate of increase in flocculation index ('Slope') give broadly similar patterns with changing coagulant dosage and coagulant type. Conventional 'alum' is seen to be the least effective and the PAC 2 sample is by far the best. The difference in the two commercial PAC products is striking in view of the relatively slight difference in their r values. It is known that increasing r (up to about 2.5) gives a higher proportion of the 'Al_{13}' species [2] and this may provide at least part of the explanation.

Effect of stirring conditions

Figure 3 shows the results of three flocculation experiments, with the same alum dosage (2.16 mg/L as Al), but different stirring conditions. In this case, the coagulant was added after 60 s, at stirring speeds of 50, 100 or 150 rpm. After 9 minutes, the stirring speed was increased to 200 rpm, and then, after a further 5 minutes, reduced to the original speed.

It is clear that floc growth is greatly influenced by the stirring speed during the first phase. At 50 rpm the FI rises to a higher value than at the more rapid stirring rates. At 150 rpm, initial floc growth is faster, but the maximum size reached is considerably less than at the lowest speed. It is well known that the maximum floc size attained in a stirred vessel is a decreasing function of the effective shear rate [8]. Furthermore, continued stirring at 150 rpm causes a noticeable decrease in FI and hence floc size. This type of behavior has been found for fairly concentrated suspensions, flocculated by polymers [9], although it is still not fully understood.

Another important feature of Figure 3 is that floc breakage at 200 rpm seems to give very similar FI values, which implies breakage to fragments of similar size, irrespective of conditions under which the flocs were formed. This suggests that a fairly simple procedure for floc strength evaluation could be developed in this way.

It is also apparent that, after returning the stirring speed to the original value, very little rise in the FI value occurs. The implication is that floc breakage is nearly irreversible. While this effect is well known and fairly well understood in flocculation by high molecular weight polymers, it is not immediately apparent why hydrolyzing coagulants should show the same behavior.

Figure 4 shows the results of similar trials with ferric sulfate as coagulant, at a dosage of 5.6 mg/L as Fe. The results are broadly similar to those with alum, but it is clear that the flocculation index reaches considerably higher values. Broadly the same floc breakage pattern and irreversibility of breakage were found.

Figures 5 and 6 show the results of similar experiments using the PAC coagulants. In these cases coagulant was added after two minutes of stirring at speeds from 50 to 200 rpm. After a further 7 minutes of stirring at the initial speed the rate was set at 200 rpm. Similar irreversibility of floc breakage was found with these coagulants, but the effect is not shown here. The results for PAC 1 (1.23 mg/L as Al) are in Figure 5 and those for PAC 2 (3.1 mg/L as Al) in Figure 6. Although the PAC 2 dosage is rather high, similar behavior was found for this product down to around 1 mg/L as Al.

The initial rate of increase in FI value is seen to be greater at the higher stirring speeds, but the maximum value attained is lower. At stirring speeds of about 100 rpm and above, FI shows a decline. Floc breakage at 200 rpm gives FI values that are nearly the same in all cases, as shown by the fact that all curves converge to the 200 rpm case.

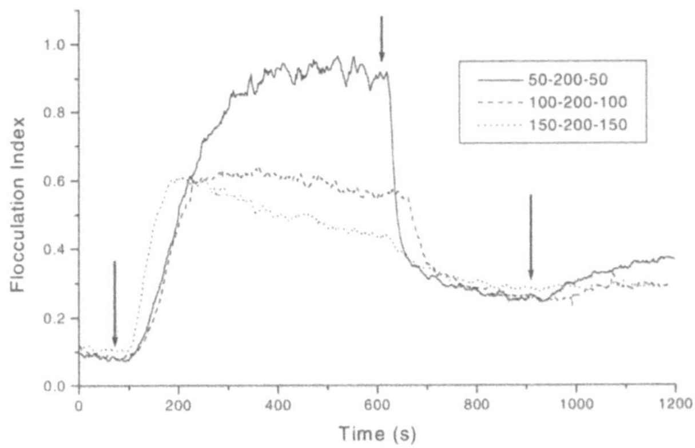

Fig. 3. Flocculation by alum (2.16 mg/L as Al) under different stirring conditions. Arrows indicate coagulant dosing and changes in stirring speed (sequence shown in the legend).

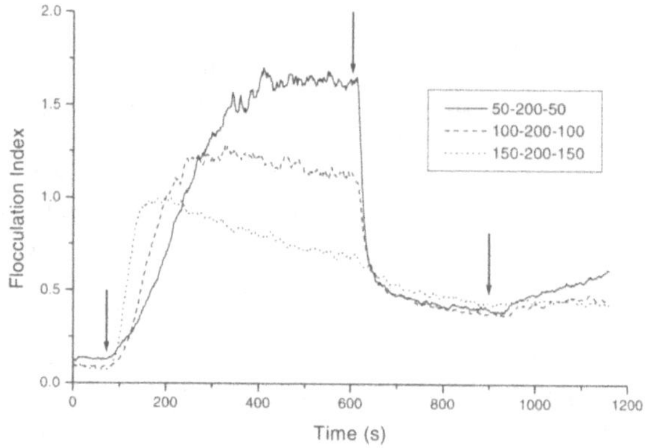

Fig. 4. As Figure 3, but with ferric sulfate as coagulant (5.6 mg/L as Fe)

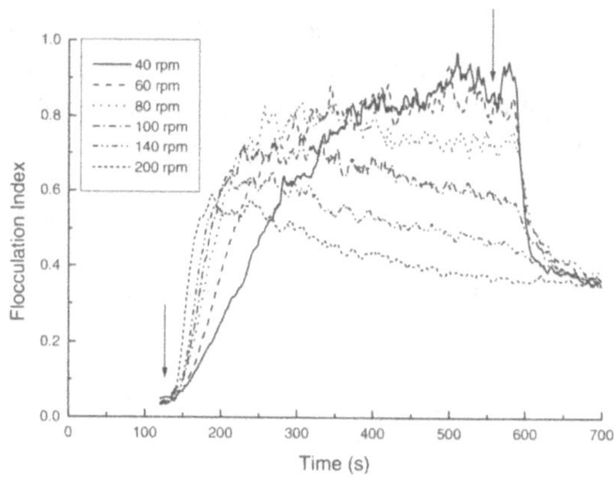

Fig. 5. Flocculation by PAC 1 at a dosage of 1.23 mg/L as Al and at different stirring speeds. Arrows indicate the point at which coagulant was added and where the stirring speed was set at 200 rpm.

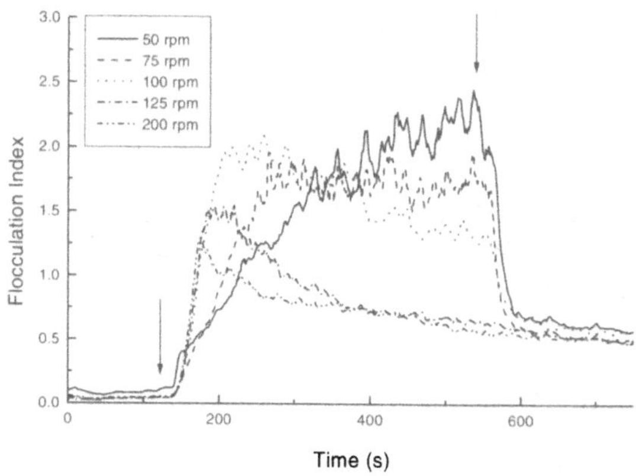

Fig. 6. As Figure 5, but with PAC 2 at a dosage of 3.1 mg/L as Al

Conclusions

The results presented here are part of a comprehensive investigation of the action of inorganic coagulants under controlled conditions, using on-line optical monitoring and a range of other techniques.

At fairly low stirring (or shear) rates, flocs grow at a certain rate and reach a limiting (plateau) size, both of which depend on the stirring rate and the type and dosage of coagulant. For higher stirring rates flocs can grow rapidly but may then show a reduction in size. By increasing the stirring rate after flocs have been formed at a lower rate, significant floc breakage occurs. For the present system it has been found that, with a given coagulant and dosage, flocs formed over a range of different stirring rates (and hence of different sizes) and then subjected to the same high stirring rate, show similar breakage behavior. This suggests that a fairly simple, standardized floc strength measurement might be developed.

With increasing coagulant dosage and fixed stirring conditions there is initially a systematic increase in the rate of floc growth, limiting floc size and floc strength. By analyzing the changing monitor response with a curve-fitting routine, it is possible to derive parameters that characterize the rate of floc formation and the ultimate size. These parameters allow realistic comparisons among different coagulants and dosages to be made. These show that conventional 'alum' is less effective than either of the two PAC products used. Furthermore, of the two PACs, the one with the slightly higher basicity ($r = 2.3$) gives markedly faster floc formation, leading to larger and stronger flocs.

This approach to flocculation monitoring gives much more detailed information than the more usual jar test procedure and is very useful in investigating the mechanism of coagulation/flocculation under typical treatment conditions. A better understanding of the mode of action of PAC and related products should enable a more rational basis for developing new materials and optimizing their use in water and wastewater treatment.

References

1. Jiang, J., Graham, N.J.D.: Pre-polymerised inorganic coagulants for treating water and wastewater. Chemistry and Industry (10) (1997) 388-391
2. Singhal, A., Keefer, K.D.: A study of aluminum speciation in aluminum chloride solutions by small angle x-ray scattering and ^{27}Al NMR. J. Mater. Res. **9**(8) (1994) 1973-1983
3. Snodgrass, W.J., Clark, M.M., O'Melia, C.R.: Particle formation and growth in dilute aluminum (III) solutions. Water Res. **18**(4) (1984) 479-488
4. Gregory, J., Duan, J.: The effect of dissolved silica on the action of hydrolysing metal coagulants. Wat.Sci.Tech. **38**(6) (1998) 113-120
5. Matsui, Y, Yuasa, A, Furuya, Y., Kamei, T.: Dynamic analysis of coagulation with alum and PACl. J. Am. Water Wks Assn. **90** (10) (1998) 96-106
6. Gregory, J., Hiller, N.: Interpretation of flocculation test data. Proc Filtech Europa/95, Karlsruhe (1995) pp 405-414
7. Gregory, J, Rossi, L.: Dynamic testing of water treatment coagulants. Water Sci. Tech., In press (2000)

8. Mühle, K.: Floc stability in laminar and turbulent flow. In Coagulation and Flocculation (Dobiás, B., ed.), Marcel Dekker, New York, pp. (1993) 355-390
9. Gregory, J., Li, G.: Effects of dosing and mixing conditions on polymer flocculation of concentrated suspensions. Chem. Eng. Comm. **108** (1991) 3-21

Floc Characterisation at Full Scale: The Case of Humic Waters

D.H. Bache*, E. Rasool*, D. Moffat* and M. Johnson**

*Department of Civil Engineering, University of Strathclyde, 107 Rottenrow, Glasgow G4 ONG, UK
d.bache@strath.ac.uk
**North West Water, Lingley Mere, Lingley Green Avenue, Great Sankey, Warrington, WA5 3LW, UK
mark@markj.demon.co.uk

Abstract

This paper describes part of a larger scale investigation designed to characterise humic flocs and focuses on those found in the flocculators at full scale. Following a section dealing with the basic parameterisation, the paper provides insight into the techniques which were employed. Insight is provided into some limitations of image analysis when attempting to measure floc size. It is shown that floc sizes measured in a sample stream withdrawn from a stirred vessel bares little resemblance to those measured in situ. From a survey carried at ten WTWs, it was apparent that the greatest influence on floc size and density was the presence of polymer. There was evidence that the presence of high alkalinity induced properties comparable with polymer addition. Little correspondence existed between floc sizes at full scale and those found in a laboratory scale mixer at identical values of velocity gradient. It was evident that DAF plants performed satisfactorily whatever the size of the incoming floc.

Introduction

Although the water industry has enormous experience of the operational aspects of coagulation at water treatment works (WTWs), there have been very few studies on the properties of the floc at full-scale. This is especially true of flocs derived from upland coloured waters. Alumino-humic are generally recognised as being small, weak and with low settling velocity. Such properties can be transformed by the addition of cationic polymers, thereby facilitating their removal during clarification and filtration. At a fundamental level, it is fair to state that the framework for describing of floc character is fairly rudimentary - tending to focus on size, density and strength. Factors, such as rheology and floc structure have received scant attention.

In the study reported in this paper, the principal aim was to gain knowledge of floc character under the conditions found at full scale. Such information is pertinant to the design and operation of clarifiers, e.g. DAF. A standardised procedure was developed to serve this purpose, a survey being carried out at ten WTWs receiving coloured, low turbidity waters. Humic flocs pose problems in measurement because of their character. Considerable effort was devoted to the development of techniques and instrumentation to allow progress.

Floc characterization

A necessary preamble is to define some of the parameters used in the survey. Size was generally expressed in terms of the parameters d_{l95} or d_{v95}. The term d_{l95} is the size below which there is 95% of the total cumulative diameter within a sample. Similarly, d_{v95} corresponds to the size below which there was 95% of the total cumulative volume within a sample.

The floc effective density (ρ_e) can be determined from its settling velocity [1]. Many studies e.g. [2,3] have shown that the size-density interdependence is well represented by the empirical relationship

$$\rho_e = Ad^{-n} \tag{1}$$

in which $\rho_e = \rho_f - \rho_w$, the terms ρ_f and ρ_w signifying the floc density and density of water respectively. Parameter A and n are fitting coefficients. Coefficient n is linked the fractal dimension $d_F = 3 - n$; the value of d_F provides insight into the packing of the solids within the floc [4].

Floc strength is generally estimated from the size of the largest flocs in a turbulent environment. When floc sizes are comparable with the smallest eddy size (represented by the Kolmogorov length, η), Bache et al [5] proposed that the strength (σ) could be estimated using the relationship

$$\sigma \sim \frac{\rho_w \varepsilon^{3/4} d}{\nu^{1/4}} \tag{2}$$

in which ε is the local rate of energy dissipation per unit mass and ν is the kinematic viscosity of the water. An immediate difficulty, is that it is not clear how to define d. Should this be linked to d_{l95} or to d_{v95} or indeed to some other reference size? Since d_{v95} and d_{l95} are different , the choice affects the calculated strength. The same dilemma arises when $d >> \eta$ or $d << \eta$ because the strength estimate based on hydrodynamic rupture is dependent on the floc size [6].

A less refined approach is to characterise strength on the basis of the behavioural pattern

$$d_{max} = CG^{-m} \tag{3}$$

in which G is the mean velocity gradient, m an index, and C a 'strength' coefficient [7]. The links between Eqs (1) and (3) were examined in [5].

From the above, it is apparent that the specification of floc character is intertwined with its size - a key factor which must be determined.

Methods

The basic framework of the survey conducted at each WTW centred on the measurement of the size-density distribution, the size distribution found in flocculators, and the response of the sampled floc to shear. Many other features were assessed, but are not reported in this paper.

The floc size distribution was measured by image analysis on the basis of video recordings using CCTV cameras and associated optics. The successful implementation of image analysis is critically dependent on object illumination, the aim being to illuminate just those flocs in the focal plane of the receiving optics. One may then discriminate between objects in focus and out of focus on the basis of their relative light intensity; this is represented by a standard grey scale with 0 representing black (no light) and 255, the maximum brightness. Past experience [3] has shown that transverse illumination using white light (or a broad band source) has been the most appropriate for this type of material. Even so, there are many imponderables. Figure1 shows the response of a floc image to different settings of the threshold setting (a selected level of brightness on the grey scale; see caption for implications). At the lowest setting (200) there is a single object, whereas at 255 (showing the brightest parts), there are six objects, even though the image frame refers to the same floc. Between these extremes the object number and size depends on the selected threshold setting. The selection of the threshold setting is a matter of judgment. Hence the measurement of size has ambiguity. The most important experimental consideration is to maintain a consistent approach (for object illumination and threshold selection) when recording measurements.

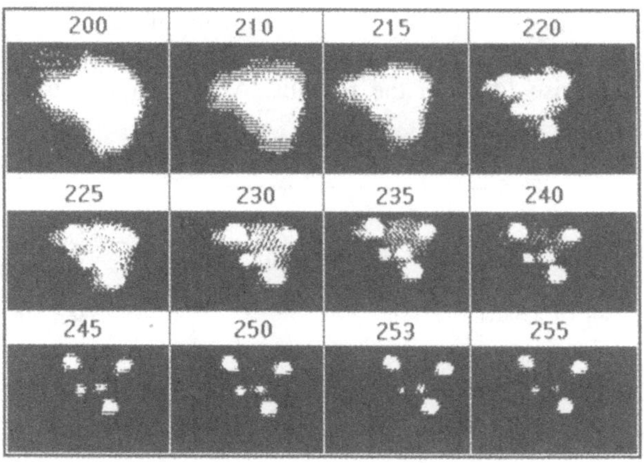

Fig. 1. Illustrating response of floc image to light threshold level; white pixels in the image correspond to part of the illuminated floc which have light intensities equal or greater than the light intensity set by the threshold level. Below the threshold, pixels are made black.

Measurements within the flocculator were achieved using an underwater camera developed by North West Water Ltd, UK. Illumination was achieved using a ring of LEDs mounted in the focal plane. The LED luminous intensity could be controlled and there were strobing facilities.

When testing the response of floc to shear, samples taken from the flocculators were placed in a two litre Phipps & Bird square beaker which was stirred using a 25 · 76 mm single bladed paddle; this is referred to as the 'square beaker test'. Following 15 min stirring at the lowest reference speed the stirrer was switched off for about 2 min to allow the fluid velocities to decrease sufficiently to obtain video recordings without image streaking. Flocs were illuminated via a slit adjacent to the inside wall. CCTV recordings were obtained using a progressive scan camera focused 50 mm above the base of the beaker. The stirrer speed was then increased, stirring for a further 15 min prior to measurement. This process was repeated until the highest speed was attained. The sample was retained for the suspended solids measurement. From knowledge of the stirring speed and temperature, the G value was obtained using a calibration.

Samples of floc were also withdrawn from an outlet located 50 mm above the base, and passed through an illuminated cuvette in accord with the procedures described in [5]. Following measurement, the sample was returned to the beaker. In both sampling methods the smallest observable floc diameter was 30 μm.

Figure 2 shows a comparison between the floc sizes by the two sampling techniques. It is evident that major discrepancies exist between the two systems of measurement, particularly at low G values. Two explanations exist. The first is that flocs are broken during their transport to the cuvette; this is likely to be more pronounced at lower speeds where the flocs are relatively large and weak (surviving in the beaker because the turbulence levels are low). A second and quite different explanation, is that during mixing, the floc size is a dynamic entity arising from continuous formation and rupture. Dynamically it may be bigger than under static conditions in which these processes are largely absent. It seems likely that the discrepancies arise for both the reasons. All subsequent data was based on remote sampling.

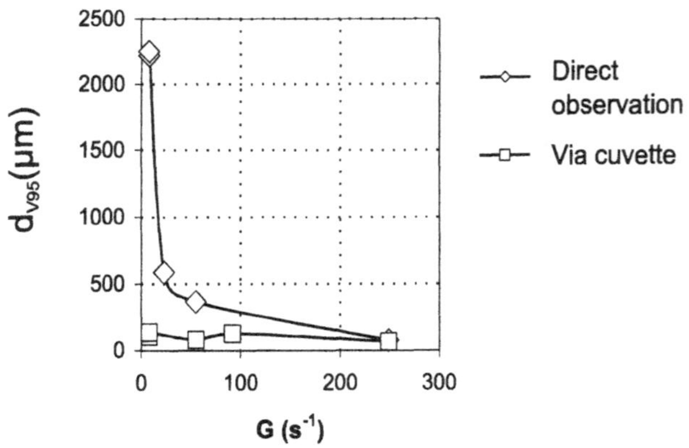

Fig. 2. Comparison of direct observation of floc size with sizes in an illuminated cuvette receiving sample stream

Survey

The principal strands of the survey included: raw water quality data; coagulation practices; estimation of mean velocity gradient; in-situ floc size measurement; floc effective density and response of floc size to G (square beaker test). All the works surveyed (see Table 1) were under the control of West of Scotland Water (WoSW) and North West Water (NWW). Features of water quality and coagulation are summarised in Table 2. Here, it is noted that the data shown in Table 2 refers to the prevailing conditions at the time of the survey. Apart from Wybersley, which treats a blended water (partly groundwater with of high calcium content), and Huntington (a river water), all remaining waters may be regarded as typical upland waters characterised by significant colour, low turbidity and low alkalinity.

Table 1. Summary of WTWs surveyed and particular features

WTW	Control authority	Clarifier	Floc aid	Special features
Neilston	WoSW	DAF		
Picketlaw		DAF		Similar design
Southmoorhouse		DAF		
Garshake		DAF		Floc barely visible
Overton		DAF	x	
Muirdykes		DAF	x	
Camphill		FBC	x	Baffled flocculator
Burncrooks		FBC	x	Baffled flocculator
Wybersley	NWW	DAF		Blended water
Huntington	NWW	Pulsator	x	Pulsator

Table 2. Water quality and coagulation data

WTW	Raw water pH	Raw water NTU	Raw water °H	Treated water pH	Treated water Coag mg/l	Treated water Poly mg/l	Treated water Flow l/s	Treated water T °C
Neilston	7.03	0.8	50	6.2	2.4		38	4.2
Picketlaw	7.1	0.77	119	6.3	5.4		99	6.7
Southm'house	7.08	1.2	142	6.4	4.7		82	5.4
Garshake	6.81	1.2	60	6.2	2.7		139	10.2
Overton	7.3	4.1	100	6.4	3.7	0.1	532	4.2
Muirdykes	7.06	3.2	110	5.9	2.5	0.1	525	10
Camphill	6.87	1.7	50	6.1	2.5	0.11	420	4.7
Burncrooks	6.51	1.73	150	6.1	3.9	0.14	486	3.9
Wybersley	6.89	3.9	83	5.0	5.0[1]		710	11.6
Huntington	7.8	3	55	6.8	4.0[2]	0.04	350	16

Notes: coagulant dose as Al apart from [1]Fe and [2]Ferral.
1 °H ≡ 1 mg Pt l^{-1} as chloroplatinic acid.

Floc size density distribution

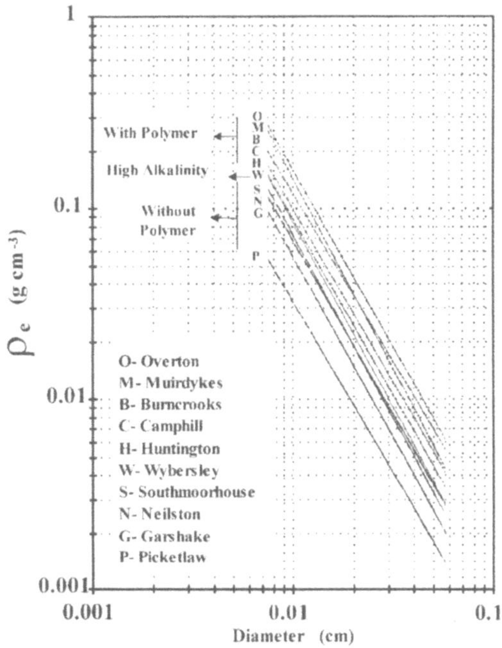

Fig. 3. Summary plot of size- density distributions at different WTWs

Trends shown in Figure 3 were fitted by Eq. (1); the values of the fitting coefficients are summarised in Table 3. The data shows that the trends are characterised by an 'n' value in the range 1.8 to 2.0. This implies that the solids within the floc are joined together in the form of a chain (or necklace structure). The highest density flocs are those associated with polymer and are characterised by a relatively 'high' A value. Where the water has high alkalinity (as at Wybersley), this yielded flocs with a density distribution which was comparable with the presence of polymer.

Floc size-G distribution

Figure 4 summarises the sensitivity of d_{V95} to G, data having been generated from the square beaker test. In all cases, it is apparent that d_{V95} decreases with increasing G. Individual trends were fitted by Eq. (3), the corresponding fitting coefficients (C and m) being shown in Table 3. It is seen that the 'm' values (representing the slope of the trends shown in Figure 4) are fairly similar - these lying in the range 0.4 to 0.8. For a given G, it is evident that floc sizes with polymer are about 4x greater than without polymer. Where there was high alkalinity (as at Wybersley), the maximum floc sizes are commensurate with those when polymer was present.

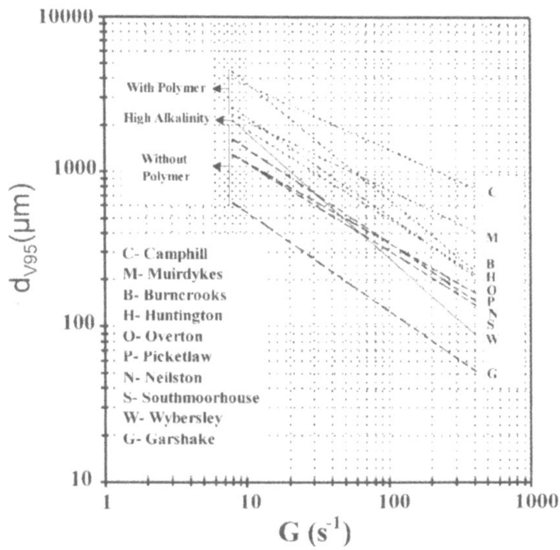

Fig. 4. Maximum floc size as a function of G and polymer presence, and one case of high alkalinity

Table 3. Summary of coefficients corresponding to trend lines shown in Figures 3 and 4. Units of A are consistent with $\Delta\rho$ in g cm^{-3} and d in cm in Eq. (1). Units of C are consistent with d_{V95} in μm in Eq.(3).

WTW	A x 10^5	n	C	m
Neilston	1.50	1.83	5784	0.61
Picketlaw	0.83	1.79	3805	0.52
Southm'house	1.00	1.93	4205	0.57
Garshake	0.88	1.90	2337	0.63
Overton	3.00	1.86	9954	0.64
Muirdykes	1.43	2.00	5875	0.44
Camphill	2.27	1.82	9415	0.41
Burncrooks	3.51	1.77	21450	0.76
Wybersley	1.11	1.94	11553	0.81
Huntington	1.75	1.85	7105	0.58

Interpretation of full scale size data

This part of the survey addressed the question of whether the square beaker data provided any useful indication of floc sizes at full scale. Comparisons were made at similar G values. Inspection of Figure 5 shows in the absence of polymer that there was no distinctive association between the large and small scale, apart from the floc being small. However, when a polymer was present, the flocs at full scale were generally much larger than encountered at bench scale.

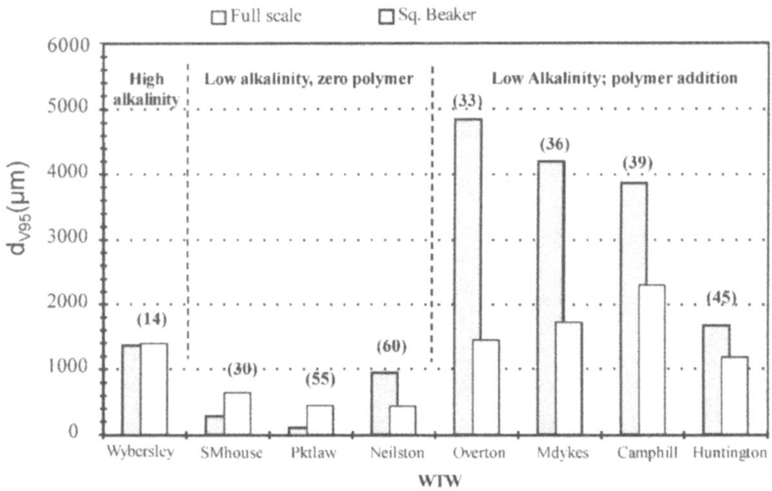

Fig. 5 Comparison of d_{V95} at full scale with square beaker data at comparable G values (shown in brackets)

Discussion

When using image analysis for measuring floc size, it must be recognised that the method is sensitive to the variations in luminous intensity across the floc. Hence the measured size is tied to the selected threshold and has an element of subjectivity. For flocs in the vicinity of the lower limit of resolution, the measurement can only be regarded as accurate to within a factor of 2 or 3 at best. For larger flocs (say ~ 1 mm), the accuracy should be within about 30 %. It was also apparent that the measured size depends on the sampling method. The comparison shown in Figure 2 serves as a warning of the dangers of taking measurements from sample streams. Given the difficulties of measuring floc size, and the imponderables over whether to use d_{I95} or d_{V95} as the representative floc size in strength calculations e.g. via Eq. (2), it is evident that estimates of strength carry all the same difficulties. They are useful for internal comparisons because of the consistency of errors, but they cannot be easily related to other studies.

The response of floc size to G (Figure 4) shows $d \propto G^{-0.6}$; this is broadly in agreement with theory [5]. The trends in Figure 5 confirm the value of polymer in increasing floc size and thereby their strength under a given shear. What is less well known, is the demonstration that the presence of hardness increases strength, presumably via Ca bonds.

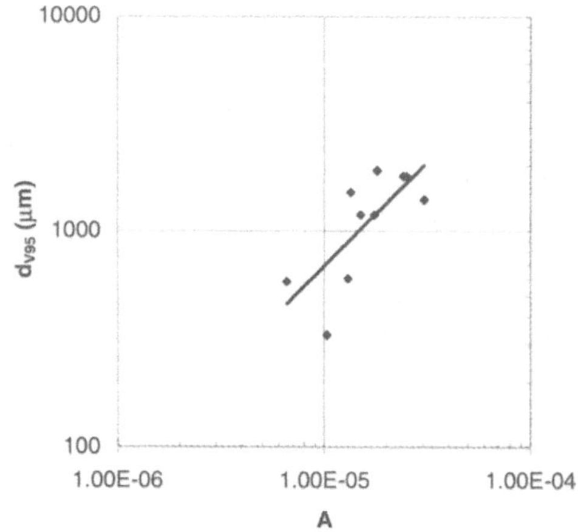

Fig. 6. Correlation between floc size at G = 32 s^{-1} and floc density distribution via parameter A in Eq. (1) . Units of A are defined in the caption of Table 3.

In Figure 3 it is seen that the presence of polymer also increases the density for a given size. This suggests that strength and density may be correlated. Since all the density trends shown in Figure 3 have much the same slope, the parameter which distinguishes their character is parameter A (see Eq.(1) and Table 3). In an attempt to show this link, the trends in Figure 3 were adjusted to a common slope (n = 1.86) and parameter A was recalculated. This reduces the error associated with variations in n, and allows one to focus on the impact of parameter A. An adjustment was also included for the influence on floc size of the variable suspended solids concentration in the stirred sample. This led to the trend shown in Figure 6 and demonstrates the coupling between density and the maximum size at a particular G value. Theory described in [5] indicates that the coupling depends on bond strengths as well as the 'A' factor.

The lack of correspondence between floc sizes at bench and full scale is well known; this is reflected in the data shown in Figure 5. What is not known is the precise reason. It is probably tied to the spatial distribution of turbulence and the relative retention times in different zones of the flocculator.

Coupled with data reported in [5], it is apparent that the controls on floc size are complex and reinforces a view expressed in [5], that it is not really feasible to use floc size as a control parameter. Given that the WTWs under scrutiny were each functioning normally, operational data indicates that the DAF units were capable of coping with the incoming floc, whatever its size. In some cases, the incoming floc was exceedingly small. This tends to support experiences reported in [8], which indicate that DAF plants perform satisfactorily when receiving a small floc. This poses questions concerning the role of polymers and the design of flocculators ahead of DAF units.

Conclusions

With appropriate illumination, the size of humic flocs can be measured using image analysis, but is subject to considerable error arising from the choice of the threshold light intensity and the lower limit of resolution of the optical system.

It is demonstrated that the presence of polymer and/or alkalinity leads to the formation of denser flocs and stronger flocs when compared with flocs in which these materials are absent.

Evidence has been provided of a coupling between the floc sizes and their response to shear, and the floc-size density distribution.

At corresponding values of velocity gradient, bench scale stirred vessels do not provide accurate insight into the behaviour at full scale, particularly when a polymeric floc aid is used.

It was evident that all the DAF plants performed satisfactorily within the process chain, whatever the size of the incoming floc. From this, one may question the necessity of the including process measures designed to enhance floc sizes ahead of DAF units.

Acknowledgements

The authors gratefully acknowledge the support of the UK Engineering and Science Research Council under grant no. GR/L/61026 together with West of Scotland Water, North West Water and WRc as collaborators in the research programme.

References

1. Bache, D.H., Rasool, E., Ali, A., McGilligan, J.F.: Floc Character: Measurement and Role in Optimum Dosing. J Wat. SRT-Aqua 44 (2) (1995) 83-92
2. Tambo, N., Watanabe, Y.: Physical Characteristics of Flocs-I. The Floc Density Function and Aluminium floc. Wat. Res. 13 (1979) 409-419
3. Bache, D.H., Hossain, M.D.: Optimum Coagulation Conditions for Coloured Water in terms of Floc Properties, J Wat. SRT-Aqua 40 (3) 1991 170-178
4. Gregory, J.: Fundamentals of Flocculation. In CRC Crit. Rev. Environ. Control. 19 (3) (1985) 185-230
5. Bache, D.H., Rasool, E., Moffat, D., McGilligan, F.J.: On the Strength and Character of Alumino-Humic Flocs. Water Sci. Tech. 40 (9) (1999), 81-88
6. Tambo, N., Hozumi, H.: Physical characteristics of flocs-II. Strength of Floc. Wat. Res. 13 (1979) 421-427
7. Parker, D.S., Warren, A.M., Kaufman, M., Jenkins, D.: Floc Breakup in Turbulent Flocculation Processes. J. San. Eng. Div., ASCE, 98 (SA1), (1972) 79-99
8. Valade, M.T., Edzwald, J.K., Tobiason, J.E., Dahlquist, J., Hedberg, T., Amato, T.: Particle removal by flotation and filtration: pretreatment effects. J AWWA 88 (12)(1996) 35-47

Mixing, Flocculation and Floc-Separation Reactors

Using Static Mixers to Mix Coagulants: CFD Modeling and Pilot-Plant Experiments

S.C. Jones*, A. Amirtharajah*, F. Sotiropoulos* and B. M. Skeens**

*Georgia Institute of Technology – School of Civil and Environmental Engineering
Atlanta, Georgia 30332–0355 – USA
cjones@ce.gatech.edu
*Jordan, Jones and Goulding – 2000 Clearview Avenue – Atlanta, Georgia 30340 – USA

Abstract

Static mixers are increasingly being used to mix coagulants like alum in water treatment plants. This study investigates static mixers for this application using two complementary approaches. In the computational approach, a state-of-the-art computational fluid dynamics model was developed to predict the turbulent fluid flow in a helical static mixer for flow rates near those found in these mixers in water treatment plants. The model predictions show that a complex, three-dimensional fluid flow is produced within the mixer. In the experimental approach, a variety of static mixers of different design and different scales were tested for mixing alum for coagulation in a pilot plant study.

Introduction

Static mixers consist of stationary mixing elements placed on the inside of a closed pipe or an open channel. Typically, coagulants are added just upstream of the elements and mix with the bulk fluid because of the complex, three-dimensional fluid motion generated by the elements. The shape of the elements determines the character of this fluid motion and thus the effectiveness of the mixer.

The advantages of static mixers include no moving parts and thus low maintenance; no external energy requirements; and at least in the context of coagulant mixing, few clogging problems [1]. One oft-quoted disadvantage of the static mixer is that the mixing is a function of the flow rate; therefore, one motivation for the CFD modeling part of this study was to explore the relationship between mixing performance and flow rate. Moreover, recent experimental evidence suggests that using a static mixer for alum mixing can result in lower chemical doses [2-4].

CFD Modeling

Model Description

The CFD model was developed by Jones [5], who extended the numerical method of Lin and Sotiropoulos [6] to laminar and turbulent flows in helical static mixers with varying numbers of elements. The model solves the steady, three-dimensional Reynolds-averaged Navier–Stokes (RANS) equations, see Brodkey [7, p. 235], with the k–ω turbulence-closure model [8]. The RANS approach was selected because it is the only practical alternative for simulating flows as complex and with Reynolds numbers as high as those encountered in most water treatment plants. The k–ω turbulence closure resolves the details of the near-wall flow without relying on the simplifying assumptions that are commonly used in commercial CFD models, the so-called wall-function approach. The RANS and turbulence-closure equations are formulated in generalized curvilinear coordinates, discretized in strong-conservation form using a finite-volume approach, and integrated in time using an explicit Runge–Kutta time-stepping algorithm. The convergence of the algorithm is enhanced with multigrid acceleration, local time-stepping, and implicit residual smoothing. Jones [5] describes the CFD model and the corresponding equations in detail.

The static mixer is discretized using a body-fitted, curvilinear grid. Figure 1 a,b shows two different two-dimensional cross-sections of this grid, which highlight the clustering of the grid nodes near the edges and sides of the mixer elements. Figure 1 a also highlights the relative length of the inlet and outlet sections of pipe in the computational domain (6 and 20 d, respectively where d is the diameter of the pipe) compared to the length of the mixer elements (1.5 d per element). Figure 1 c shows the three-dimensional shape of the mixer elements. The number of grid nodes primarily depends on the number of mixer elements and the Reynolds number (Re) of the flow. For the results presented in this paper ($Re = 10^5$), the number of nodes was 1,035,125 or 245×65×65 in the longitudinal direction and cross-sectional directions. All of the grid parameters including the length of the inlet and outlet pipe sections and the number of grid nodes were chosen after careful numerical experiments to minimize the numerical error in the predicted flowfield [5]. The method was validated using the recently published experimental data of the flow in a pipe bend by Sudo et al. [9]. The flow in a pipe bend is a good substitute for the flow in a static mixer because both flows are complex, three-dimensional, internal shear flows that have strong secondary (transverse) motions. The measurements and predictions of mean pressure and velocity components were in good agreement [5].

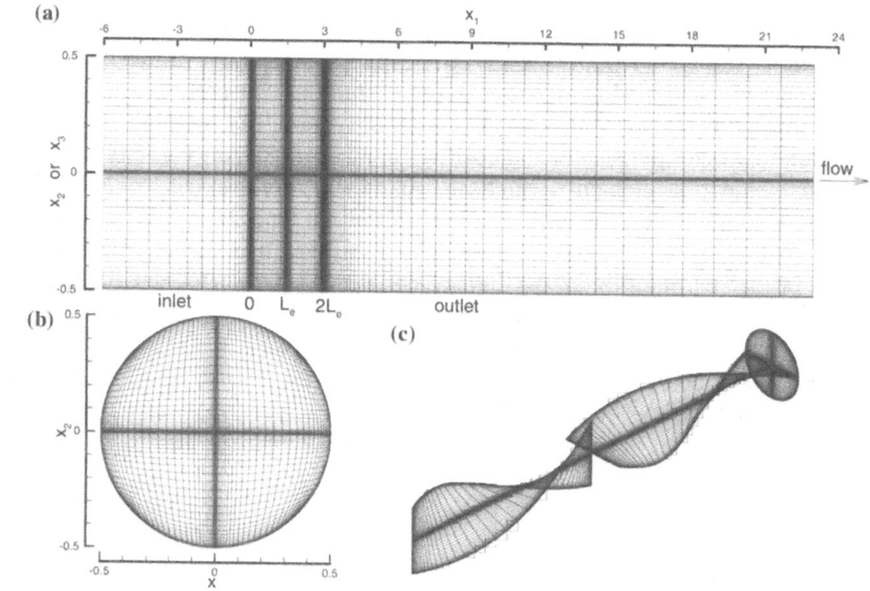

Fig. 1. Three views of the computational grid for a two-element static mixer

Flowfield Predictions

From the perspective of understanding the mechanisms that cause mixing to occur within a static mixer, the most important and immediately useful information produced by a CFD model is the velocity field. Figure 2 shows contours of the predicted velocity field within the second element of the static mixer. (All velocities are normalized by the bulk velocity U in the inlet section of the pipe.) In this set of cross-sections, the second element twists in a clockwise direction. The first row of Figure 2 contains contours of the mean longitudinal velocity at four locations along the element: (2 a) the intersection of the first and second elements, (2 b) approximately 20% along the length of the second element, (2 c) 50% along the second element, and (2 d) 80% along the second element. The second row of Figure 2 contains vector plots of the transverse velocities at the same locations, and the third row contains contour plots of the mean longitudinal vorticity. The mean longitudinal vorticity is a measure of the strength of the transverse flow. Positive vorticity indicates counterclockwise rotation; negative vorticity indicates clockwise rotation.

In Figure 2, the second element appears to split the approach flow into four regions of high mean longitudinal momentum. As the flow progresses through the element, the high-momentum pockets that were not separated by the element merge into one pocket of high-momentum fluid through the action of the transverse flow. Regions of very low momentum and even negative mean longitudinal velocity occur at the intersection of the element with the pipe wall. For this particular Reynolds number ($Re = 10^5$), the volume occupied by these reversed-flow regions is relatively small. Simulations at lower Re indicate that the size of these regions strongly depends on Re [5]. A closer look at the mean transverse velocity vectors reveals some of the features of this flow, in particular, for cross-sections in Figure 2 b two vortices can be seen in each half of the pipe. The mean longitudinal vorticity plots show the positive (counterclockwise) vorticity that was produced in the first element (which also has a counterclockwise twist) being transported from the element surface to the wall of the pipe. Negative (clockwise) vorticity is produced on the second element and this negative vorticity determines the twist of the swirling flow in the outlet section of the pipe. Although not shown here, the complexity of the flow leaving the mixer (see Figure 2 d) persists for many diameters into the outlet section and exhibits a very complex recovery to fully developed pipe flow. In fact, the predictions suggest that the flow has still not recovered even 20 d downstream of the elements.

Characterizing Mixing

Camp [10] developed the root-mean-squared G-value to characterize mixing in flocculation basins, and the G-value is now used to characterize mixing in a wide variety of processes in Environmental Engineering. Several convincing theoretical arguments have shown that the original derivation of the G-value is flawed [11,12] and thus should not be used. Even two decades ago [13], the G-value concept was recognized to be a gross oversimplification of the mixing process because it attempts to represent a complex flowfield with multiple, widely varying length and time scales with one single number. However, the G-value is thoroughly entrenched in the Environmental Engineering literature, and it is likely to continue to be used for design purposes in the near future. With these facts motivating us, we investigated the G-value using the CFD model for static mixers.

For a static mixer (or any other in-line mixer for that matter), the G-value is calculated as

$$G = \sqrt{\frac{P}{\mu V}} = \sqrt{\frac{\gamma Q \Delta h}{\mu V}} = \sqrt{\frac{f U^3}{2 \nu d}} \qquad (1)$$

where P is the power dissipated by the mixer, μ is the dynamic viscosity, V is the volume of water, Q is the volumetric flow rate, γ is the specific weight of water.

$\Delta h = f\left(L/d\right)\left(U^2/2g\right)$ is the head loss in the mixer,

where f is the friction factor, L is the length of the mixer, d is the diameter of the pipe, g is the acceleration due to gravity, U is the bulk velocity in the pipe, and ν is the kinematic viscosity.

Using the G-value for in-line mixers is difficult because the actual volume over which mixing takes place is not known. In the following, we consider the volume corresponding to the length of the two elements and exclude any volume downstream of the elements. The predicted friction factor for the two elements at a $Re = 10^5$ is 0.96.

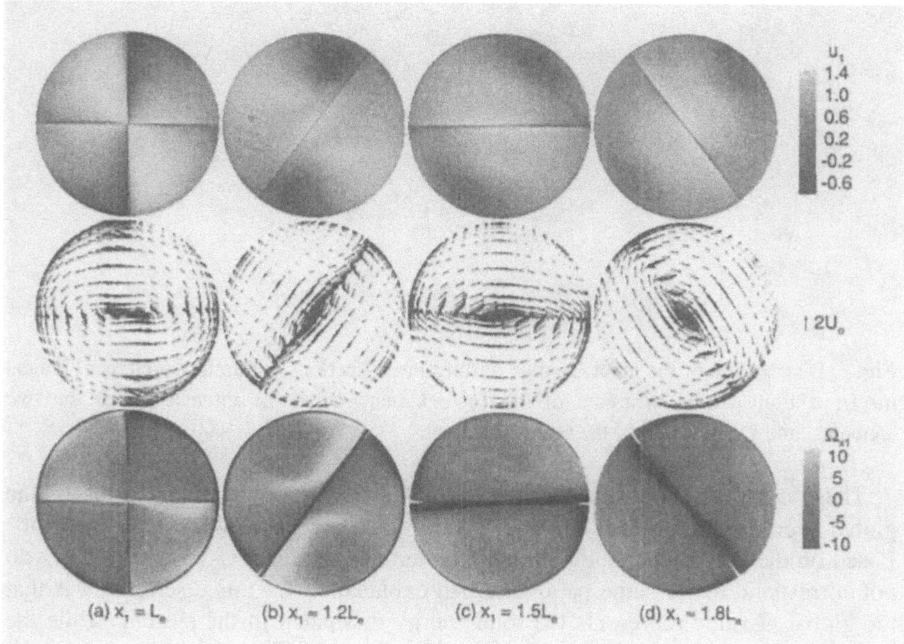

Fig. 2. Predicted flowfield within the second element of a two-element helical mixer at a $Re = 10^5$. The top row shows contours of mean longitudinal velocity; the middle row shows vectors of mean transverse velocity; and the bottom row shows contours on the mean longitudinal vorticity.

Assuming a pipe diameter of 50 mm and a bulk velocity of 2 m s^{-1} (a typical design velocity for a raw water pipeline [1, p. 61]) gives a *G*-value of 8800 s^{-1} – a value several times larger than those suggested for design [13]. (For reference, this *G*-value corresponds to $\Delta h = 0.5$ m at $Q = 4$ L s^{-1}.) If we include the pipe downstream of the mixer to calculate the *G*-value, the calculated *G*-value will decrease, but this decrease is countered by the larger head loss in the extra length of pipe.

To improve the ability of the *G*-value concept to characterize a spatially varying flowfield, a local *G*-value is sometimes used [14]. This local *G*-value is commonly defined as

$$G = \sqrt{\varepsilon / \nu} \tag{2}$$

where ε is the turbulence energy dissipation in the mixer. The turbulence closure equations in the CFD model provide an estimate of the turbulence energy dissipation; thus, the local *G*-value can be easily calculated from the predicted flowfield. Figure 3 shows the distribution of the local *G*-value at three cross-sections within the mixer: (3 a) at the leading edge of the first element, (3 b) at the intersection of the two elements, and (3 c) at the trailing edge of the second element.

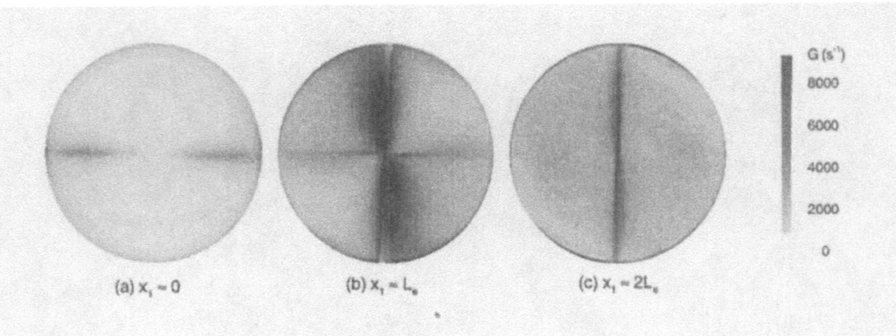

Fig. 3. Distribution of the local *G*-value at three cross-sections within a two-element helical mixer: **a**) immediately upstream of the first element, **b**) at the intersection of the two elements, and **c**) at the end of the second element.

These contours show that the local *G*-value only approaches the value of the global average *G*-value of 8800 s^{-1} in small regions of the flow near the elements. Based on these predictions, the global *G*-value and the local *G*-value obviously do not correspond to the same parameter. An explanation for this discrepancy is that the global *G*-value represents the total energy dissipated in the system, while the local *G*-value represents energy dissipated by the turbulence only. At present, the role of these different fractions of energy dissipation in the mixing process is not well understood. However, this inconsistency in the *G*-value concept suggests that other quantities (for example, head loss and intensity of segregation) should also be used to evaluate the efficacy of static mixers for water treatment.

Pilot Plant Experiments

Materials and Methods

The pilot plant is owned by the City of Atlanta, Georgia, USA. The plant consists of two parallel treatment trains each of which have a rapid mix unit (containing the static mixer), three stages of tapered flocculation (*G*-value = 60, 40, or 20 s^{-1}), sedimentation with inclined plate settlers, and two dual-media 100-mm (4-inch) diameter filters. One filter on each train consists of anthracite and sand; the other consists of granular activated carbon and sand. A constant filtration rate of 10 m h^{-1} (4 gpm ft^{-2}) was used for all tests. Raw water for the plant comes directly from the Chattahoochee River, which is a low organic (~2.5 mg L^{-1} as total organic carbon) and precipitation-sensitive water with turbidity often near 100 ntu following rainstorms. Most of the time, however, the turbidity was 15–30 ntu. Experiments were performed at flow rates (Q) of 0.12 or 0.25 L s^{-1} (2 or 4 gpm) for each train. The coagulant was either alum or polyaluminum chloride[1], and doses were set using a streaming current detector. All experiments were performed under charge

[1] PAX-XL70 provided by Kemwater North America Co., Savannah, Georgia, USA

neutralization conditions as defined by plotting the alum dose and coagulation pH on the alum coagulation diagram [15,16]. Turbidity was measured from grab samples or from on-line instruments whose output was transmitted directly to a computer.

Results and Discussion

Effect of the Number of Elements. Experiments were performed with a two- and six-element helical static mixer (d = 25 mm) from Chemineer.[2] The measured head loss in the mixers at $Q = 0.12\ \mathrm{L\ s^{-1}}$ was 80 and 240 mm for the two and six elements, respectively. A typical result from these experiments is shown in Figure 4 a, which shows very little difference in the performance of the two trains. In fact, little difference was observed for any of the criteria measured: settled-water turbidity, filtered-water turbidity, or filtered-water particle counts. The additional mixing provided by the four elements had no effect on the performance of the pilot plant.

Scale-up for Several Types of Static Mixer. Experiments were performed to compare the performance of a given static mixer to an empty-pipe mixer of the same size (both length and diameter). An empty-pipe mixer simply consists of a pipe with no static mixer. This experimental design was motivated by observations from bench-scale experiments [4] that showed very little difference in the performance of a static mixer compared to an empty-pipe mixer – a result that was theorized to be due to the small scale of the bench-scale experiments (d = 12 mm). To investigate this further, similar experiments were performed at the pilot plant but with diameters between 25 and 50 mm (1 and 2 inches). Four static mixers were used in these tests: a Kenics helical static mixer, a TAH[3] Stata-tube, a TAH Spiral mixer, and Koch[4] (Sulzer) SMV mixer. In each experiment, the alum dose for each mixer was optimized independently allowing the optimum alum dose for each mixer to be determined. Figure 4 b compares the optimum alum dose for the empty pipe mixer and static mixer at three diameters for the Kenics static mixer. As the figure indicates, as the diameter increases, the static mixer needs less alum than the empty-pipe mixer. This trend also held for the TAH Spiral and Stata-tube mixers. This trend was not observed, however, for the SMV static mixers and both the empty-pipe and SMV mixers had the same optimum alum dose at each diameter tested. Based on the performance of the different static mixers compared to the empty-pipe mixer, we judged that observing performance differences using the current experimental methodology would be unlikely when comparing mixers from different manufacturers in side-by-side tests, and thus side-by-side comparisons of the mixers from different manufacturers were not pursued.

---

[2] Chemineer, Inc., Dayton, Ohio, USA, http://www.chemineer.com

[3] TAH Industries, Inc., Robbinsville, New Jersey, USA, http://www.tah.com

[4] Koch-Glitsch, Inc., Wichita, Kansas, USA, http://www.koch-glitsch.com; Sulzer-Chemtech, Ltd., Winterthur, Switzerland, http://www.sulzerchemtech.com

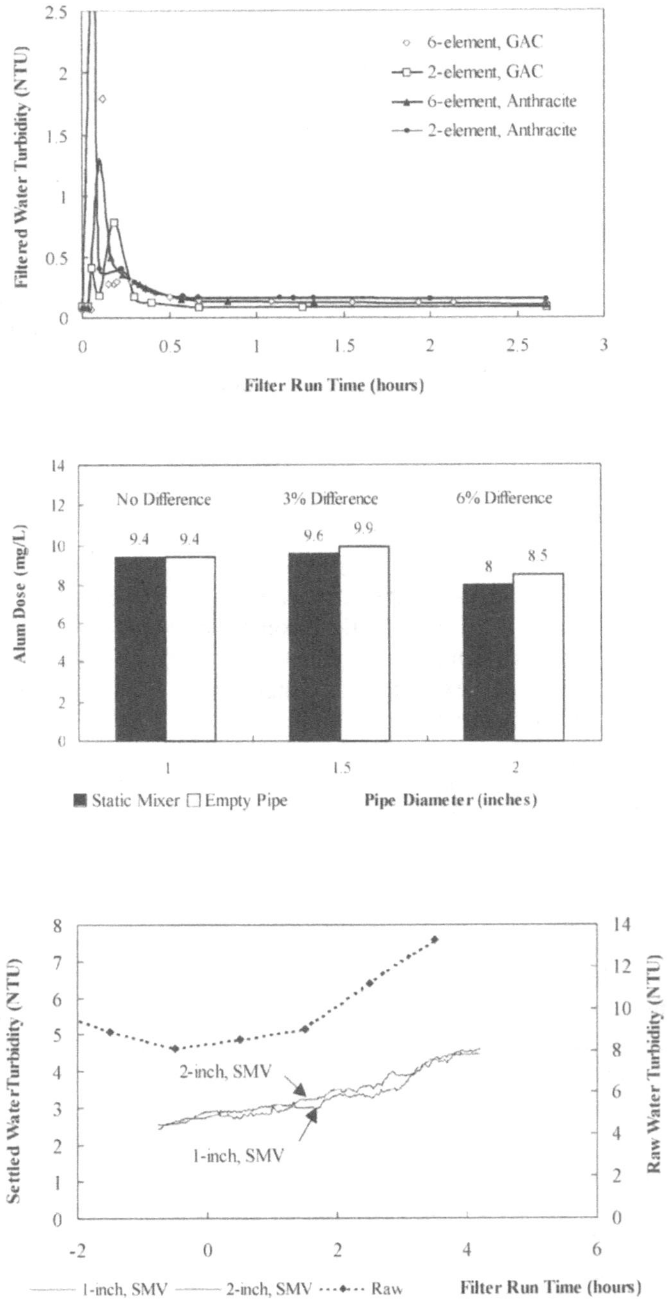

Fig. 4 Results of the pilot-plant experiments (from top to bottom): **a**) comparison of filtered-water turbidity between Kenics mixers with two and six elements, **b**) comparison of optimum alum dose for Kenics mixer and empty-pipe mixer at three diameters, and **c**) comparison of settled-water turbidity for SMV mixer at two diameters.

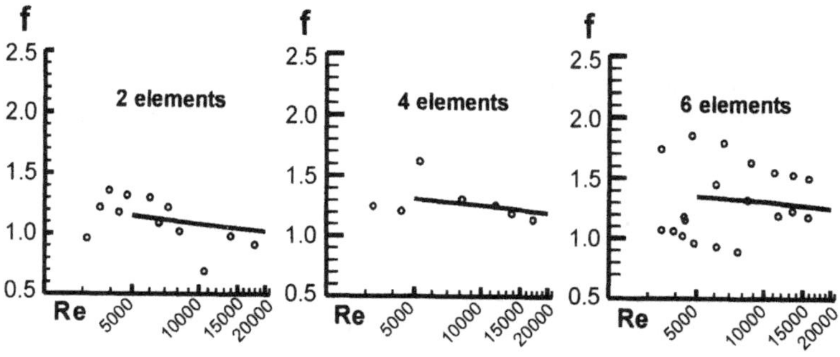

Fig. 5. Comparison of the head loss measurements at the pilot-plant (circles) with the model predictions (lines)

Direct Comparisons of Different Diameter Mixers. The Kenics and SMV mixers were used to compare the performance of different diameter mixers at a $Q = 0.25$ L s^{-1}. However, no consistent trends were observed based on the optimum alum dose or the performance of the pilot plant. As an example, Figure 4 c shows one of the larger differences in settled-water turbidity for these tests; however, based on the repeatability of the tests, this difference is not significant.

Comparison of Computations and Experiments. Experiments were also performed at the pilot plant to measure the head loss across the Kenics helical static mixer and compare the head loss to that predicted by the CFD model. Shown in Figure 5, the agreement between the predicted and measured friction factor (directly related to the head loss) is satisfactory. However, further experimental validation of the model should include detailed experimental measurements of the velocity field and turbulence quantities.

Summary and Conclusions

A CFD model was developed to predict the turbulent flow in a helical static mixer. The predicted flowfield is extremely complex and contains regions of reversed mean flow and a transverse flow that is dominated by the interaction of longitudinal vortices produced by the mixer elements. Analysis of the predicted flowfield shows that common concepts for characterizing mixing, the global and local *G*-values, are inconsistent. Pilot-plant experiments showed little difference between the performance of static mixers with different numbers of elements under the charge-neutralization mechanism of coagulation. Although not conclusive, the pilot-plant experiments did show that as the diameter increases static mixers require slightly less chemicals than the empty-pipe mixer – a mixer that provides only the mixing present in the turbulent flow in a pipe at a moderate Reynolds number.

Acknowledgments

This work was supported by the American Water Works Association Research Foundation and the City of Atlanta. Mixers were provided by Chemineer, Inc., TAH Industries, Inc., and Koch-Glitsch, Inc.

References

1 Kawamura, S.: Integrated design of water treatment facilities. John Wiley, New York (1991)
2 Clark M.M., Srivastava R.M., Lang J.S., Trussell R.R., McCollum L.J., Bailey D., Christie J.D., Stolarik G.: Selection and design of mixing processes for coagulation. AWWA Research Foundation, Denver (1994)
3 Latimer R., Amirtharajah A.: Pilot scale comparison of static mixers and backmix reactors for water treatment. In: Proceedings of the 1998 Annual Conference, AWWA, Denver (1998)
4 Amirtharajah A., Jones S.C.: Mixing for coagulation: organic polymers, static mixers, and modeling. In: Hahn HH, Hoffmann E, and Ødegaard H (eds) Chemical water and wastewater treatment IV. Springer, Berlin, (1996) pp 3–15
5 Jones S.C.: Static mixers for water treatment: a computational fluid dynamics model. PhD Dissertation, Georgia Institute of Technology (1999)
6 Lin F.-B., Sotiropoulos F.: Strongly-coupled multigrid methods for 3-d incompressible flows using near-wall turbulence closures. ASME J Fluids Eng 119 (1997) 314–324
7 Brodkey R.S.: The phenomena of fluid motions. Dover, New York (1995)
8 Wilcox D.C.: Turbulence modeling for CFD. DCW Industries (1994)
9 Sudo K., Sumida M., Hibara H.: Experimental investigation on turbulent flow in a circular-sectioned 90-degree bend. Exp in Fluids 25 (1998) 42–49
10 Camp T.R., Stein P.C.: Velocity gradients and internal work in fluid friction. J Boston Soc Civil Eng 30 (1947) 219–237
11 Clark M.M.: Critique of Camp and Stein's RMS velocity gradient. ASCE J Environmental Engineering 111 (1985) 741–754
12 Graber S.D.: A critical review of the use of the *G*-value (RMS velocity gradient) in Environmental Engineering. Dev Theoretical and Applied Mech 17 (1994) 533–556
13 Amirtharajah A.: Design of rapid mix units. In: Sanks RL (ed) Water treatment plant design. Butterworths-Heinemann, Boston (1978) pp 131–147
14 Amirtharajah A., Trusler S.L.: Destabilization of particles by turbulent rapid mixing. ASCE J Envir Eng 112 (1986) 1085–1108
15 Skeens B.M.: Pilot scale evaluation and comparison of static mixers for coagulation in water treatment, Master's Thesis, Georgia Institute of Technology (1999)
16 Amirtharajah A., Mills K.M.: Rapid mix design for mechanisms of alum coagulation J AWWA 74 (1982) 210–216

Using CFD in the Study of Mixing in Coagulation and Flocculation

J. Korpijärvi*, E. Laine* and H. Ahlstedt**

*Kemira Chemicals Oy, Oulu Research Centre, P.O. Box 171, FIN-90101 Oulu, Finland
jarmo.korpijarvi@kemira.com, eija.laine@kemira.com
**Tampere University of Technology, Energy and Process Engineering, P.O. Box 589, FIN-33101 Tampere, Finland
ahlstedt@cc.tut.fi

Abstract

Mixing has always been acknowledged to be one of the key issues in producing a good quality treated water in the chemical water treatment process. In the rapid mixing stage, where the inorganic coagulant is first mixed into the raw water, the charge neutralization step occurs in less than a second. This being the case, charge neutralization and adsorption are very mixing-sensitive processes. A lack of effective mixing when using inorganic coagulants results in chemical overdosing and decreased efficiency as sweep coagulation becomes the dominant mechanism for floc formation. On the other hand, in the flocculation stage excess velocity gradients should be avoided in designing a flocculator. This work highlights the possible use of computational fluid dynamics, CFD, in numerical modeling of mixing in water treatment processes.

Background

Computational fluid dynamics, CFD, has increasingly been used in the chemical process industry as a tool for simulating, troubleshooting, and designing reactors, for example. This is partly due to the increased computational power of cheap desktop computers and the recent availability of commercial CFD codes for these platforms. The programs themselves have also become easier to use with graphical user interfaces and improved grid generation tools. However, there are several shortcomings in the numerical solutions of the Navier-Stokes equations coupled with turbulence closures in particular. The user must have knowledge of these shortcomings and of fundamental fluid flow to be able to apply CFD to new areas.

The above mentioned Navier-Stokes equations govern any Newtonian fluid flow in any geometry. Together with turbulence models, such as the standard k-ε model, they form a closed group of equations, which can be solved numerically to engineering accuracy.

Navier-Stokes Equations

Commercial CFD package CFX-4 [1] was used in this project to solve the flow field. Time-averaged Navier-Stokes equations are discretised and solved together with turbulence closures at each node point inside the computational domain. The following set of equations is the time-dependent Reynolds-averaged Navier-Stokes equations derived in three dimensions. For the conservation of mass over a small fluid element, i.e., the continuity equation, one can write:

$$\frac{\partial U}{\partial x}+\frac{\partial V}{\partial y}+\frac{\partial W}{\partial z}=0 \tag{1}$$

For the conservation of momentum in the x, y and z directions the following three equations apply:

$$\begin{aligned}\frac{\partial U}{\partial t}+U\frac{\partial U}{\partial x}+V\frac{\partial U}{\partial y}+W\frac{\partial U}{\partial z}&=-\frac{1}{\rho}\frac{\partial p}{\partial x}+\nu\left(\frac{\partial^2 U}{\partial x^2}+\frac{\partial^2 U}{\partial y^2}+\frac{\partial^2 U}{\partial z^2}\right)\\&+B_x-\frac{\partial \overline{u^2}}{\partial x}-\frac{\partial \overline{uv}}{\partial y}-\frac{\partial \overline{uw}}{\partial z}\end{aligned} \tag{2}$$

$$\begin{aligned}\frac{\partial V}{\partial t}+U\frac{\partial V}{\partial x}+V\frac{\partial V}{\partial y}+W\frac{\partial V}{\partial z}&=-\frac{1}{\rho}\frac{\partial p}{\partial y}+\nu\left(\frac{\partial^2 V}{\partial x^2}+\frac{\partial^2 V}{\partial y^2}+\frac{\partial^2 V}{\partial z^2}\right)\\&+B_y-\frac{\partial \overline{uv}}{\partial x}-\frac{\partial \overline{v^2}}{\partial y}-\frac{\partial \overline{vw}}{\partial z}\end{aligned} \tag{3}$$

$$\begin{aligned}\frac{\partial W}{\partial t}+U\frac{\partial W}{\partial x}+V\frac{\partial W}{\partial y}+W\frac{\partial W}{\partial z}&=-\frac{1}{\rho}\frac{\partial p}{\partial z}+\nu\left(\frac{\partial^2 W}{\partial x^2}+\frac{\partial^2 W}{\partial y^2}+\frac{\partial^2 W}{\partial z^2}\right)\\&+B_z-\frac{\partial \overline{uw}}{\partial x}-\frac{\partial \overline{vw}}{\partial y}-\frac{\partial \overline{w^2}}{\partial z}\end{aligned} \tag{4}$$

We now have four equations, namely, the continuity equation and three momentum equations, and 10 unknowns:

- three velocities (*U*, *V* and *W*)
- six Reynolds stresses,
- $(\overline{u^2}, \overline{v^2}, \overline{w^2}, \overline{uv^2}, \overline{uw^2}, \overline{vw^2})$
- and pressure *p*.

To close the above equations turbulence models are needed. The k-ε model is based on the eddy viscosity hypothesis by Boussinesq. The eddy viscosity concept is analogous to the movement of molecules. Turbulent eddies are thought of as fluid particles, which collide and exchange energy like the molecules obeying the kinetic gas theory. The eddy viscosity model simply states that the Reynolds stresses in Eqs. (2)-(4) are related to the local shear via eddy viscosity ν_t as follows:

$$-\overline{u_i u_j} = \nu_t \left(\frac{\partial U_i}{\partial x_j} + \frac{\partial U_j}{\partial x_i} \right) - \frac{2}{3} k \delta_{ij} = 2\nu_t s_{ij} - \frac{2}{3} k \delta_{ij} \tag{5}$$

In the k-ε model the eddy viscosity ν_t can be calculated from

$$\nu_t = C_\mu \frac{k^2}{\varepsilon} \tag{6}$$

The transport equation for turbulence kinetic energy k according to the standard k-ε model by Jones and Launder [2] is as follows:

$$\frac{\partial k}{\partial t} + \nabla \cdot (\mathbf{U}k) = \nabla \cdot \left[\left(\nu + \frac{\nu_t}{\sigma_k} \right) \nabla k \right] + P - \varepsilon \tag{7}$$

where P is the production of the turbulence kinetic energy due to shear. Turbulence kinetic energy is dissipated, that is, the turbulent eddies are dampened by the viscous forces, which ultimately transform turbulence into heat. Dissipation has its own transport equation, hence the name k-ε, and it can be calculated as follows:

$$\frac{\partial \varepsilon}{\partial t} + \nabla \cdot (\mathbf{U}\varepsilon) = \nabla \cdot \left[\left(\nu + \frac{\nu_t}{\sigma_\varepsilon} \right) \nabla \varepsilon \right] + C_1 \frac{\varepsilon}{k} P - C_2 \frac{\varepsilon^2}{k} \tag{8}$$

In order to solve turbulent flow, all we have to do is to give values for the five coefficients, which were originally derived from measurements and computer optimisation. Table I presents the values recommended by Launder and Spalding [2].

Table 1. Values of the coefficients in the k-ε model.

C_μ	C_1	C_2	σ_k	σ_ε
0.09	1.44	1.92	1.0	1.3

In addition, scalar transport equations are needed to model the transport of tracer concentration, for example, in turbulent flows. A general scalar transport equation for constant density flow can be written as

$$\frac{\partial \Phi}{\partial t} + \nabla \cdot (\mathbf{U}\Phi) = \nabla \cdot (\Gamma \nabla \Phi - \overline{\mathbf{u}\phi}) + S \tag{9}$$

Here, Γ is the molecular diffusion and S is a source or a sink term corresponding to the creation or destruction of Φ.

Traditionally, the average G value for a mixing vessel is calculated from the dissipation per unit mass in the vessel, ε, by Camp and Stein [3] as follows:

$$\bar{G} = \sqrt{\frac{\varepsilon}{\nu}} \qquad \textbf{(10)}$$

In order to evaluate the local impact of the mean flow and turbulence on the flocs one can calculate local G values as suggested by Luo [4]:

$$G = \left(\frac{\mu_{eff}}{\mu}\right)\left\{\begin{matrix} 2\left[\left(\frac{\partial U}{\partial x}\right)^2 + \left(\frac{\partial V}{\partial y}\right)^2 + \left(\frac{\partial W}{\partial z}\right)^2\right] \\ + \left(\frac{\partial U}{\partial y} + \frac{\partial V}{\partial x}\right)^2 + \left(\frac{\partial V}{\partial z} + \frac{\partial W}{\partial y}\right)^2 + \left(\frac{\partial U}{\partial z} + \frac{\partial W}{\partial x}\right)^2 \end{matrix}\right\}^{1/2} \qquad \textbf{(11)}$$

Flocculation

Flocculation is extensively used in raw and wastewater treatment to remove unwanted compounds, particles and debris from the treated water by means of flocculation chemicals. In the first stage, the coagulation chemicals, usually dissolved aluminium or iron salts and also various polymers, are rapidly mixed into the raw water stream by means of an inline mixer or immersed pumping system.

Usually pH has to be adjusted beforehand to about 8 for aluminium and 5-6 for iron chemicals. Aluminium sulphate reacts with water and forms aluminium hydroxides within 1-7 seconds upon addition of the chemical. This is the mixing critical stage, where the intensity of turbulence is of great importance to the outcome. The positive aluminium ions (Al^{3+}), aluminium hydroxy ions and aluminium hydroxides adsorb to negative particle surfaces (e.g. clay) in the raw water. This all happens very fast, within 0.2-1 second, depending on the extent of mixing [3]. The flocculation itself takes place in a more quiescent part of the process. After the coagulants have been mixed in, the raw or wastewater is fed into a large mixing tank or a baffled reservoir where mixing is very gentle. Here, the particles with positive and negative sites form flocs or aggregates due to mutual collisions. Once these aggregates have grown enough in size, they can easily be removed by sedimentation or flotation.

Kemira Kemwater's jar test device, a mini-flocculator, was used as a first target for model building and measurements. The jar test is a standard procedure used for determining the required amount of chemical to be added in a particular raw or wastewater. The device consists of a supported paddle mixer and a one-litre beaker as well as a console, which allows adjustment of the mixer rotation speed. The device is shown in Figure 1. The mini-flocculator was chosen because of its optimal size for measurements as well as for modeling.

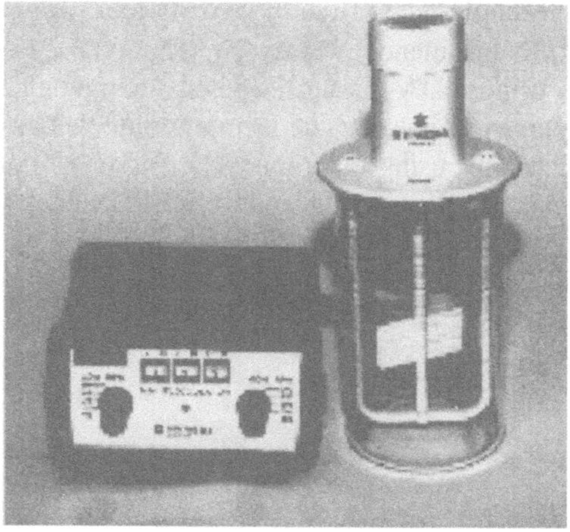

Fig. 1. Kemira jar test device, mini-flocculator

Modeling

CFD was used to model the flow field inside the jar test device. A rotating grid or sliding grid approach was chosen to mimic the behaviour of the mixer rotating in the actual mini-flocculator. Another possibility would have been to use time independent calculation, where the mixer is modelled as a source term for momentum equations. The advantages of the chosen method are that no experimental data are required and that the accuracy of the calculations is better than that using the source term method. The disadvantage of the approach is that the calculation of the flow field tends to be quite lengthy. The basic idea of the sliding grid is illustrated in Figure 2. The time steps in the sliding calculations are chosen so that the advancement of the rotating grid is one cell length per time step. This means that with a finer grid the time steps are smaller than those with a coarser grid.

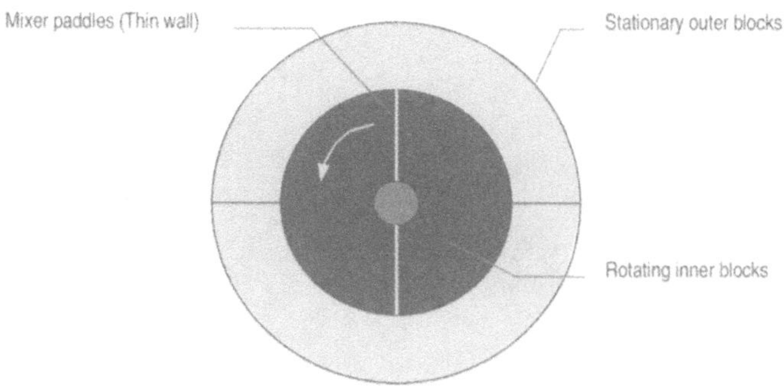

Fig. 2. Simplified illustration of the sliding (rotating) grid

In the first stage, the single-phase (liquid) velocity field was calculated by using k-ε, RNG k-ε and RSM turbulence models [5]. The axial and tangential velocities were validated by using LDA (Laser Doppler Anemometry) and PIV (Particle Image Velocimetry) measurements. The local absolute velocity gradient, G values, can also be calculated by using equation (11). Figure 4 illustrates some of the values that can be calculated and visualized from CFD calculations.

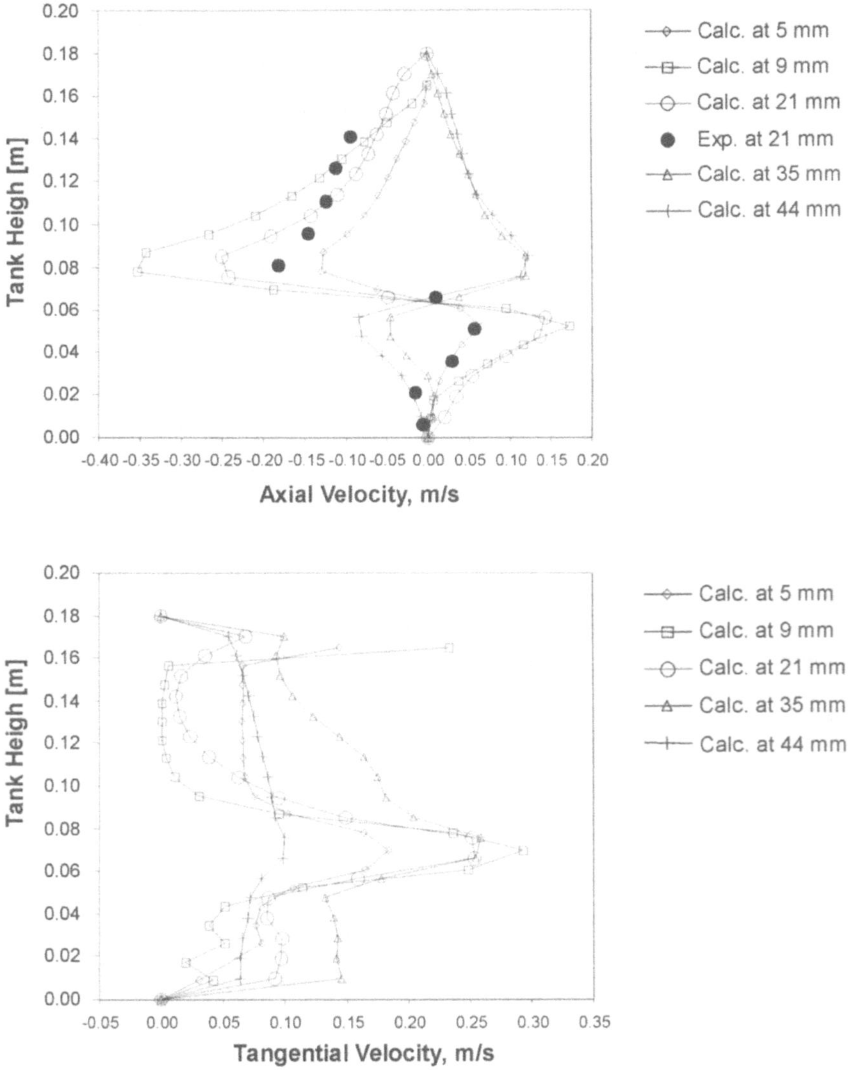

Fig. 3. Calculated and measured axial (top) and tangential (below) velocity profiles

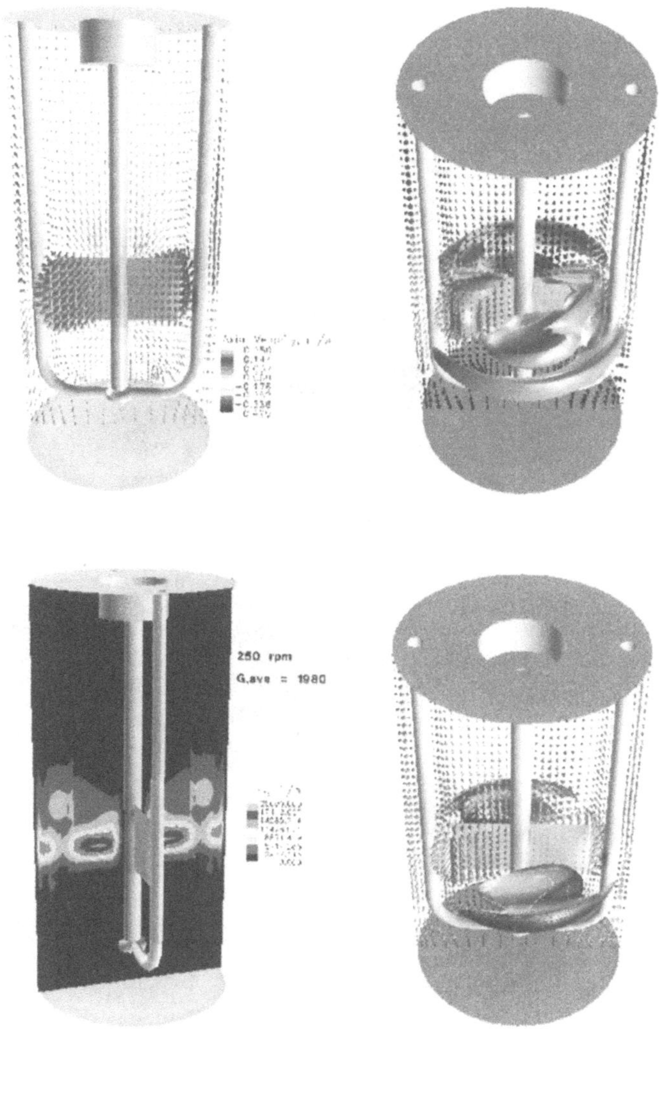

Fig. 4. Calculated velocity field at 250 rpm (top left). Iso-surface of the maximum G values (top right). Local G values at the mini flocculator mid-plane (lower left). Maximum turbulence values, turbulence kinetic energy (lower right).

Figure 4 shows that the largest G values are concentrated near the paddles and particularly in the turbulent area behind the paddle all the way to the wall in a radial direction. There is also a very large dispersion of G values throughout the vessel, as can be expected. The average G value at the mid-plane is less than 2000 1/s, which seems quite a high value compared to the values reported in the literature [6], [7], and [8] for a similar vessel. However, the average G value in Figure 4 is only for a single plane, not the whole vessel volume.

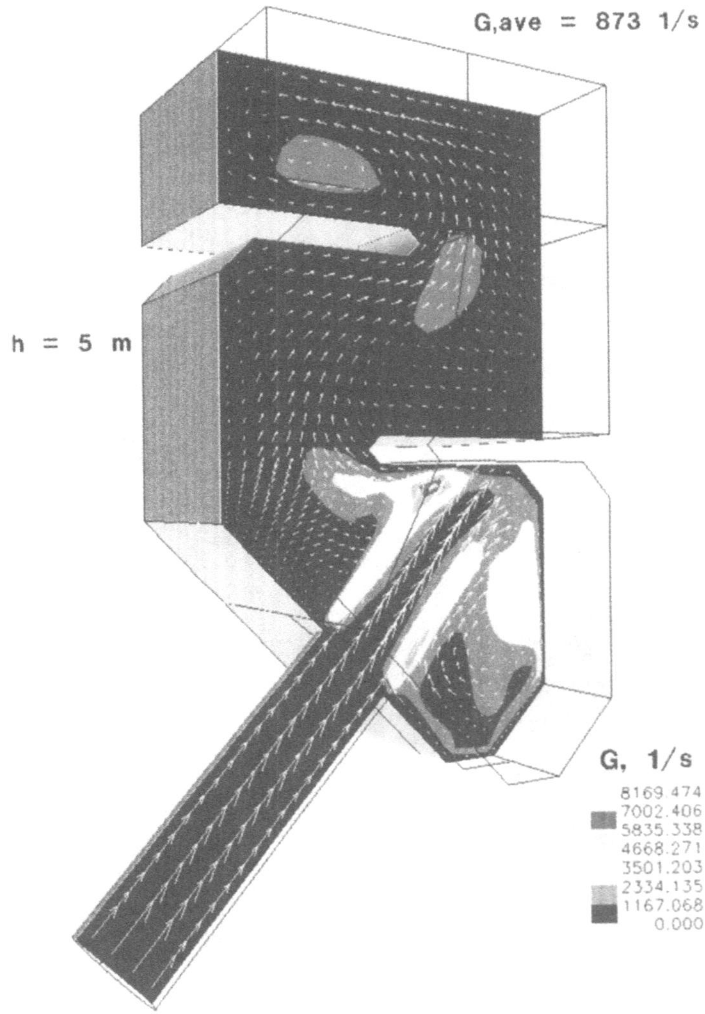

Fig. 5. The average and local G values for a large-scale rapid mixer at a drinking water treatment plant

Figure 5 presents a 5-meter tall, large-scale rapid mixing tank, which is used to mix aluminum coagulant into raw water at a drinking water treatment plant. From the analysis of this particular device it could be concluded that the mixing of the coagulant was most likely inefficient when the chemical was fed at the top of the large inlet pipe (as it was). The recommendation was to feed the coagulant below the pipe, from where it more likely enters the mixing vortex, where the largest G values are present. In general, this type of mixer is not recommended due to the wide distribution of residence times and strong dependence of the mixing behavior on inlet flow rates.

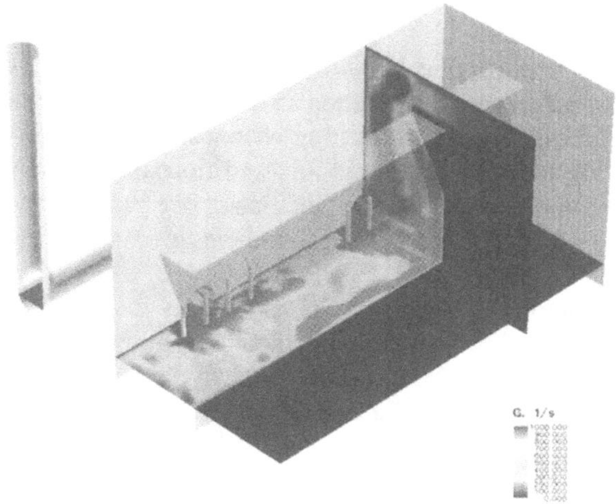

Fig. 6. Calculated local G values in a DAF tank inlet section. The color spectrum was scaled between 0 and 2000 1/s

The example in Fig. 6 presents the calculated G values at the inlet section of a DAF tank. The inlet section is located behind a baffle, which has been made transparent in this figure for better visualization. For simplification, only liquid flow was modeled in the tank. Fig. 7 shows the flow streamlines for liquid entering the tank.

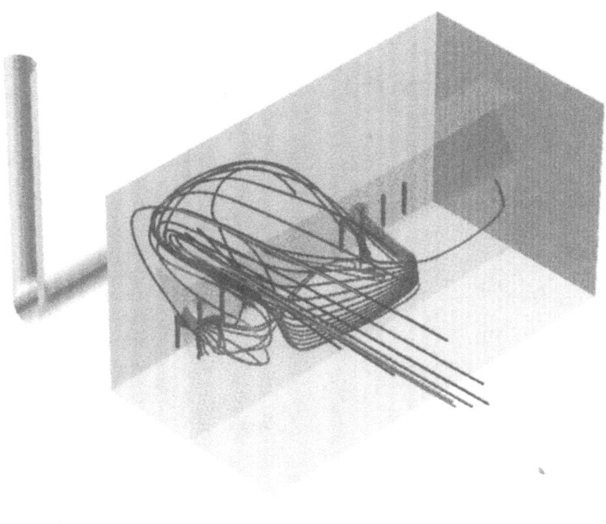

Fig. 7. Flow streamlines in the DAF tank

Conclusions

This work dealt with the applicability of CFD in the study of mixing in coagulation and flocculation processes. Coagulation in water treatment is extremely mixing-sensitive due to infinitely fast competing chemical reactions. By using CFD, one can study the effect of different rapid mix and flocculator designs and determine the extent of mixing required. On the other hand, one can also use flow simulation and visualization to avoid the velocity gradients that are likely to lead to floc breakage.

The next phase in the modeling would be to incorporate the interactions between flocs and the fluid into the CFD models. There are several possibilities to do this, ranging from full Eulerian-Eulerian two-phase modeling into one-phase modeling, assuming that the influence of the floc phase on the liquid phase is negligible. In addition, one needs to address the chemical interactions between different species in order to move towards a more general CFD flocculation model.

Symbols

B	body force	N/kg
C_1	turbulence model coefficient	-
C_2	turbulence model coefficient	-
C_μ	turbulence model coefficient	-
G	local absolute velocity gradient	1/s
$\overline{G}$	average velocity gradient	1/s
k	turbulence kinetic energy	J/kg
P	stress production by shear	W/kg
t	time	s
U	velocity in the x-direction	m/s
u_i	fluctuating velocity component	m/s
V	velocity in the y-direction	m/s
W	velocity in the z-direction	m/s
	Greek symbols:	
δ_{ij}	Kronecker delta	-
ε	turbulence dissipation	W/kg
Φ	scalar quantity	-
ϕ	scalar fluctuation	-
μ	dynamic laminar viscosity	Pa·s
μ_{eff}	effective viscosity	Pa·s
ν	kinematic viscosity	m^2/s
ν_t	turbulent viscosity	m^2/s
ρ	density	kg/m^3
σ_k	turbulence model coefficient	-
σ_ε	turbulence model coefficient	-

References

1. CFX-4.2: Solver, AEA Technology plc, 1997
2. Launder, B.E., Spalding, D.B.: Lectures in Mathematical Models of Turbulence. Academic Press, London, England (1972)
3. Camp, T.R., P.C. Stein: Velocity Gradients and Internal Work in Fluid Motion. Jour. Boston Soc. Civil Engrg. 30 (10) (1943) 217
4. Luo, C. : Distribution of Velocities and Velocity Gradients in Mixing and Flocculation vessels: Comparison between LDV data and CFD prediction. Ann Arbor, Mich. UMI Dissertation Services (1997)
5. Korpijärvi, J., Ahlstedt, H., Saarenrinne, P., Reunanen, J.: Modelling of flow field in the Mini-flocculator. In: Fluid Mixing 6 , H. Benkreira (Ed) IChemE 146, UK, (1999) 361-372
6. Amirtharajah A., Tambo, N.: Mixing in Water Treatment. In: Mixing in Coagulation and Flocculation, Amirtharajah A., Clark, M.M., Trussel, R.R. (Eds) American Water Works Research Foundation (1990) 3 - 34
7. Mhaisalkar V.A., Paramasivam R., Bhole A.G.: An Innovative Technique for Determining Velocity Gradient in Coagulation-Flocculation Process. Wat. Res. 20(10) (1986) 1307-1314
8. Rossini M., Garcia Garrido J., Galluzzo M.: Optimization of the Coagulation-Flocculation Treatment: Influence of Rapid Mix Parameters. Wat. Res. 33 (8) (1999) 1817-1826

Flocculation Kinetics of Colloidal Suspensions: Effects of Metallic Coagulant Dosage and Primary Particle Concentration on the Breakup and Aggregation Constants

S.S.F. Filho*, I. Hespanhol* and H.A. Moreira*

Departamento de Engenharia Hidráulica e Sanitária, Escola Politécnica da Universidade de São Paulo, Avenida Professor Almeida Prado 271 – Prédio de Engenharia Civil, Bairro: Cidade Universitária – Cep: 04514-030, São Paulo – Capital – Brazil
ssffilho@usp.br; ivanhesp@usp.br

Abstract

The scientific knowledge of the flocculation process has progressed substantially since the experimental work of Argaman and Kaufman (1970). However several aspects, particularly the ones associated with process mathematical modeling, still remain controversial. One of these controversies surrounds the effect of coagulant dosage and the concentration of primary particles on the aggregation and breakup kinetic constants, K_A and K_B, respectively. In this study, the experimental data collected to evaluate this effect showed that: (a) The dosage of metallic coagulants influences considerably the flocculation kinetics of colloidal suspensions, reflecting directly on K_A and K_B ; (b) For short detention times, in order to allow for adequate aggregation of colloidal particles, the coagulant dosage should be the one which produces the largest possible K_A; for relatively longer detention times, the dosage can be smaller, leading to the sweep coagulation mechanism. In this case the coagulant dosage is actually defined by flocculation and not by coagulation.

Introduction

Flocculation is the single most important process in water treatment. While its qualitative aspects have been well understood since the thirties, important quantitative advances have been achieved only recently.

The pioneer work performed by several researchers [1,2,3] provided the scientific basis for the development of a flocculation kinetics framework, allowing for evaluation of process behavior under different hydraulic conditions.

While the scientific knowledge of the process has progressed substantially since the experimental work of Argaman and Kaufman [3], there are still a few aspects associated with process mathematical modeling that remain controversial. For instance, to what extent are the kinetic constants K_A, the aggregation constant, and K_B, the breakup constant, affected by the metallic coagulant dosage and by the concentration of primary particles in the raw water?

Even though three basic mechanisms can be associated with the flocculation of colloidal suspensions, orthokinetic flocculation is the most important one as far as process engineering is concerned, and that is why it is the mechanism of choice for mathematical modeling.

Assuming that the kinetics of colloidal suspensions are associated with aggregation and breakup phenomena, Argaman and Kaufman [3] demonstrated that the overall process can be described by the following equation:

$$\left(\frac{dn_p}{dt}\right) = -\left(\frac{dn_p}{dt}\right)_a + \left(\frac{dn_p}{dt}\right)_r = -\frac{4}{3}\cdot\alpha\cdot n_p\cdot n_f\left(R_p+R_f\right)^3\cdot\frac{du}{dz}+\frac{Cd\cdot\overline{U}^2\cdot n_f\cdot 4\cdot\pi\cdot}{2} \quad (1)$$

where

$(dn_p/dt)_a$ = rate of change of the concentration of primary particles with time, resulting from the aggregation of colloidal particles ($ML^{-3}\,T^{-1}$),

$(dn_p/dt)_r$ = rate of change of the concentration of primary particles with time, resulting from the breakup of flocs ($ML^{-3}\,T^{-1}$),

n_f = concentration of flocs resulting from precipitation of metallic hydroxides ($ML^{-3}\,T^{-1}$),

R_p = radius of primary particles (L),

R_f = radius of flocs (L),

α = collision aggregation ratio,

du/dz = local laminar velocity gradient (T^{-1}),

C_d = drag coefficient,

U^2 = mean square velocity fluctuation (L^2T^{-2}).

Equation 1 can be simplified to read:

$$\left(\frac{dn_p}{dt}\right) = -K_a\cdot\overline{G}\cdot n_p + K_b\cdot n_0\cdot\overline{G}^2 \quad (2)$$

$$K_A = \frac{K_1\cdot K_2\cdot\phi}{\pi} \quad (3)$$

$$K_B = \frac{3\cdot C_d\cdot K_2\cdot\phi}{2\cdot K_3\cdot n_0} \quad (4)$$

Where

n_0 = concentration of primary particles entering the flocculation system (ML^{-3})

ϕ = floc volume fraction (volume of particles per unit volume of suspension)

G = Overall root-mean square velocity gradient (T^{-1})

K_1, K_2 and K_3 = constants, function of the performance and physical arrangement of the flocculation system

K_A = general aggregation constant or overall flocculation constant (T)

K_B = breakup constant (T)

Both, the aggregation and breakup constants are dependent on the collision aggregation ratio and the concentration of primary particles entering the flocculation system. This implies that the coagulant dosage has a direct influence on flocculation kinetics and, consequently, on process performance.

Objectives

The basic objective was to investigate the effect of coagulant dosage and the concentration of primary particles in the raw water on K_A and K_B. Three synthetic waters prepared with different turbidity values were tested in a "jar test" apparatus, with the goal of determining K_A and K_B for different dosages of a metallic coagulant.

Methods of analysis and experimental apparatus

Water from the public supply system taken at the Civil Engineering Laboratory, University of São Paulo in São Paulo, was utilized for the preparation of the raw water. The concentration of primary particles was associated with turbidity measurements and was adjusted to 15 NTU, 50 NTU and 100 NTU by adding kaolin, to represent waters of low, medium and high turbidity, respectively.

Each water was tested using different coagulant dosages, mean velocity gradients, and flocculation times. The "jar-test" apparatus utilized as batch reactors consisted of 12 individual jars, each one with a volume of 2.0 liters, allowing for simultaneous operation of a large range of detention times.

The metallic coagulant utilized was aluminum sulfate, $Al_2(SO_4)_3.18\ H_2O$. The water pH was maintained between 6.0 and 6.5 by adding 0.1 M sodium hydroxide or 0.1 M hydrochloric acid solutions.

The basic procedure was to perform a group of tests with the same raw water and coagulant dosages but changing the velocity gradient to pre-established values. For each velocity gradient, different detention times were allowed for flocculation, taking advantage of the large number of jars utilized.

Table 1. Coagulant dosages and velocity gradients

Raw water Turbidity NTU	Coagulant dosage mg/l as Al	Velocity Gradient s^{-1}
15	0.81; 1.62; 2.43; 3.24; 4.86; 6.49	20, 30, 40, 60, 80, 100
50	1.62; 2.43; 3.24; 4.86; 6.49; 8.11	20, 30, 40, 60, 80, 100
100	1.62; 2.43; 3.24; 4.86; 6.49; 8.11	20, 30, 40, 60, 80, 100

Once the coagulant dosage was determined, the test was performed a total of 6 runs with velocity gradients of 20, 30, 40, 60, 80 and 100 s^{-1}. The stirring device utilized is made of two flat blades measuring 2.0 cm high by 7.0 cm long, located at opposite sides of the shaft. The velocity gradient was calculated according to the procedures recommended in the literature [4]. Table 1 indicates the dosages and velocity gradients utilized for each one of the turbidities adopted for the raw water

For each velocity gradient, flocculation times of 2.5, 5.0, 7.5, 10.0, 12.5, 15.0, 17.5, 20, 25, 30, 40, 50, and 60 minutes were established. Afterwards, the suspension was allowed to settle for 2 or 4 minutes, which represented settling rates of 27 m^3/m^2·day and 54 m^3/m^2·day, respectively. Samples were also collected after 30 minutes to account for infinite detention time.

The numerical values of the aggregation and breakup constants were calculated through the values of n/n_0 as a function of the flocculation times. By feeding the experimental data into the computational program "SOLVER", it was possible to determine the values of K_A and K_B that minimized a previously established objective function. The computation of the constants K_A and K_B as a function of n_0/n was avoided in order to minimize errors associated with the variation of the settled water turbidity at long detention times.

Results and discussion

Figures 1 and 2 show typical results of the coagulation-flocculation tests for a fixed value of velocity gradient, settling velocity of 54 m/day, and different coagulant dosages for raw waters with turbidity of 15 NTU and 50 NTU respectively.

Regarding the importance of coagulant dosage on process efficiency, it can be observed from Figures 1 and 2 that, for relatively large flocculation times, the lowest settled water turbidity values were obtained with lower values of coagulant dosage.

On the other hand, for the same velocity gradient and increased coagulant dosages there was a decrease in the flocculation efficiency since higher values of n_p/n_0 were observed for the same detention time. This trend was observed for the three levels of water turbidity tested, and for both settling rates, 27 and 54 m/day.

This experimental evidence is very important as far as process engineering is concerned, since the establishment of a coagulant dosage is dependent on detention time as well as on the hydraulic characteristics of the flocculation system.

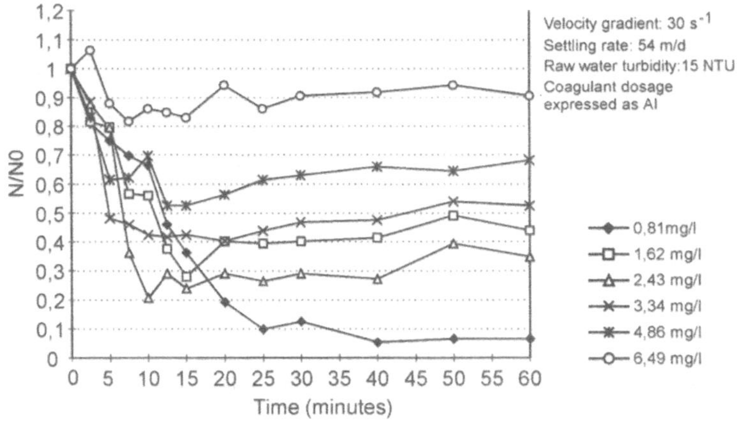

Fig. 1. Ratio of settled water turbidity to raw water turbidity as a function of flocculation time and coagulant dosage. Raw water turbidity: 15 NTU

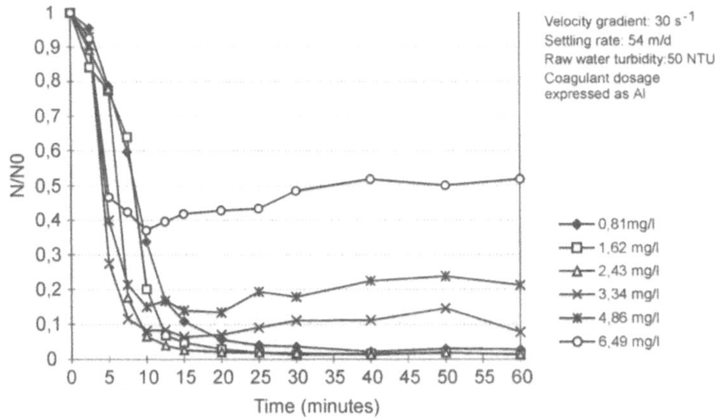

Fig. 2. Ratio of settled water turbidity to raw water turbidity as a function of flocculation time and coagulant dosage. Raw water turbidity: 50 NTU

When a water treatment plant is overloaded, and consequently operating with short detention times, it would be necessary to add higher coagulant dosages to provide higher rates of colloidal particle aggregation. In this case, dosage increment is not associated with the coagulation process (destabilization of colloidal suspensions) but with flocculation, allowing for a higher aggregation rate which will increase the K_A value. If the plant's flocculation system was designed with a longer detention time, the coagulant dosages could be smaller with no hampering of the unit operation.

Figures 3 to 6 show typical values for the aggregation and breakup constants depending on coagulant dosage, velocity gradient and settling rates for the three raw waters investigated.

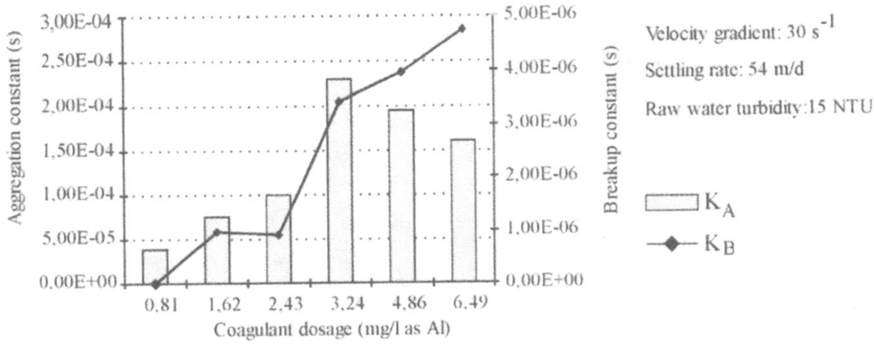

Fig. 3. Aggregation and breakup constants as function of coagulant dosage

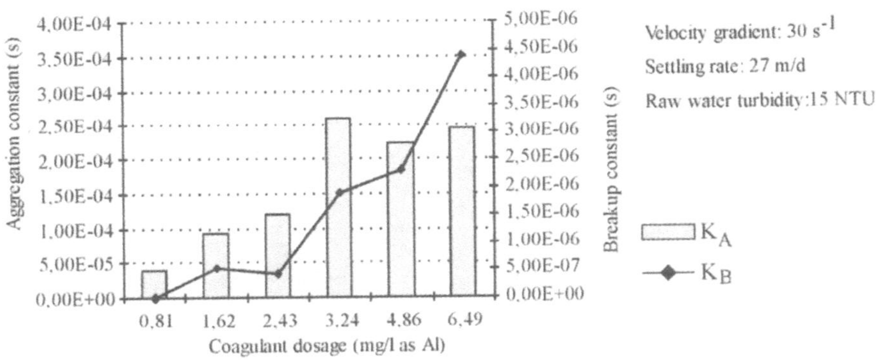

Fig. 4. Aggregation and breakup constants as a function of coagulant dosage

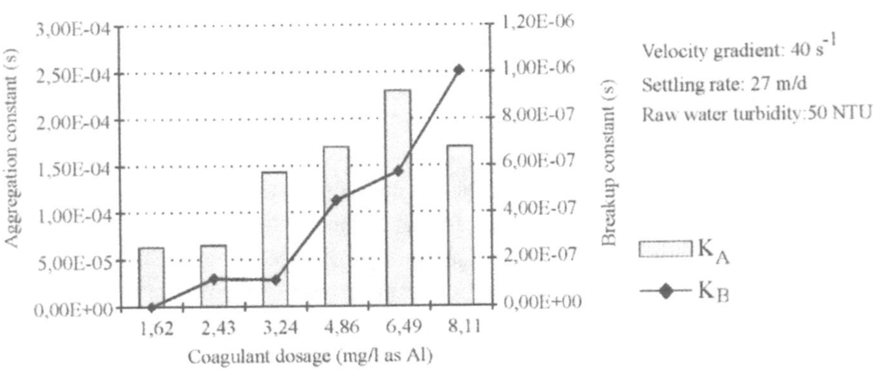

Fig. 5. Aggregation and breakup constants as a function of coagulant dosage

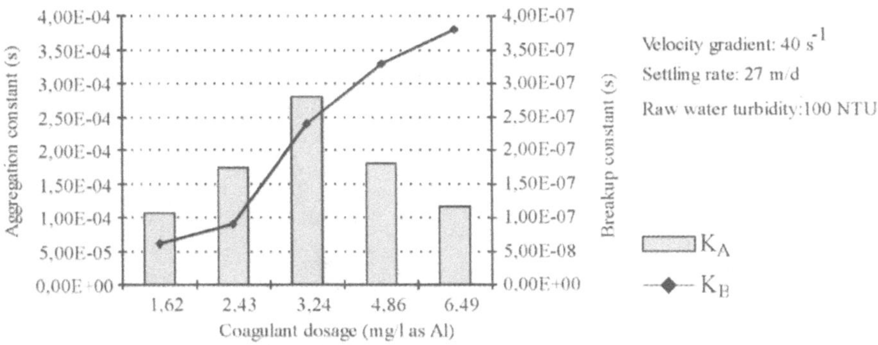

Fig. 6. Aggregation and breakup constants as a function of coagulant dosage

It can be seen for all waters that when the velocity gradient and the settling rate are kept constant, an increase in dosage results in a continuous increase in the aggregation constant. This expected result was actually observed during routine operation of water treatment plants. However, increasing the dosage above a certain value causes a reduction in the aggregation constant, reducing the flocculation efficiency.

Under the same conditions it was observed that increasing the dosage increases the breakup coefficient, indicating that the larger the amount of metallic hydroxide in the floc, the less would be its resistance to shearing tension. This was demonstrated by comparing the breakup constants for the three raw waters: at the same coagulant dosage, the less the turbidity, and the higher the breakup constant.

For short detention times increasing the coagulant dosage to a certain value, the effect is beneficial, since it leads lower breakup constant values and higher aggregation constants, improving flocculation efficiency. For longer detention times the effect happens to be detrimental, since floc breakup starts to be more significant when flocculation is too extensive.

For fixed velocity gradients, increased coagulant dosages resulted in higher values for both the aggregation and the breakup constants. As far as the breakup constant is concerned, the effect is enhanced over longer detention times, since its tendency is to increase faster than the aggregation constant.

Figures 7 and 8 show typical values of the aggregation and breakup constants as a function of turbidity and at the same coagulant dosages, velocity gradients and settling rates.

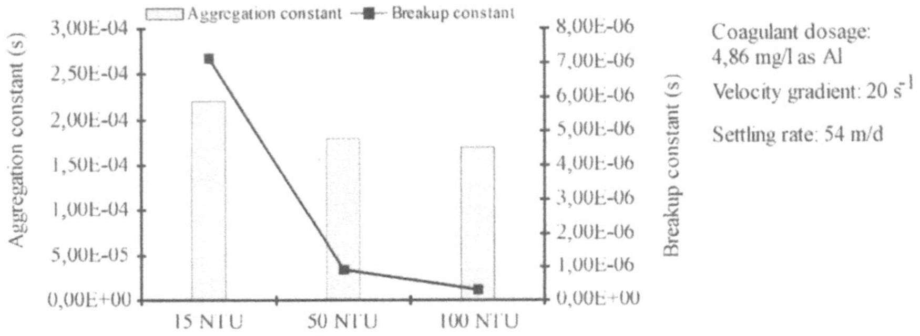

Fig. 7. Aggregation and breakup constants as a function of the primary particle concentration

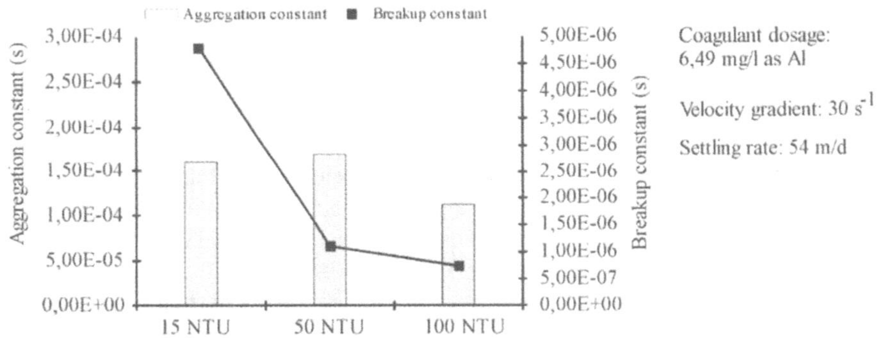

Fig. 8. Aggregation and breakup constants as a function of the primary particle concentration

Increased raw water turbidities led to a significant reduction in the breakup constant, regardless of the coagulant dosage used. The same effect was observed for the aggregation constant, but the variation was very small compared to that for the breakup constant.

It was also observed that the breakup constant does not decrease linearly as raw water turbidity increases. In fact, the largest breakup constant variation rate was observed when the raw water turbidity went from 15 NTU to 50 NTU, while just a marginal decrease was noticed for turbidity values ranging from 50 NTU to 100 NTU. These results suggest that, while the breakup constant is a function of turbidity, the correlation is not linear, as suggested by Argaman and Kaufman [3].

Conclusions

The following conclusions can be drawn from the experimental data:

- The metallic coagulant dosage has a significant effect on the flocculation kinetics of colloidal suspensions, and this is reflected directly in the numerical values of the aggregation and breakup constants.
- The aggregation constant increases with coagulant dosage. However, beyond a certain dosage, the aggregation constant only increased marginally.
- The rate of change in the breakup constant was close to linear as a function of coagulant dosage. This reflects the increased floc fragility with increasing metallic hydroxide content.
- In order for satisfactory flocculation of colloidal particles to occur at short flocculation times, the coagulant dosage has to be close to the one that produces the highest aggregation constant.
- For relatively longer detention times, the dosage can be lower, leading to the sweep coagulation mechanism. In this case, the coagulant dosage is actually defined by flocculation and not by coagulation.

Acknowledgements

The authors are pleased to acknowledge the financial support of Fundação de Amparo à Pesquisa do Estado de São Paulo (FAPESP - Process # 1997/00345-3) and the assistance of Centro Tecnológico de Hidráulica (CTH-DAEE) in constructing the experimental apparatus and for logistical support.

References

1. Hudson, H.E.:. Physical aspects of flocculation, Journal of American Water Works Association, 57 (7) (1965) pp. 885-892
2. Harris, H.S., Kaufman, W.J., Krone, R.B:. Orthokinetic flocculation in water purification. Journal of the Sanitary Engineering Division, 92 (6)(1966) pp. 95-111
3. Argaman, Y. A. , Kaufman, W.J.. Turbulence and flocculation. Journal of the Sanitary Engineering Division, 96 (2) (1970) pp. 223-241
4. Kawamura, S: Integrated design of water treatment facilities. John Wiley & Sons, New York (1991)

Oxidation Processes

Broadening the Scope of Modified Photo-Fenton Processes in Water and Wastewater Treatment through Ferric Complex Design

T.D. Waite, A.J. Feitz and R. Aplin

School of Civil and Environmental Engineering, The University of New South Wales
Sydney, NSW 2052, Australia
d.waite@unsw.edu.au

Abstract

The modified photo-Fenton process is an interesting variant of the standard Fenton reagent that enables degradation of selected contaminants at higher rates and higher pHs than achievable by the thermal process. The effectiveness of the process is variable, however, and its application should be used with care. In some instances, it appears that addition of oxalate actually inhibits contaminant degradation as a result of the formation of non-photoactive ternary oxalato complexes. Scope appears to exist for investigation of alternative Fe(III) chromophores that may retain their photoactivity at even higher pH (and possibly with more effectiveness) than is achieved by ferrioxalate.

Introduction

Fenton's reagent is a well-known process that generates hydroxyl radicals ($OH^{\bullet}$) by the reaction of Fe^{2+} with H_2O_2:

$$Fe^{2+} + H_2O_2 \rightarrow Fe^{III}OH^{2+} + OH^{\bullet} \tag{1}$$

Fe acts as a catalyst, with Fe(II) being regenerated by the reaction of Fe(III) with H_2O_2, though this process is relatively slow with the result that Fe is largely present as Fe(III) [1]. Fenton's reagent is an example of an advanced oxidation process (AOP) that produces $OH^{\bullet}$ to degrade organic contaminants. Hydroxyl radicals are highly reactive and are able to oxidise many organic compounds [2] There has been considerable study of waste treatment using Fenton's reagent and other AOPs such as ozonation, UV/H_2O_2 and UV/TiO_2 [3,4].

Fenton's reagent is most effective at a pH between 2 and 4, with an optimum pH of approximately 3. At higher pHs, iron flocs and precipitates may form [5,6].

The inactivity of Fenton's reagent at pH > 4 has also been attributed to faster decomposition of H_2O_2 to water and oxygen [6] ,oxidation of Fe^{2+} by an oxidant other than H_2O_2 or the reaction between Fe^{2+} and H_2O_2, not producing hydroxyl radicals [7] in this pH region.

A modified form of Fenton's reagent was recently developed [8,9,10], which potentially overcomes some of the problems with the standard Fenton's reagent. The production of $OH^\bullet$ and hence the degradation of organic contaminants is accelerated by the addition of UV light and a ligand such as oxalate to the Fe/H_2O_2 system. Oxalate forms photo-active Fe(III) complexes and also forms Fe(II) complexes. The major reactions that occur between iron, oxalate and H_2O_2 are as follows [11,12] (reactions are shown for the major species present under the conditions of this study, namely, $Fe^{III}(C_2O_4)_3^{3-}$, Fe^{2+} and $HC_2O_4^-$, with similar reactions occurring for related species such as $Fe^{III}(C_2O_4)_2^-$, $Fe^{II}C_2O_4$ and $C_2O_4^{2-}$). Both complexed and uncomplexed Fe(II) react with H_2O_2 to produce $OH^\bullet$ radicals (reaction 1). These radicals can oxidise either oxalate, H_2O_2 or Fe(II) (reactions 2-4).

$$HC_2O_4^- + OH^\bullet \rightarrow C_2O_4^{\bullet-} + H_2O \tag{2}$$

$$H_2O_2 + OH^\bullet \rightarrow HO_2^\bullet + H_2O \tag{3}$$

$$Fe^{2+} + OH^\bullet \rightarrow Fe^{III}OH^{2+} \tag{4}$$

Fe(III)-oxalato complexes undergo a photo-induced ligand to metal charge transfer, producing Fe(II) and an oxalate radical (reaction 5). The oxalate radical can either reduce another Fe(III)-oxalato complex or react with O_2 to produce $O_2^{\bullet-}$ radicals (reactions 6-7). The oxalate radical can also decompose to form $CO_2^{\bullet-}$ radicals, which react similarly.

$$Fe^{III}(C_2O_4)_3^{3-} + h\nu \rightarrow Fe^{II}(C_2O_4)_2^{2-} + C_2O_4^{\bullet-} \tag{5}$$

$$Fe^{III}(C_2O_4)_3^{3-} + C_2O_4^{\bullet-} \rightarrow Fe^{II}(C_2O_4)_3^{4-} + 2CO_2 \tag{6}$$

$$C_2O_4^{\bullet-} + O_2 \rightarrow 2CO_2 + O_2^{\bullet-} \tag{7}$$

The addition of oxalate and UV light to Fenton's reagent therefore has a number of effects [12,13]. The reaction of Fe(II)-oxalato complexes with H_2O_2 is faster than the corresponding reaction of Fe^{2+}(aq), while Fe(III) is reduced to Fe(II) by the photolysis of Fe(III)-oxalato complexes, which is faster than the reduction of Fe(III) in the standard Fenton's process. The presence of the oxalate ligand increases Fe solubility, decreasing the formation of iron flocs and precipitates and possibly allowing the process to be effective over a wider pH range than Fenton's reagent.

In this paper, we examine the effectiveness of the modified Fenton's process in degrading a variety of contaminants (trichlorethylene, a reactive dye, and phenol).

We then assess the scope for improving the effectiveness of the modified Fenton's process by detailed consideration of the role of the oxalate ligand.

Methods

Experiments were performed in a water-jacketed reactor with a quartz window for illumination. Appropriate amounts of sodium oxalate, ferrous sulphate and (if required) a contaminant (TCE, dye, or phenol) were dissolved in Milli-Q (Millipore Corp.) water and 0.1 M HNO_3 to give either 1.2 L or 250 mL solution at the required pH. A 1.2 M H_2O_2 stock solution was made from 30 wt % H_2O_2. To start the experiment, the required volume of either 30 wt % or 1.2 M H_2O_2 was added and the solution was irradiated by an Oriel 200 W Hg arc lamp source. For studies of TCE and oxalate degradation, an ABB 4600 pH/redox transmitter was used to control pH by adding either 1 M HNO_3 or 1 M NaOH as needed to keep the pH of the solution within 0.15 of the required value. The solution was stirred using a magnetic stirrer bead, and water was circulated through the water-jacket to maintain the solution temperature constant. A neutral density filter (Omega Optical) was used to reduce the light intensity by 60% in some instances.

At regular intervals a sample was taken from the reactor and added to a $Na_2S_2O_3$ solution to destroy any remaining H_2O_2. Analysis of organic constituents was then undertaken in various ways. In studies in which oxalate was the only added organic reagent, samples were analysed using a Shimadzu TOC-5000A total organic carbon analyser. The TOC results were then converted to oxalate concentrations, as oxalate was the only source of organic carbon in these samples. In studies of reactive dye degradation, the pH of samples was adjusted to 7 by adding phosphate buffer solution, and dye concentration was analysed using a Hewlett Packard 1100 HPLC system with detection at 525 nm. HPLC was also used in studies of phenol degradation with detection at 280 nm. Trichloroethylene (TCE) was analysed following preconcentration by solid phase micro-extraction. A Hewlett Packard 6890 gas chromatograph with a Hewlett Packard 5973 mass selective detector was used for analysis.

Results

Results from various studies of the application of Fenton's and modified photo-Fenton's reagents are given below to show the possibilities and drawbacks of the light-assisted process. Selected results of more detailed investigations of the mechanisms underlying the photo-Fenton's process are then presented.

Reactive dye degradation

Selected results of the degradation of the mono-azo dye Reactive Red 235 by conventional Fenton's reagent and by a modified photo-Fenton's process at various values of initial pH are shown in Figures 1 and 2, respectively.

Rates of degradation of the azo dye by Fenton's reagent are observed to be slow, particularly at pH 6 where very little degradation is observed over the 90-minute duration of the experiment. Even at pH 3, over 90 minutes was needed for complete degradation of the dye. Much more rapid degradation was observed in the modified photo-Fenton process. At pH 3, complete degradation was achieved within 20 minutes and even at pH 6 more than 80% of the contaminant was degraded in 90 minutes.

In addition to pH, many other system variables might be expected to influence the efficiency of dye degradation including the dye, oxalate, hydrogen peroxide and iron concentrations, and light intensity. Preliminary examination of some of these factors yielded the following selected findings.

Increase in oxalate concentration had little effect on dye degradation rate. This is presumably because (at the added concentration) there is sufficient oxalate to bind all Fe present. In addition, while oxalate can be oxidised by hydroxyl radicals, the reaction of hydroxyl radicals with the dye ($k \cong 9{\cdot}10^9\ M^{-1}s^{-1}$) is presumably considerably faster than with oxalate (discussed in more detail below). Increase in dye concentration (and other reactants proportionately) resulted in a decrease in rate of dye degradation. This effect may simply be a result of the zero order nature of the Fe(III) photodegradation process (where the extent of Fe(II) formation is related to light intensity but not reactant concentration).

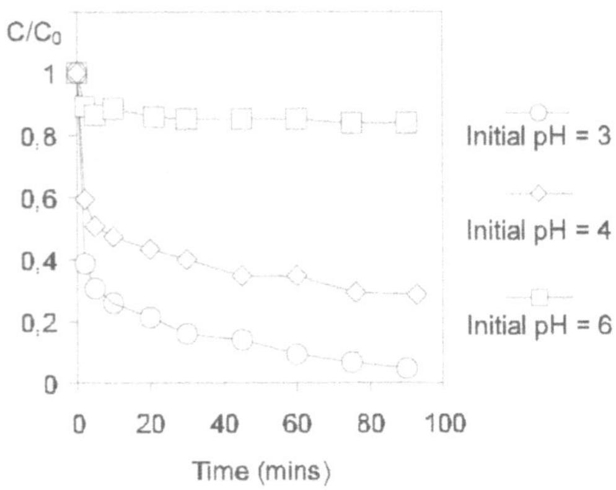

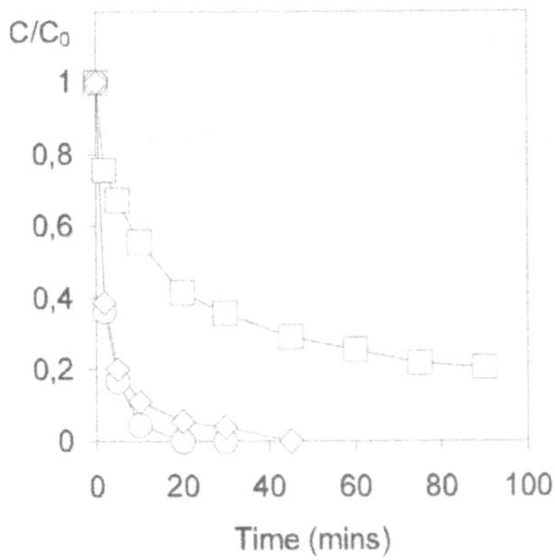

Fig. 1. Degradation of Reactive Red 235 (Co = 0.02 g/L) by modified photo-Fenton's process at various initial pH. [Fe]o= 0.02 mM, [Oxalate]o = 0.06 mM, $[H_2O_2]$o= 0.2 mM.

Fig. 2. Degradation of Reactive Red 235 (Co = 0.02 g/L) by Fenton's reagent at various initial pH values. [Fe]o = 0.02 mM, $[H_2O_2]$o = 0.2 mM.

This effect may also be related to a lower efficiency of hydroxyl radical generation due to a decrease in light absorption by ferrioxalate as a result of absorption by the dye. The dye might also compete with oxalate in the formation of Fe(III) complexes. Such complexes in some instances are themselves light active initiating ligand to metal charge transfer reactions which regenerate Fe(II) [14].

TCE degradation

TCE was degraded rapidly (typically within 5 minutes) by Fenton's reagent at pH ≅ 3 (Figure 3) but the rate of degradation dropped considerably at higher pHs (Figure 4).

In comparison, rapid TCE degradation was maintained up to pH ≅ 6.5 upon using a modified photo-Fenton system. Figure 4 shows selected results for degradation of TCE showing the proportion of initial TCE remaining after 5-minute reaction with both Fenton's reagent and modified photo-Fenton's reagent.

While the dark Fenton degradation of TCE could be satisfactorily modeled using constants within the range of typical literature values (Appendix), a model of the significantly more complex modified photo-Fenton process has yet to be developed.

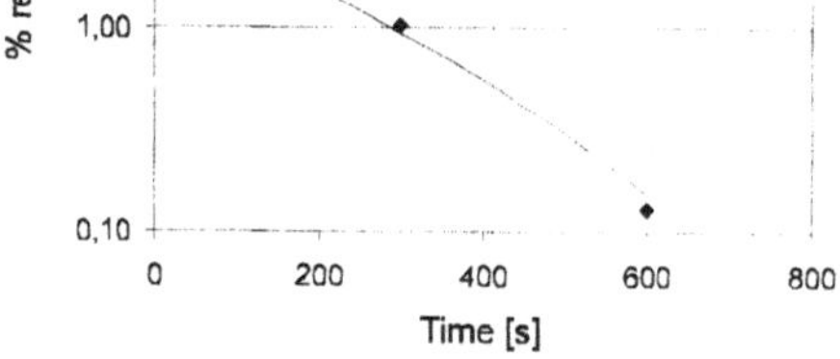

Fig. 3. Degradation of 0.076 mM TCE at pH 3 using Fenton's reagent with [H O] = 1.47 mM and $[H_2O_2]/[Fe^{2+}]$ = 11/1. Model given in Appendix 1.

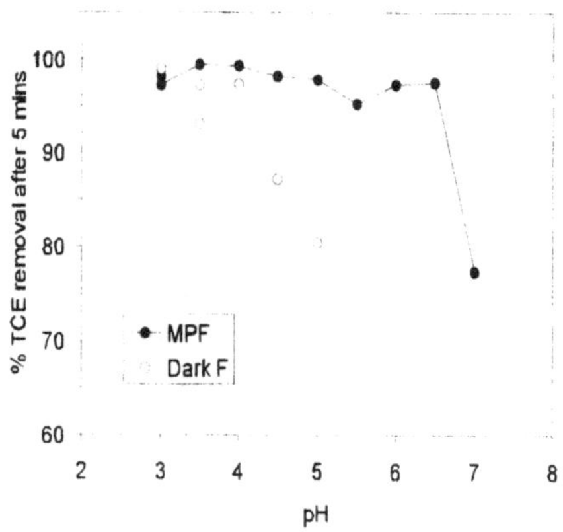

Fig. 4. Percentage of TCE remaining in solution after 5-minute exposure to either standard Fenton's reagent (Dark F) or modified photo-Fenton's reagent (MPF) at various initial pH values. Conditions as in Figure 3 but with $[H_2O_2]/[C_2O_4^{2-}]$ = 10/3 in modified photo-Fenton studies.

Phenol degradation

Selected results of studies onto the degradation of phenol using both a light-assisted Fenton ("photo-Fenton") reagent and a modified photo-Fenton's reagent are shown in Figures 5 and 6.

As can be seen from Figure 5, the ability of an Fe(II)/H_2O_2 mixture to catalyse the breakdown of phenol is dramatic with a strong dependency of the rate of degradation on the hydrogen peroxide concentration. On introduction of oxalate, however, significantly diminished rates of degradation are obtained (Figure 6). While the reason for the lack of effectiveness of the oxalate-modified photo-Fenton reagent in the case of phenol is unclear, it may be related to the tendency of Fe(III) to form relatively non-photoreactive mixed ligand oxalato-phenolate complexes [15]. The lower quantum yields of Fe(II) formation for such mixed ligand complexes have been attributed to the greater number of possibilities for radiationless deactivation and to inner filter effects [16].

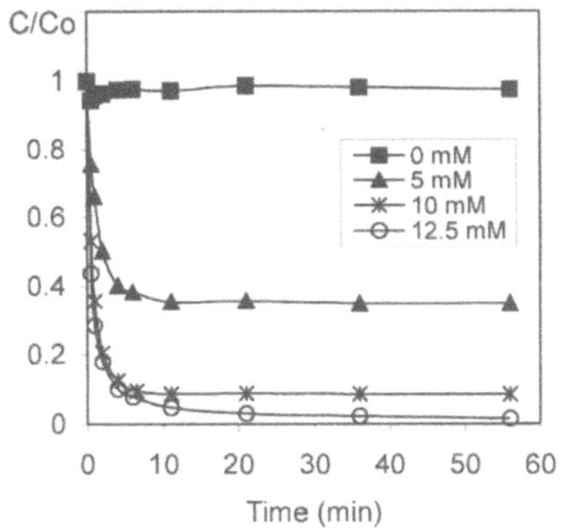

Fig. 5. Effect of H_2O_2 concentration on degradation of phenol using photo-Fenton's reagent at pH 3. [Phenol]o = 2.5 mM and [Fe(II)]o = 0.5 mM.

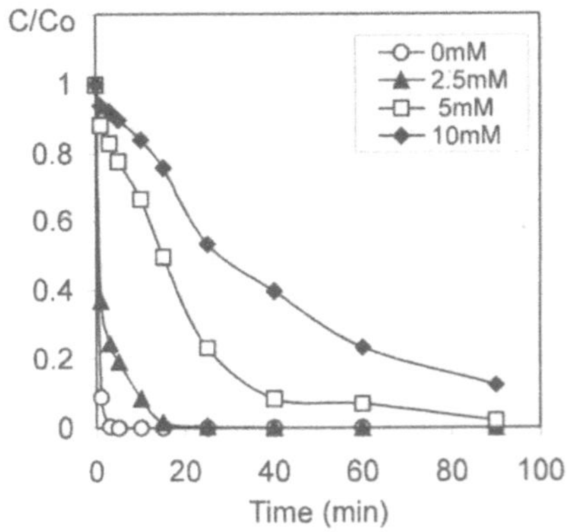

Fig. 6. Effect of oxalate concentration on degradation of phenol using modified photo-Fenton's reagent at pH 3. [Phenol]o = 2.5 mM, $[H_2O_2]$o = 15 mM and [Fe(II)]o = 1 mM.

Oxalate degradation

In order to better understand the mechanism underlying the photo-Fenton process, we investigated the degradation of oxalate in the absence of other hydroxyl radical scavengers. Such a process may be considered to be a photo-Fenton degradation of oxalate and should enable us to a) develop a kinetic model of use in underpinning modeling efforts in more complex systems (involving added radical scavengers), and b) gain added insight into the factors critical in determining the effectiveness of the photo-Fenton process.

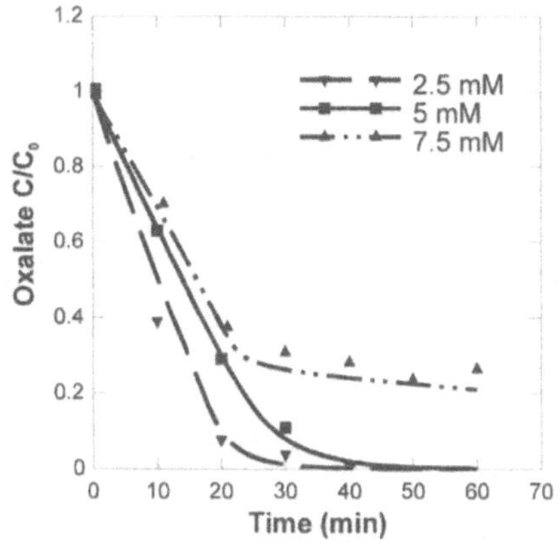

Fig. 7. Effect of initial oxalate concentration on oxalate degradation.

In each case, $[H_2O_2]o = 5$ mM, $[Fe] = 0.5$ mM, pH = 3 and a neutral density filter was used. Symbols represent experimental results and lines represent model results (see Appendix).

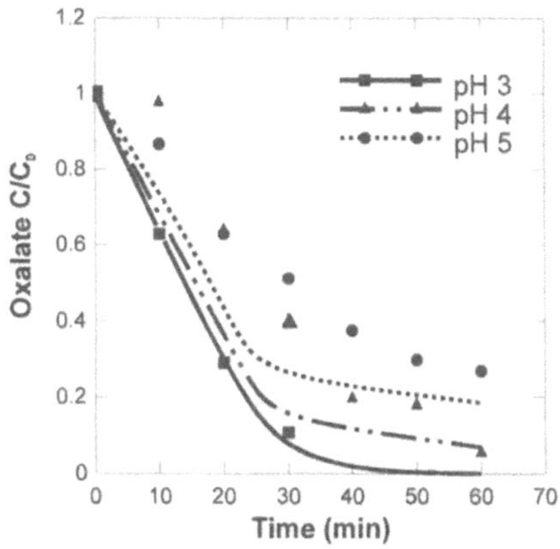

Fig. 8. Effect of pH on oxalate degradation.

In each case, [oxalate]o = 5 mM, $[H_2O_2] = 5$ mM, $[Fe] = 0.5$ mM and a neutral density filter was used. Symbols represent experimental results and lines represent model results (see Appendix).

Results showing the effect of initial oxalate concentration and pH on oxalate degradation are shown in Figures 7 and 8.

As the initial oxalate concentration was increased, the degradation of oxalate became relatively slower, though the absolute degradation rate was increased as a result of oxalate consuming a larger fraction of $OH^{\bullet}$ radicals (reactions 2 in Appendix, Table 2). When the initial concentration of oxalate was greater than that of H_2O_2, the degradation of oxalate became very slow once all H_2O_2 was consumed (reaction 1 in Appendix, Table 2 no longer effective).

The decrease in rate of oxalate degradation on increase in pH may be attributed to a number of factors though Fe(III) hydrolysis as complete Fe complexation by oxalate is not predicted from equilibrium calculations. Thermodynamic calculations indicate that there is an increase in the concentration of the trioxalato species $Fe^{III}(C_2O_4)_3^{3-}$ at the expense of the dioxalato species $Fe^{III}(C_2O_4)^{2-}$ on increase in pH from 3 to 5. We have found the lower dentate oxalate complexes to be somewhat more effective Fe(II) generators than the higher dentate complexes. However, this effect is unlikely to be significant enough to account for the observed decrease in oxalate degradation rate (as shown by the underprediction of the model in Figure 8). Abrahamson et al. [17] investigated the photoreduction of a range of Fe(III) carboxylate complexes and also found a reduction in Fe(II) quantum yield (0.65 to 0.30) on increasing the pH of ferrioxalate solutions from

2.7 to 4.0 though, again, speciation changes (and the difference in photoreactivity of these species) would seem inadequate to account for his observations.

Discussion

The above results provide a range of insights into the behaviour of the modified photo-Fenton process and provide possible avenues for refinement. An important first step, however, is to confirm that the process is effective. While the results for contaminants such as reactive dyes and TCE are promising, other contaminants (such as phenol) appear to be unsuitable candidates for a ferrioxalate-driven degradation process.

The advantages in adopting such a process should be clear prior to implementation. Thus, the results of studies on Reactive Red 235 suggest advantages in using the modified photo-Fenton process both at low pHs (where a standard Fenton reagent might otherwise have been used) and also at higher (near neutral) pH where the conventional process is inoperative. In comparison to the results for the reactive dye, however, TCE is seen to be rapidly degraded by standard Fenton reagent (with little enhancement through use of the modified photo-Fenton process) at low pH. The modified photo-Fenton process does, however, have significant advantages in enabling effective degradation of TCE up to pH 6.5.

The results of experimental and computational studies of oxalate degradation also provide useful insights into optimal operating conditions and possible choice of complexant. For example, it appears that whilst operation at near neutral pH is possible, the effectiveness of the ferrioxalate-mediated process is lower than may be achievable with other chromophores. As mentioned above, the quantum yield for Fe(II) production on photolysis of ferrioxalate drops from 0.65 to 0.40 on increase in pH from 2.7 to 4.0. This reduced quantum yield may account for the observed reduction in rate and extent of oxalate degradation at higher pH. In comparison, the quantum yields for production of Fe(II) on photolysis of L(+)-tartaric acid and citric acid increase from 0.40 to 0.58 and 0.28 to 0.45, respectively, on increasing the solution pH from 2.7 to pH 4.0. Thus, while other factors doubtless play a role, alternate complexes with Fe(III) in place of ferrioxalate may yield a greater rate of Fe(II) production and, subsequently, hydroxyl radical production. As so elegantly summarised some years ago by Balzani and Carassiti [18] and more recently by Sima and Makanova [19], a wide variety of photolabile ferric complexes exist and considerable scope for development exists in examining some of the more likely systems.

A word of caution should be aired in that whilst photolability of the ferric complex is important, so too is the relative insensitivity of the complex (or free ligand) to attack by hydroxyl radicals. As shown above, in the absence of a hydroxyl radical scavenger, oxalate itself is degraded. Whilst a more reactive contaminant may render any competition from the complexing ligand insignificant, this issue should be borne in mind as alternative Fe(III) ligands are assessed.

References

1. Pignatello J. J.: Dark and photoassisted Fe^{3+}-catalyzed degradation of chlorophenoxy herbicides by hydrogen peroxide. Environ. Sci. Technol., 26 (5) (1992) 944-951
2. Buxton G. V., Greenstock C. L., Helman W. P., Ross A. B.: Critical review of rate constants for reactions of hydrated electrons, hydrogen atoms and hydroxyl radicals ($\cdot OH/\cdot O^-$) in aqueous solution. J. Phys. Chem. Ref. Data, 17 (2) (1988) 513-886
3. Venkatadri R., Peters R. W.: Chemical oxidation technologies: ultraviolet light/hydrogen peroxide, Fenton's reagent, and titanium dioxide-assisted photocatalysis. Haz. Waste Haz. Mat., 10 (2) (1993) 107-149
4. Rice R. G. : Applications of ozone for industrial wastewater treatment – a review. Ozone Sci. Eng., 18 (6) (1996) 477-515
5. Kiwi J., Pulgarin C., Peringer P., Grätzel M.: Beneficial effects of homogeneous photo-Fenton pretreatment upon the biodegradation of anthraquinone sulfonate in waste water treatment. Applied Catal. B: Environ., 3 (1993) 85-99
6. Pulgarin C., Kiwi J.: Overview on photocatalytic and electrocatalytic pretreatment of industrial non-biodegradable pollutants and pesticides. Chimia, 50 (3) (1996) 50-55
7. Gallard H., De Laat J., Legube B. : Influence du pH sur la vitesse d'oxydation de composés organiques par $FeII/H_2O_2$: mécanismes réactionnels et modélisation. New J. Chem., 22 (3) (1998) 263-268
8. Safarzadeh-Amiri A. : Photocatalytic method for treatment of contaminated water. U.S. Patent No. 5, 266, 214 (1993)
9. Sun Y., Pignatello J. J.: Activation of hydrogen peroxide by iron(III) chelates for abiotic degradation of herbicides and insecticides in water. J. Agric. Food Chem., 41 (2) (1993) 308-312
10. Safarzadeh-Amiri A., Bolton J. R., Cater S. R.: Ferrioxalate-mediated photodegradation of organic pollutants in contaminated water. Water Res., 31 (4) (1997) 787-798
11. Zuo Y., Hoigné J.: Formation of hydrogen peroxide and depletion of oxalic acid in atmospheric water by photolysis of iron(III)-oxalato complexes. Environ. Sci. Technol., 26 (5) (1992) 1014-1022
12. Faust B. C., Zepp R. G.: Photochemistry of aqueous iron(III)-polycarboxylate complexes: roles in the chemistry of atmospheric and surface waters. Environ. Sci. Technol., 27 (12) (1993) 2517-2522
13. Balmer M. E., Sulzberger B. : Atrazine degradation in irradiated iron/oxalate systems: effects of pH and oxalate. Environ. Sci. Technol., 33 (14) (1999) 2418-2424
14. Herrera F., Kiwi J., Lopez A., Nadtochenko V.: Photochemical decoloration of Remazol Brilliant Blue and Uniblue A in the presence of Fe^{3+} and H_2O_2. Environ. Sci. Technol. 33 (18) (1999) 3145-3151
15. Horvath O., Stevenson K. L.: Charge Transfer Photochemistry of Coordination Compounds. VCH, New York (1993)
16. Roewer G., Kempe G.: Photolyse von ternären Fe(III)-komplexen mit oxalat- and phenolat-liganden. J. Prakt. Chem. 323 (6) (1981) 864-868
17. Abrahamson H. B., Rezvani A. B., Brushmiller J. G.: Photochemical and spectroscopic studies of complexes of iron(III) with citric acid and other carboxylic acids. Inorg. Chim. Acta, 226 (1994) 117-127
18. Balzani V., Carassiti V.: Photochemistry of Coordination Compounds. Academic Press, New York (1970)
19. Sima, J., Makanova, J.: Photochemistry of iron(III) complexes. Cord. Chem. Rev. 160 (1997) 161-189

20. Tang W. Z., Huang C. P. : Stoichiometry of Fentons reagent in the oxidation of chlorinated aliphatic organic pollutants. Environ. Technol., 18 (1) (1997) 13-23
21. Chen R. Z., Pignatello J. J.: Role of quinone intermediates as electron shuttles in Fenton and photoassisted Fenton oxidations of aromatic compounds. Environ. Sci. Technol., 31 (8) (1997) 2399-2406
22. Sedlak D. L., Hoigné J. : The role of copper and oxalate in the redox cycling of iron in atmospheric waters. Atmos. Environ., 27A (14) (1993) 2173-2185
23. Bielski B. H. J., Cabelli D. E., Arudi R. L., Ross A. B.: Reactivity of HO_2/O_2^- radicals in aqueous solution. J. Phys. Chem. Ref. Data, 14 (4) (1985) 1041-1100
24. Huston P. L., Pignatello J. J.: Reduction of perchloroalkanes by ferrioxalate-generated carboxylate radical preceding mineralization by the photo-Fenton reaction. Environ. Sci. Technol., 30 (12) (1996) 3457-3463
25. Goldstein S., Czapski G., Cohen H., Meyerstein D.: Formation and decomposition of iron-carbon σ-bonds in the reaction of iron(II)-poly(amino carboxylate) complexes with CO_2^- free radicals. A pulse radiolysis study. J. Am. Chem. Soc., 110 (1988) 3903-3907
26. Morel F. F., Hering J. G.: Principles and Applications of Aquatic Chemistry. Wiley, New York (1993)
27. Aplin R., Waite T. D.: Degradation of oxalate in a modified photo-Fenton (UV/ferrioxalate/H_2O_2) system. Environ. Sci. Technol. (2000) (submitted).

Appendix

Table 1. Model of Fenton reagent degradation of TCE

Reaction	k or K	Ref.
(1) $ClCH{=}CCl_2 + OH^\bullet \rightarrow {}^\bullet CCl_2CHClOH$	$k = 4.2 \times 10^9\ M^{-1}s^{-1}$	2
(2) $Fe(II) + H_2O_2 \rightarrow Fe(III) + OH^\bullet$	$k = 51\ M^{-1}s^{-1}$	20
(3) $Fe(II) + OH^\bullet \rightarrow Fe(III)$	$k = 4.3 \times 10^8\ M^{-1}s^{-1}$	1
(4) $Fe(II) + HO_2^\bullet\ (+H^+) \rightarrow Fe(III) + H_2O_2$	$k = 1.2 \times 10^6\ M^{-1}s^{-1}$	11
(5) $Fe(II) + O_2^{\bullet-}\ (+2H^+) \rightarrow Fe(III) + H_2O_2$	$k = 1 \times 10^7\ M^{-1}s^{-1}$	11
(6) $Fe(III) + HO_2^\bullet \rightarrow Fe(II) + O_2 + H^+$	$k = 2.7 \times 10^5\ M^{-1}s^{-1}$	*b*
(7) $Fe(III) + H_2O_2 \rightarrow Fe(II) + HO_2^\bullet$	$k = 0.02\ M^{-1}s^{-1}$	1
(8) $Fe(III) + O_2^{\bullet-} \rightarrow Fe(II) + O_2$	$k = 1.5 \times 10^8\ M^{-1}s^{-1}$	11
(9) $H_2O_2 + OH^\bullet \rightarrow HO_2^\bullet + H_2O$	$k = 2.7 \times 10^7\ M^{-1}s^{-1}$	20
(10) $HO_2^\bullet + HO_2^\bullet \rightarrow H_2O_2 + O_2$	$k = 8.3 \times 10^5\ M^{-1}s^{-1}$	11
(11) $HO_2^\bullet + O_2^{\bullet-}\ (+H^+) \rightarrow H_2O_2 + O_2$	$k = 9.7 \times 10^7\ M^{-1}s^{-1}$	11
(12) $OH^\bullet + OH^\bullet \rightarrow H_2O_2$	$k = 5.3 \times 10^9\ M^{-1}s^{-1}$	20
(13) $HO_2^\bullet \leftrightarrow O_2^{\bullet-} + H^+$	$K = 1.6 \times 10^{-5}\ M$	11

b Fitted value.

Table 2. Model of photo-Fenton reagent degradation of oxalate *

Reaction	k or K	Ref.
Major reactions		
(1a) $Fe^{2+} + H_2O_2 \rightarrow Fe^{III}OH^{2+} + OH^{\bullet}$	$k = 76\ M^{-1}s^{-1}$	21
(1b) $Fe^{III}(C_2O_4)_m^{2-2m} + H_2O_2 \rightarrow Fe^{II}(C_2O_4)_m^{3-2m} + OH^- + OH^{\bullet}$	$k = 3.1 \times 10^4\ M^{-1}s^{-1}$	22
(2a) $H_2C_2O_4 + OH^{\bullet} \rightarrow C_2O_4^{\bullet -} + H_2O + H^+$	$k = 1.4 \times 10^6\ M^{-1}s^{-1}$	2
(2b) $HC_2O_4^- + OH^{\bullet} \rightarrow C_2O_4^{\bullet -} + H_2O$	$k = 4.7 \times 10^7\ M^{-1}s^{-1}$	2
(2c) $C_2O_4^{2-} + OH^{\bullet} \rightarrow C_2O_4^{\bullet -} + OH^-$	$k = 7.7 \times 10^6\ M^{-1}s^{-1}$	2
(2d) $Fe^{III}(C_2O_4)_n^{3-2n} + OH^{\bullet} \rightarrow Fe^{III}(C_2O_4)_{n-1}^{5-2n} + C_2O_4^{\bullet -} + OH^-$	$k = 1.5 \times 10^8\ M^{-1}s^{-1}$	
(3) $H_2O_2 + OH^{\bullet} \rightarrow HO_2^{\bullet} + H_2O$	$k = 2.7 \times 10^7\ M^{-1}s^{-1}$	2
(4a) $Fe^{2+} + OH^{\bullet} \rightarrow Fe^{III}OH^{2+}$	$k = 4.3 \times 10^8\ M^{-1}s^{-1}$	2
(4b) $Fe^{II}(C_2O_4)_m^{2-2m} + OH^{\bullet} \rightarrow Fe^{III}(C_2O_4)_m^{3-2m} + OH^-$	$k = 1 \times 10^{10}\ M^{-1}s^{-1}$	
(5) $Fe^{III}(C_2O_4)_n^{3-2n} + h\nu \rightarrow Fe^{II}(C_2O_4)_{n-1}^{4-2n} + C_2O_4^{\bullet -}$		
(6a) $Fe^{III}(OH)_p^{3-p} + C_2O_4^{\bullet -} \rightarrow Fe^{II}(OH)_p^{2-p} + 2CO_2$	$k = 8 \times 10^8\ M^{-1}s^{-1}$	
(6b) $Fe^{III}(C_2O_4)_n^{3-2n} + C_2O_4^{\bullet -} \rightarrow Fe^{II}(C_2O_4)_n^{2-2n} + 2CO_2$	$k = 8 \times 10^9\ M^{-1}s^{-1}$	
(7) $C_2O_4^{\bullet -} + O_2 \rightarrow 2CO_2 + O_2^{\bullet -}$	$k = 2.4 \times 10^9\ M^{-1}s^{-1}$	22
Other reactions		
(8a) $Fe^{2+} + HO_2^{\bullet}\ (+H^+) \rightarrow Fe^{3+} + H_2O_2$	$k = 1.2 \times 10^6\ M^{-1}s^{-1}$	23
(8b) $Fe^{II}(C_2O_4)_m^{2-2m} + HO_2^{\bullet}\ (+H^+) \rightarrow Fe^{III}(C_2O_4)_m^{3-2m} + H_2O_2$	$k = 1 \times 10^7\ M^{-1}s^{-1}$	
(9a) $Fe^{2+} + O_2^{\bullet -}\ (+2H^+) \rightarrow Fe^{3+} + H_2O_2$	$k = 1 \times 10^7\ M^{-1}s^{-1}$	23
(9b) $Fe^{II}(C_2O_4)_m^{2-2m} + O_2^{\bullet -}\ (+2H^+) \rightarrow Fe^{III}(C_2O_4)_m^{3-2m} + H_2O_2$	$k = 1 \times 10^7\ M^{-1}s^{-1}$	
(10) $Fe^{III}(OH)_p^{3-p} + H_2O_2 \rightarrow Fe^{II}(OH)_p^{2-p} + HO_2^{\bullet} + H^+$	$k = 0.01\ M^{-1}s^{-1}$	21
(11a) $Fe^{III}(OH)_p^{3-p} + HO_2^{\bullet} \rightarrow Fe^{II}(OH)_p^{2-p} + O_2 + H^+$	$k = 3.1 \times 10^5\ M^{-1}s^{-1}$	23
(11b) $Fe^{III}(C_2O_4)_n^{3-2n} + HO_2^{\bullet} \rightarrow Fe^{II}(C_2O_4)_n^{2-2n} + O_2 + H^+$	$k = 1 \times 10^2\ M^{-1}s^{-1}$	
(12a) $Fe^{III}(OH)_p^{3-p} + O_2^{\bullet -} \rightarrow Fe^{II}(OH)_p^{2-p} + O_2$	$k = 1.5 \times 10^8\ M^{-1}s^{-1}$	21
(12b) $Fe^{III}(C_2O_4)_n^{3-2n} + O_2^{\bullet -} \rightarrow Fe^{II}(C_2O_4)_n^{2-2n} + O_2$	$k = 1 \times 10^3\ M^{-1}s^{-1}$	
(13) $C_2O_4^{\bullet -} + H_2O_2 \rightarrow 2CO_2 + OH^- + OH^{\bullet}$	$k = 7.3 \times 10^5\ M^{-1}s^{-1}$	24
(14) $C_2O_4^{\bullet -} \rightarrow CO_2 + CO_2^{\bullet -}$	$k = 2 \times 10^6\ s^{-1}$	24
(15) $CO_2^{\bullet -} + CO_2^{\bullet -} \rightarrow C_2O_4^{2-}$	$k = 5 \times 10^8\ M^{-1}s^{-1}$	25
(16) $HO_2^{\bullet} + HO_2^{\bullet} \rightarrow H_2O_2 + O_2$	$k = 8.5 \times 10^5\ M^{-1}s^{-1}$	23
(17) $HO_2^{\bullet} + O_2^{\bullet -}\ (+H^+) \rightarrow H_2O_2 + O_2$	$k = 9.7 \times 10^7\ M^{-1}s^{-1}$	23
Equilibria		
(18) $H_2C_2O_4 \leftrightarrow HC_2O_4^- + H^+$	$K = 5.62 \times 10^{-2}\ M$	12
(19) $HC_2O_4^- \leftrightarrow C_2O_4^{2-} + H^+$	$K = 5.37 \times 10^{-5}\ M$	12
(20) $Fe^{3+} + OH^- \leftrightarrow Fe^{III}OH^{2+}$	$K = 6.31 \times 10^{11}\ M^{-1}$	26
(21) $Fe^{III}OH^{2+} + OH^- \leftrightarrow Fe^{III}(OH)_2^+$	$K = 3.16 \times 10^{10}\ M^{-1}$	26
(22) $Fe^{III}(OH)_2^+ + OH^- \leftrightarrow Fe^{III}(OH)_3$	$K = 3.16 \times 10^{16}\ M^{-1}$	26
(23) $Fe^{3+} + C_2O_4^{2-} \leftrightarrow Fe^{III}C_2O_4^+$	$K = 2.51 \times 10^9\ M^{-1}$	12
(24) $Fe^{III}C_2O_4^+ + C_2O_4^{2-} \leftrightarrow Fe^{III}(C_2O_4)_2^-$	$K = 6.31 \times 10^6\ M^{-1}$	12
(25) $Fe^{III}(C_2O_4)_2^- + C_2O_4^{2-} \leftrightarrow Fe^{III}(C_2O_4)_3^{3-}$	$K = 1.58 \times 10^4\ M^{-1}$	12
(26) $Fe^{2+} + C_2O_4^{2-} \leftrightarrow Fe^{II}C_2O_4$	$K = 2.00 \times 10^4\ M^{-1}$	12
(27) $Fe^{II}C_2O_4 + C_2O_4^{2-} \leftrightarrow Fe^{II}(C_2O_4)_2^{2-}$	$K = 126\ M^{-1}$	12
(28) $HO_2^{\bullet} \leftrightarrow O_2^{\bullet -} + H^+$	$K = 1.58 \times 10^{-5}\ M$	12

* See Aplin and Waite [27] for details

Modelling Full-Scale Advanced Micropollutant Oxidation

M. Boller, U. von Gunten, R. Pianta and L. Solcà

Swiss Federal Institute for Environmental Science and Technology (EAWAG),
CH-8600 Duebendorf, Switzerland
Boller@eawag.ch

Abstract

Advanced oxidation has proven to be an efficient process to increase micropollutant removal compared to conventional ozonation and to lower considerably the load to subsequent activated carbon treatment. Laboratory tests to determine the most important reaction rate constants involved in conventional and advanced oxidation and the OH radical/ozone ratio in the specific water to be treated can provide the basic parameters to simulate direct ozone and OH radical reactions. The combination of tracer based hydraulic reactor and chemical reaction modelling made it possible to simulate full-scale reactor performance. Model extrapolations used to optimise the peroxide dosing point for maximum micropollutant removal showed that disinfection for Cryptosporidium may severely be impaired.

Introduction

Ozonation is a water treatment process usually applied in combination with other processes for potable water production from surface waters or sometimes as a single step for the treatment of contaminated spring and groundwater. Ozone (O_3), primarily introduced as a powerful disinfectant, reacts with all kinds of substances present in the water to be treated. Positive effects are the oxidation of NOM to biodegradable compounds and the removal of certain micropollutants. The possible production of oxidation by-products, above all the formation of bromate, has to be mentioned as a negative point. The oxidation reactions are governed by the concentrations of O_3 and OH radicals. The latter species is produced during the decay of O_3, which is strongly dependent on the water constituents. In order to enhance the formation of OH radicals, some treatment plants combine O_3 with H_2O_2 (hydrogen peroxide) in a so-called advanced oxidation process.

Treatment plant operators are often concerned with optimising dose and dosing point for the two chemicals involved. With the complex nature of the advanced oxidation process, only enormous analytical efforts may lead to optimised operation on an empirical basis. With the potential carcinogenicity of bromate, the aim of process optimisation is to minimise the formation of bromate (molecular ozone and OH radical reactions) while ensuring efficient disinfection (mainly molecular ozone reactions) and micropollutant oxidation (molecular ozone and OH radical reactions, [1]). In view of the many goals to be attained and the many chemical reactions involved in advanced oxidation, process modelling may be a useful tool to gain insight into the relative importance of the reactions and into hydraulic reactor performance [5].

In order to allow reality-consistent modelling, kinetic data such as O_3 decay, the molar concentration ratio of OH radical/O_3 (R_{ct}-value), and the reaction constants of O_3 and OH radicals with various micropollutants and microorganisms have to be known or elaborated in the laboratory. In addition, empirical information on the hydraulic reactor configuration in the form of tracer studies is necessary to describe reactor performance mathematically.

The modelling concept was investigated as a case study at the treatment plant of Porrentruy, Switzerland. The raw water contains many trace organics of which trichlorethene (TCE), tetrachlorethene (PCE) and atrazine are considered to be major contaminants. The study is therefore focused on the removal efficiency of these compounds by advanced oxidation.

Materials and Methods

Raw water characteristics

The experimental studies were performed with karstic spring water originating from calcite rock formations in Western Switzerland. The major quality problems are caused by natural turbidity from soil particles, hygienic contaminants, taste and odour, frequently elevated levels of ammonia, nitrates and pesticides from agriculture, chlorinated hydrocarbons and other synthetics from leachates of solid waste deposits. Most of the quality parameters depend strongly on weather conditions and fluctuate in an unpredictable way. Average and maximum raw water concentrations of various quality parameters are depicted in Table 1.

Chemical Analysis and Laboratory Tests

DOC, alkalinity and pH were measured according to standard methods. The ozone concentration was determined by the indigo method [3]. The OH radical concentration cannot be measured directly but was determined from laboratory experiments in which the R_{ct}-value (ratio of O_3/OH radical) was measured. From this value the OH radical concentration in the reactor can be calculated [4,5]. TCE and PCE were analysed by head space GC-ECD with a detection limit < 1 μg/l. Atrazine was measured by HPLC-UV after preconcentration with solid phase extraction, with a detection limit of 10-20 ng/l.

Table 1. Raw water characteristics of the spring Betteraz in Porrentruy, Switzerland

Parameter		Average concentration range	Maximum value
pH		7.1 – 7.5	
Conductivity	µS/cm	502 - 573	
Turbidity	FTU	0.7 – 4.9	300
DOC	mgC/l	0.6 – 1.8	4
Nitrate	mgNO_3/l	26 - 30	40
Hardness	mg$CaCO_3$/l	310 - 340	
Calcium	mgCa/l	110 - 120	
Magnesium	mgMg/l	3	
Chloride	mgCl/l	14	32
Sulfate	mgSO_4/l	21	28
Bromide	µgBr/l	17	
Trichlorethene	µg/l	10	30
Tetrachlorethene	µg/l	10	70
Atrazine	ng/l	50-200	500
E. coli	Number/100ml	0-200	5000
HPC	Number/ml	0-1600	10000
Cryptosporidium	Number/20l	2	5

Laboratory experiments with the raw water were carried out in order to (1) determine the first-order rate constant for ozone decomposition, (2) measure the ratio of OH radical and ozone concentration (R_{ct}-value) during the ozonation experiment, and (3) determine the rate constants for the reaction of the different micropollutants with ozone and OH radicals as a function of temperature.

Treatment Plant

The spring water is treated in a multi-stage plant including flocculation, coarse filtration, advanced ozonation, fine filtration, layered upflow GAC filtration, fixed bed GAC filtration and final disinfection with ClO_2. The hydraulic design capacity is 260 m^3/h. The ozone reactor is placed between the two contact filter steps and consists of 4 chambers in series. Ozone is dosed through a gas diffuser into the first chamber and H_2O_2 is added as approx. 15% solution in the last chamber usually in a molar ratio of 0.3-0.5 mol H_2O_2/mol O_3. The reactor scheme in Figure 1 shows the dosing points for the two oxidants and the sampling points used in this investigation. The theoretical detention time in the total reactor volume is 47.4 min at the normal operational flow of 200 m^3/h.

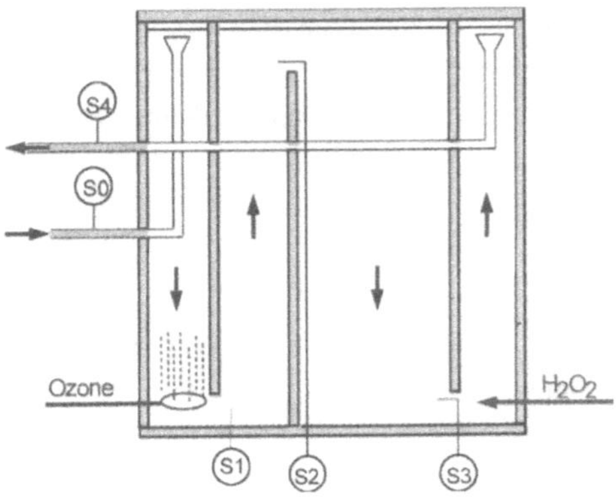

Fig.1. Ozone reactor with 4 chambers and indication of sampling points S0 to S4

Tracer Experiments and Hydraulic Model

In order to determine the hydraulic characteristics of the ozone reactor, three tracer studies at different flow capacities of 160 m^3/h, 200 m^3/h and 260 m^3/h were performed. The NaCl tracer solution was dosed to the first reactor until a steady state concentration in the effluent was observed. Thus, two conductivity signals during rising and declining tracer concentrations could be evaluated. Figure 2 shows the tracer curves at 4 sampling points and the corresponding modelled breakthrough curves. The hydraulic model was set up empirically using a series of completely stirred tank reactors (CSTR) with partial back-mixing, taking into account considerable dispersion and dead zones. In order to simulate satisfactorily the tracer at each sampling point, the reactor compartments were characterised hydraulically as follows: (1) the first aerated tank as 1 CSTR, (2) the second tank as a series of 5 CSTRs, (3) the largest third tank as a series of 6 CSTRs of unequal size with back-mixing and (4) the last tank with a series of 3 CSTRs of unequal size and back-mixing. The model was established with the help of the computer code AQUASIM [6] and was used as basic reactor configuration for the simulation of the chemical reactions.

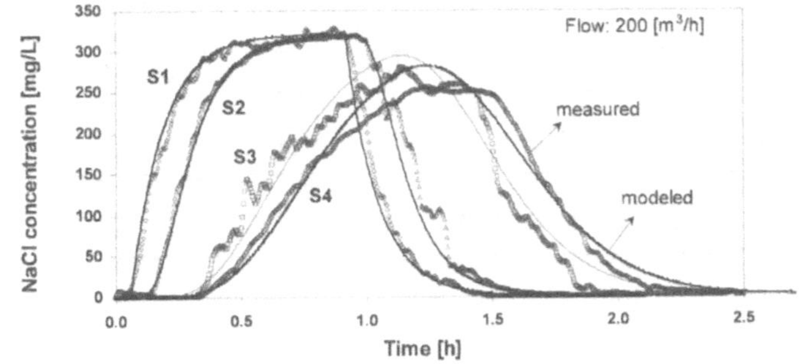

Fig.2. Tracer curves and simulated tracer breakthrough curves at 4 sampling points of the ozone reactor at a flow rate of 200 m^3/h

Laboratory Experiments

Ozone Decay Rate

In order to simulate mathematically the various oxidation reactions, kinetic data concerning the ozone decay and the R_{ct} value are required, and have to be determined experimentally for the specific spring water. The stability of O_3 in natural water depends strongly on the water constituents. Temperature, pH, DOC, and alkalinity may influence the ozone decay. Detailed kinetic information on the effects of the various parameters may be avoided by direct measurement of the O_3 decay under the prevailing water composition. Acero and von Gunten [5] measured the ozone decay for the Betteraz water at 11°C. Depending on the DOC concentration of the raw water they found the first order rate constant to be in the range of $k = 1.5 \cdot 10^{-4}$ to $1.0 \cdot 10^{-3}$ s^{-1}. The O_3 decay in advanced oxidation is faster and the rate constants were determined to be $k = 1.4 \cdot 10^{-3}$ to $4.4 \cdot 10^{-3}$ s^{-1} for DOC concentrations of 1 mg/l and 2.2 mg/l, respectively. The data reveal that a spontaneous O_3 decay phase of a few minutes exists, which deviates from pure first order kinetics. The faster decay at the beginning is compensated in the model by introducing a higher rate constant in the first reactor.

OH Radical Concentration

The concentration of OH radicals cannot be measured directly. It has to be determined via the decay of a standard substance reacting only with OH. In accordance with investigations by Elovitz and von Gunten [4], p-chlorobenzoic acid (pCBA) was used in this study because it reacts very slowly with O_3 ($k_{O3/pCBA}$ = 0.15 $M^{-1}s^{-1}$) but very fast with OH radicals ($k_{OH/pCBA}$ = 5·109 $M^{-1}s^{-1}$). The experimentally measured decay of pCBA in the Betteraz water was used to determine the corresponding R_{ct} values as model parameters [5]. The following set of equations (1) to (4) leads to the important relationship between the ozone exposure (time X concentration) and pCBA decay from which the OH radical concentrations may be calculated. The experimental values expressed according to equation (4) resulted in R_{ct} values of $7.9 \cdot 10^{-10}$ to $1.6 \cdot 10^{-9}$ with a higher R_{ct} of $4.1 \cdot 10^{-9}$ over the first exposure of about 0.01 M s. When hydrogen peroxide was dosed to the same water, the R_{ct} values amounted to $1.4 \cdot 10^{-8}$ in the low DOC concentration range and $2.1 \cdot 10^{-8}$ in the higher.

$$\frac{-d\,[pCBA]}{dt} = k_{OH/pCBA}\,[pCBA]\cdot[OH] \qquad (1)$$

$$\ln\left(\frac{[pCBA]_t}{[pCBA]_o}\right) = -k_{OH/pCBA}\int_o^t [OH]\cdot dt \quad (2)$$

$$R_{ct} = \frac{\int [OH]\cdot dt}{\int [O_3]\cdot dt} \quad (3)$$

$$\ln\left(\frac{[pCBA]_t}{[pCBA]_o}\right) = -k_{OH/pCBA} R_{ct} \int_o^t [O_3]\cdot dt \quad (4)$$

The low R_{ct} value reveals that OH radical concentrations are indeed very low at O_3 concentrations usually applied in drinking water treatment. The findings make clear that low OH radical levels are important for certain reactions with organic micropollutants but are of secondary importance for disinfection. In order to come up with realistic data on the O_3 decay and OH radical concentrations with H_2O_2 dosing, experiments were carried out to determine the R_{ct} values under different O_3 and H_2O_2 dosing conditions. The results are depicted in Table 2, and were used for modelling the reactors with and without H_2O_2, respectively.

Chemical Oxidation and Disinfection Model

The water specific information from lab experiments was combined with kinetic data on micropollutant oxidation reactions with O_3 and OH radicals and with the hydraulic reactor performance. In this study the micropollutants of interest were tri- (TCE), tetrachlorethene (PCE) and atrazine. The following equations, which have to be solved simultaneously, were thus set up.

Ozone decay

$$\frac{dO_3}{dt} = -k_{O3}[O_3] \quad (5)$$

OH-radical concentration

$$[OH] = R_{ct}[O_3] \quad (6)$$

Micropollutant oxidation

$$-\frac{ds}{dt} = k_{s,O_3}[O_3]\cdot[s] + k_{s,OH}[OH]\cdot[s] \quad (7)$$

where s = micropollutant

Disinfection

$$\frac{dN}{dt} = -k_{d,N}[N]\cdot[O_3] \quad (8)$$

with N = number of organisms/ml.

The reaction rate constants used for the simulation are summarised in Table 2.

Table 2. First-order reaction rate constants and R_{ct} values used in the model (T = 11°C)

Reaction Rate Constants	Units	Value	Reference
k_{O3}	s^{-1}	$1.6 \cdot 10^{-4}$	This study
k_{O3} advanced	s^{-1}	$2.2 \cdot 10^{-3}$	This study
$k_{TCE,O3}$	$M^{-1} s^{-1}$	8.6	This study
$k_{TCE,OH}$	$M^{-1} s^{-1}$	$4.15 \cdot 10^{9}$	[7]
$k_{PCE,O3}$	$M^{-1} s^{-1}$	0.05	This study
$k_{PCE,OH}$	$M^{-1} s^{-1}$	$2.1 \cdot 10^{9}$	[7]
$k_{atrazin,O3}$	$M^{-1} s^{-1}$	4.0	This study
$k_{atrazin,OH}$	$M^{-1} s^{-1}$	$2.3 \cdot 10^{9}$	This study
$k_{d,Crypto\ parvum}$	$M^{-1} s^{-1}$	154	[8]
$k_{d,Bacillus\ subtilis}$	$M^{-1} s^{-1}$	438	[9]
$k_{d,E.coli}$	$M^{-1} s^{-1}$	$2 \cdot 10^{5}$	[9]
R_{ct} without H_2O_2			
1st reactor	-	$2.4 \cdot 10^{-9}$	[5]
other reactors		$1.6 \cdot 10^{-9}$	
R with H O /O = 0.4 (molar)	-	$1.5 \cdot 10^{-8}$	[5]

Full-scale Experiments

Six experiments were carried out with the full-scale ozone reactor at a constant flow of 200 m^3/h. Ozone and peroxide dose were varied according to the values given in Table 3. The ozone concentration was measured at all sampling points throughout the reactor and H_2O_2 in the last compartment. At each point TCE, PCE, and atrazine were measured, as well as bicarbonate, ammonium, DOC, pH and temperature at sampling point S0.

Table 3. Experimental conditions during full-scale ozonation

Experiment No.	Flow m^3/h	O_3 dose g/m^3	H_2O_2 dose g/m^3
1	200	0.95	0
2	200	1.48	0
3	200	1.95	0
4	200	1.02	0.13
5	200	1.55	0.25
6	200	2.02	0.43

Results

Ozone decay

Taking the decay rates from the lab experiments in the range between 1.1 to $2 \cdot 10^{-4}$ s^{-1}, it was possible to simulate accurately the measured O_3 decay in chambers 2, 3, and 4 following the ozonation chamber (Figure 3). In the experiments with H_2O_2, the higher decay rates measured by Acero and von Gunten [5] in the range between 1.4 to $2.5 \cdot 10^{-3}$ s^{-1} lead to a rapid O_3 decay in the last reactor chamber where H_2O_2 was dosed.

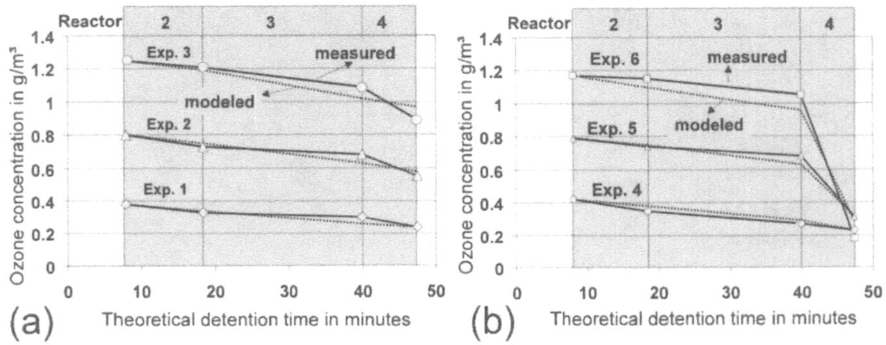

Fig. 3. (a) Measured and modelled O_3 decay in the ozone reactor without H_2O_2 dosing (b) Measured and modelled O_3 decay in the ozone reactor with H_2O_2 dosing

Micropollutant oxidation

The measured micropollutant concentrations throughout the ozone reactor are shown for experiments 3 and 6 in Figure 4 for atrazine and Figure 5 for TCE and PCE. The results show that atrazine, TCE and PCE are removed 28.9%, 44.2%, and 7.1%, respectively, without H_2O_2 dosing. When 0.4 gH_2O_2/m^3 was dosed, the respective removal rates amounted to 64.0%, 85.2%, and 49.0%. Obviously, micropollutant removal was increased significantly by introducing advanced oxidation. The removal of the micropollutants was calculated by introducing the reaction rate constants for atrazine, TCE and PCE with O_3 and OH radicals (Table 2), respectively. The modelled results confirmed that the set of kinetic equations and corresponding rate constants in combination with the hydraulic model gave a consistent description of conventional ozonation. However, the R_{ct} values analysed in the laboratory in the presence of H_2O_2 led to an underestimation of the OH radical oxidation capacity in the full-scale reactor. In order to meet the measured micropollutant concentrations, R_{ct} would have to be increased by a factor of 2 over the measured values. The simulated results are compared with the measurements in Figures 4 and 5. In the full-scale treatment plant advanced oxidation is followed by activated carbon filters.

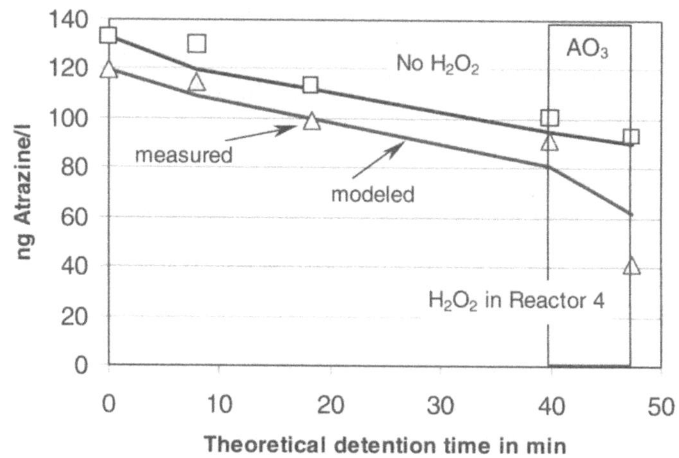

Fig. 4. Measured (symbols) and modelled (lines) removal of atrazine in conventional and advanced ozonation (AO_3 =advanced ozonation in the last chamber)

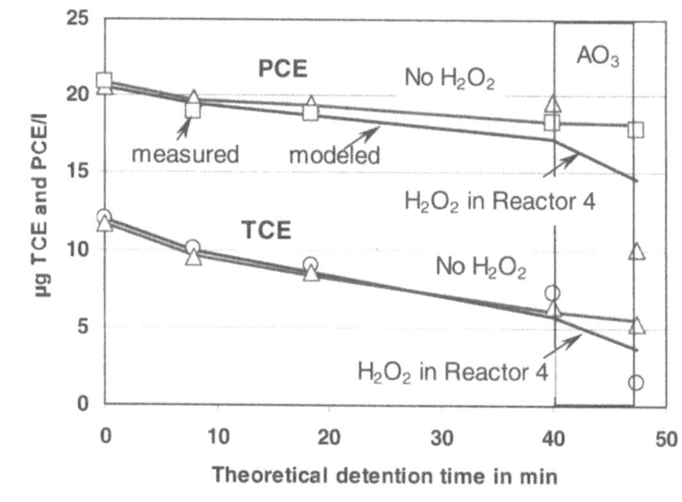

Fig.5. Measured (symbols) and modelled (lines) removal of TCE and PCE in conventional and advanced ozonation (AO_3 =advanced ozonation in the last chamber)

Due to the additional micropollutant removal in advanced oxidation, the carbon is supposedly loaded with lower concentrations of chlorinated hydrocarbon compounds. In reality, atrazine is oxidised to stable products, which may still lead to a loss of carbon adsorption capacity. Nevertheless, a cost comparison reveals that the possible decrease in carbon use compensates for the H_2O_2 dosing to the ozone reactor [10].

Model Calculations

The model was also used to simulate the performance of advanced oxidation under different operational conditions. While earlier dosing of H_2O_2 may increase the oxidation of micropollutants via OH radical reactions, this will reduce the efficiency of direct O_3 reactions such as the disinfection of microorganisms. Therefore, several model calculations were performed in order to optimise the dosing point for H_2O_2, aiming at both micropollutant oxidation and sufficient disinfection. A 2-log removal for *Cryptosporidium parvum* was set as the major criterion for disinfection. Figure 6 summarises the results for atrazine, TCE and PCE for three different operational conditions: (1) no H_2O_2 was added (Exp. 3), (2) H_2O_2 was dosed to the second reactor chamber at a molar ratio of about H_2O_2/O_3 = 0.8, and (3) H_2O_2 was dosed to the last reactor chamber at a ratio of H_2O_2/O_3 = 0.4 (Exp. 6). The results indicate that maximum removal efficiency can be expected for all micropollutants if H_2O_2 is dosed to the second reactor. The fast OH radical reactions are practically complete within the third chamber. However, the disinfection reactions modelled according to Chick-Watson for *Cryptosporidium parvum* and other microorganisms shown in Figure 7 reveal that disinfection is insufficient under these conditions. For a 2-log removal of *Cryptosporidium parvum* and *Bacillus subtilis* the dosing point has to be shifted towards chamber 3. This would still allow maximum micropollutant removal, but increase disinfection efficiency. In the present case, the irregular size of the contact chambers, with the largest volume in chamber 3, does not offer much flexibility for setting the H_2O_2 dosing point in an optimal position. The example shows that a subdivision of the ozone reactor into a series of at least five contact chambers using multiple baffles is especially advantageous for advanced oxidation.

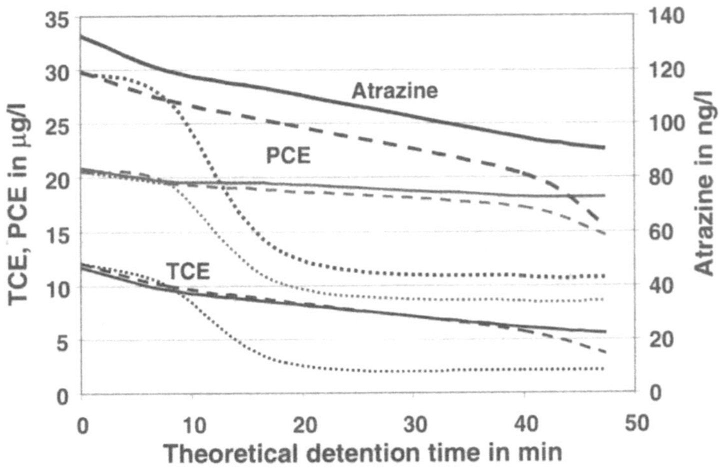

Fig. 6. Simulated oxidation of atrazine, TCE and PCE along the ozone reactor with no H_2O_2 addition (solid lines), dosing H_2O_2 to the last reactor chamber (dashed lines), and dosing H_2O_2 to the second reactor chamber (dotted lines)

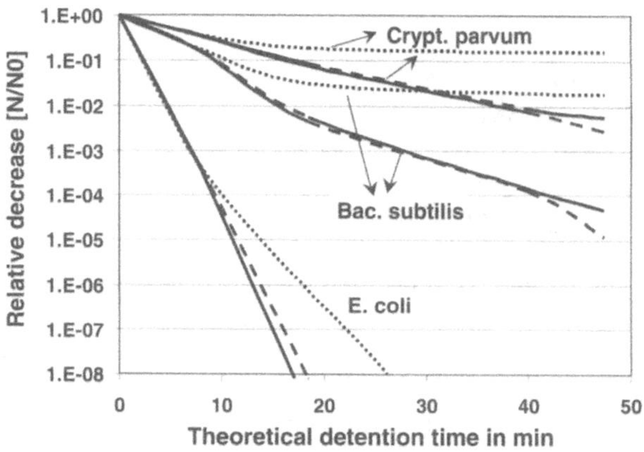

Fig. 7. Simulated disinfection of *Cryptosporidium parvum*, spore-forming *Bacillus subtilis*, and *Escherichia coli* across the ozone reactor with no H_2O_2 addition (solid lines), dosing H_2O_2 to the last reactor chamber (dashed lines), and dosing H_2O_2 to the second reactor chamber (dotted lines)

Conclusions

Known reaction rate constants for direct O_3 and OH radical reactions with atrazine, tri- and tetrachlorethene from literature data, and analysis of the O_3 decay and the molar OH/O_3 ratio (R_{ct} value) for the water in question are basic information needed to simulate ozone reactions with respect to micropollutant removal. The combination of reactor hydraulics, characterised with the help of tracer studies in an existing reactor or assumed in a designed reactor, with the reaction equations turned out to be a useful tool to model consistently the performance of full-scale ozonation reactors. The addition of H_2O_2 to the ozone reactor for advanced oxidation promotes OH radical reactions with the micropollutants and considerably improves the performance of the ozonation step and subsequent activated carbon treatment. Modelling advanced oxidation is somewhat more delicate. More research is needed to increase the information on the exact R_{ct} values and their possible variation under the prevailing water composition. Modelling conventional and advanced oxidation seems to be a beneficial tool to assess the performance of existing ozone reactors and to optimise the contradictory requirements between micropollutant removal and disinfection with regard to H_2O_2 dosing. Further studies would also have to consider bromate formation under different operational conditions. Model calculations indicate that advanced ozone reactors for micropollutant removal and disinfection have to be designed and constructed as multiple stage baffled tanks with a high flexibility for setting the H_2O_2 dosing point.

References

1. Von Gunten, U., Bruchet A., Costentin, E.: Bromate formation in advanced oxidation processes, JAWWA, Vol. 88 (6) (1996) 53-65
2. Von Gunten, U., Elovitz, M., Kaiser, H.P. : Calibration of full-scale ozonation systems with conservative and reactive tracers, J Water SRT – Aqua 48(6) (1999) 250-256
3. Hoigné, J., Bader, H. : Determination of ozone in water by the indigo method, Wat. Res. 15 (1981) 449-456
4. Elovitz, M.S., von Gunten, U.: Hydroxyl Radical/Ozone Ratios during Ozonation Processes, The Rct concept, Ozone Sci. and Engrg. 21(1999) 239-260
5. Acero, J.L., von Gunten, U.: From ozonation to the hydrogen peroxide induced AOP O_3/H_2O_2: characterization of oxidation processes, submitted to JAWWA (2000)
6. Reichert, P.: Concepts underlying a Computer Program for the Identification and Simulation of Aquatic Systems (Manual of the Computer Code AQUASIM), Schriftenreihe EAWAG, No. 7 (1994) EAWAG, Duebendorf, Switzerland
7. Buxton, G.V., Greenstock, C.L., Helman, W.P., Ross, A.B.: Critical review of rate constants for reactions of hydrated electrons, hydrogen atoms and hydroxyl radicals in aqueous solution, J. Phys. Chem. Ref. Data 17 (1988) 513-886
8. Rennecker, J.L., Marinas, B.J., Owens, J.H., Rice, E.W.: Inactivation of cryptosporidium parvum oocysts with ozone, Wat. Res., Vol. 33 (11) (1999) 2481-2488
9. Kaiser, H.P., von Gunten, U., Elovitz, M.: Die Bewertung von Ozonreaktoren, Gas-Wasser-Abwasser No. 1 (2000) 50-61
10. Pianta, R., Boller, M.: Cost evaluation of conventional water treatment and membrane filtration, in preparation (2000)

Advanced Oxidation Processes in Water Treatment

A. Latifoglu* and M.D. Gürol**

*Environmental Engineering Department, Hacettepe University, 06532, Ankara.
**Civil and Environmental Engineering Department, San Diego State University, CA, 92182-1324.
aysegull@hacettepe.edu.tr

Abstract

H_2O_2/UV and O_3/UV oxidized nitrobenzene in water with an initial nitrobenzene concentration of 0.15 mM, inorganic carbon concentration of 0.8 mM and pH 7 at a rate of $1.13 \cdot 10^{-8}$ and $2.68 \cdot 10^{-8}$ M/sec, respectively. Oxidation by ozonation offered a comparable rate of $2.02 \cdot 10^{-8}$ M/sec with O_3/UV under similar operating conditions. At higher ozone dosages, ozonation also resulted in appreciable nitrobenzene removal rates. Nitrobenzene promoted radical reactions in the ozonation process. On the other hand, these radical-promoting reactions were strongly inhibited at an inorganic carbon concentration of 4 mM. For O_3/UV treatment, the nitrobenzene removal rate decreased when the inorganic carbon concentration was increased from 0.8 to 4 mM. This decrease, however, was not as great in O_3/UV as in ozonation, since the nitrobenzene-promoted radical reactions are less important in O_3/UV. An increase in pH in the range of 5 to 10 adversely affected the nitrobenzene removal for the same process.

Introduction

There has been considerable interest in advanced oxidation processes that include combinations of traditional oxidants with various catalysts for the treatment of water contaminated with pollutants. The effectiveness of these processes is attributed to the generation of highly reactive, non-selective oxidants of hydroxyl radicals [$OH^•$]. The reaction rate constant of the [$OH^•$] with many synthetic aromatic and aliphatic organic compounds has been shown to be in the range of 10^8 to 10^{10} $M^{-1}sec^{-1}$ [1].

Ozonation or hydrogen peroxide in the presence of ultraviolet radiation (O_3/UV and H_2O_2/UV) are examples of advanced oxidation processes. O_3/UV was developed in the early 1970s for treatment of cyanide-containing wastewaters [2]. Since then, the process has been primarily used for the oxidation of various synthetic organic compounds in water: aliphatic carboxylic acids (oxalic, acetic, succinic, propionic and malonic acids) and aliphatic alcohols (methyl and ethyl alcohol) are effectively destroyed in water by O_3/UV [3]. On-site oxidation of various volatile chlorinated organic compounds, including trichloroethane, trichloroethylene, tetrachloroethylene and chloroform, was also found to be successful in treating groundwater [4]. Substituted nitrogen heterocyclics and aromatic amines in oil shale retort water were transformed by O_3/UV into compounds that are more easily biodegradable [5]. The O_3/UV process was also used to eliminate natural organic matter in water in order to reduce subsequent formation of trihalomethanes during disinfection by means of chlorination [6,7]. On the other hand, H_2O_2/UV has been shown to be effective in the treatment of phenols [8], aliphatic acids of formic, acetic and propionic acids [9], and aromatic compounds such as toluene and chlorobenzene [10].

Besides the oxidizing hydroxyl radicals [$OH^{\bullet}$], O_3, H_2O_2 and UV light may have an individual affect on the decomposition of the targeted organic compound. Although highly selective in their reactions, O_3 and H_2O_2 alone can serve as oxidizing agents with oxidation potentials of 2.07 and 1.77 volts, respectively. Depending on the type of compound, photodegradation by direct photolysis can also be successful for some nitroaromatic compounds like 2,4,6-trinitrotoluene [11].

Oxidation rate of organic compounds by these processes may depend on several factors: 1) nature and concentration of organic pollutants, 2) operational conditions, 3) water quality characteristics. This paper focuses mainly on the oxidation of organic contaminants by H_2O_2/UV, O_3/UV, and ozonation processes. Nitrobenzene was selected as a model organic contaminant, which was introduced into the water resources as a result of improper disposal of waste from industrial facilities. Nitrobenzene can only react with $OH^{\bullet}$ and is essentially nonreactive towards O_3 and H_2O_2. It does not directly photolyze under UV radiation. This study investigated the processes involved in the removal of nitrobenzene at different nitrobenzene concentrations, dosages of the oxidants, and water quality characteristics.

Materials and methods

The experiments were conducted on a bench-scale reactor containing 2 liters of solution. Nitrobenzene was added to the solutions at high and low concentrations, 0.15 mM and 0.006 mM, respectively. Ozone was sparged into the reactor containing solution through a bubble diffuser. The solution was exposed to ultraviolet light using a photochemical reactor. Concentration of ozone in gas and liquid were measured using the potassium iodide [12] and indigo methods [13], respectively. The concentration of hydrogen peroxide in the solution was determined using the horseradish peroxidase method [14]. UV light intensity (I_o) was measured using ferrioxalate actinometry [15]. Nitrobenzene concentration in the solution was measured using high pressure liquid chromotagraphy (HPLC). Millipore-filtered milli-Q water was used for the preparation of all aqueous solutions. The pH of the solutions was adjusted with diluted H_2SO_4 and NaOH solutions. Inorganic carbon concentrations (C_T) were adjusted by using certified ACS sodium bicarbonate in the solutions.

Results and discussion

For the first set of experiments, the solutions contained initial nitrobenzene concentrations of 0.15 mM, and were adjusted to pH = 7 and C_T = 0.8 mM. Ozone gas at a dosage of 3 mg/l was used for ozonation and O_3/UV. Solutions with an initial hydrogen peroxide concentration of 0.1 mM were prepared for H_2O_2/UV. $I_o=11\cdot10^{-6}$ Einstein/L-sec were used for the experiments. UV photolysis and stripping (to remove nitrobenzene) were controlled by running the experiments under UV radiation and oxygen gas sparging. Our control experiments showed that nitrobenzene was quite stable against direct photolysis by UV light and stripping due to gas sparging. The removal of nitrobenzene by O_2/UV, O_3, O_3/UV and H_2O_2 /UV is shown in Figure 1.

Zero-order removal rates of nitrobenzene for each treatment were calculated from the slopes of the decomposition profiles. The results are given in Table 1.

The removal profiles as well as the rates indicated that O_3/UV is more effective for nitrobenzene removal than is H_2O_2/UV. The removal rate was calculated to be more than two times faster in O_3/UV than in H_2O_2/UV. The results also demonstrated that nitrobenzene is removed in appreciable amounts by ozonation only, although the direct reactivity of ozone with nitrobenzene has been found to be as low as 0.09 $M^{-1}sec^{-1}$ [16]. Nitrobenzene removal by ozonation was also

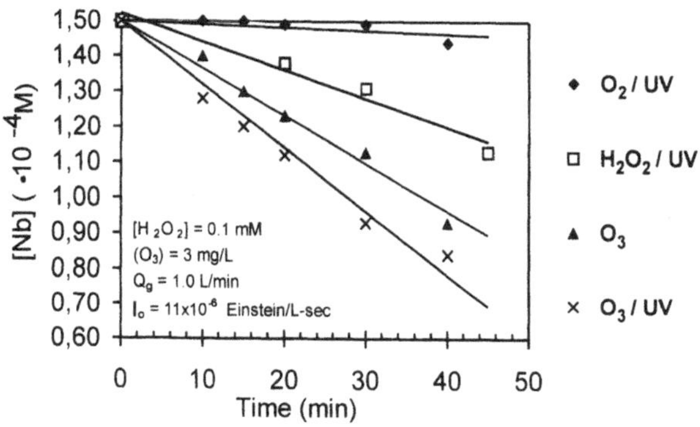

Fig. 1. Removal of nitrobenzene by O_2/UV, ozonation, O_3/UV and H_2O_2 /UV

Table 1. Removal rates of nitrobenzene by ozonation, H_2O_2/UV and O_3/UV

Type of Treatment	Reaction Rate (M/sec)
O_3	$2.02 \cdot 10^{-8}$
H_2O_2/UV	$1.13 \cdot 10^{-8}$
O_3/UV	$2.68 \cdot 10^{-8}$

appreciable at higher ozone dosages in this study. At an ozone dosage of 6 mg/l, the removal rates of nitrobenzene were calculated to be $3.48 \cdot 10^{-8}$ and $4.27 \cdot 10^{-8}$ M/sec for ozonation and O_3/UV, respectively. At a higher ozone dosage of 9 mg/l, the calculated rates were $4.26 \cdot 10^{-8}$ and $5.80 \cdot 10^{-8}$ M/sec for ozonation and O_3/UV, respectively. There is evidence that nitrobenzene promotes hydroxyl radical production in the ozonation process at low concentrations of carbonates [7]. This may be why decomposition reactions of nitrobenzene were promoted in the ozonation process at the low C_T of 0.8 mM.

The effect of carbonates (HCO_3^- and $CO_3^=$) on O_3/UV was investigated in the C_T range of 0.8 to 4 mM at a pH = 7. Carbonates are the natural inorganic carbon content of the water resources. The results of the experiments are given in Figure 2. The figure shows that an increase in carbonate concentration decreased the removal rate of nitrobenzene. Previous work [1] has shown that HCO_3^- and $CO_3^=$ react with $OH^\bullet$ at a reaction rate constant of $3.6 \cdot 10^7$ and $4.2 \cdot 10^8$ $M^{-1}sec^{-1}$, respectively. Thus it is likely that nitrobenzene removal was retarded through the consumption of $OH^\bullet$ by HCO_3^- and $CO_3^=$ species.

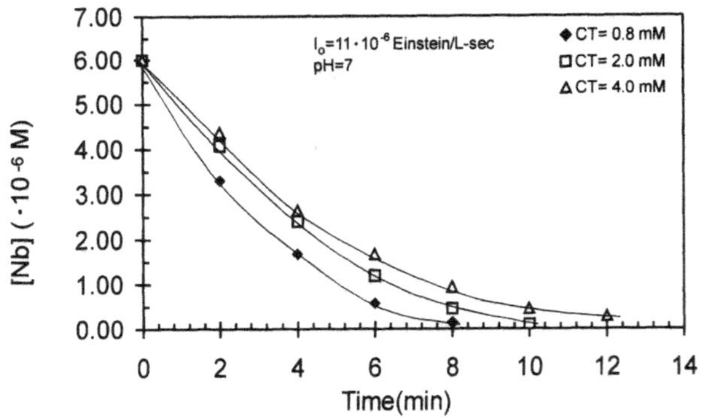

Fig. 2. The effect of C_T on the removal of nitrobenzene by O_3/UV

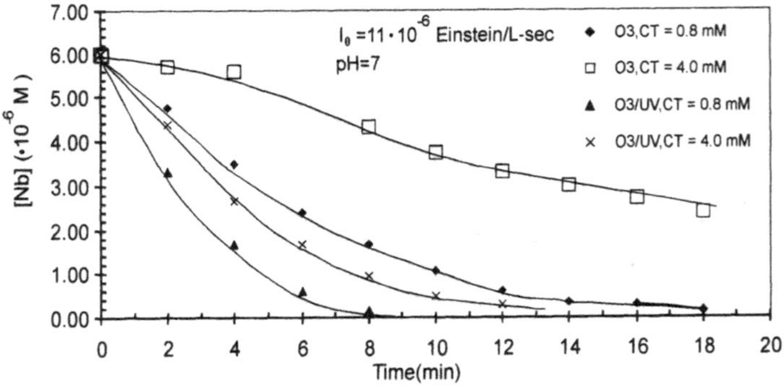

Fig. 3. Comparison of ozonation and O_3/UV for nitrobenzene removal at high and low C_T

Figure 3 shows the difference in nitrobenzene removal for high and low C_Ts for both ozonation and O_3/UV. The results demonstrated that increasing C_T hindered nitrobenzene oxidation more with ozone treatment than with O_3/UV. In the ozonation process, the reaction of nitrobenzene with radicals is the main producer of $OH^{\bullet}$. Hence, consumption of the $OH^{\bullet}$ significantly retarded the nitrobenzene removal rate. On the other hand, photolysis of ozone is the major producer of hydroxyl radicals in O_3/UV treatment, and the chain promoting character of nitrobenzene is not as pronounced in O_3/UV as in the ozonation process.
The effect of pH on nitrobenzene removal for O_3/UV treatment is shown in Figure 4. The rate of nitrobenzene oxidation by O_3/UV was adversely affected with increasing pH in the range of 5 to 10. Formation of $OH^{\bullet}$ through the reactions of ozone is base-catalyzed as indicated in the summary of the reactions for O_3/UV below:

$$O_3 + H_2O \xrightarrow{\text{UV Light}} H_2O_2 + O_2 \qquad k_1 \quad \mathbf{(1)}$$

$$H_2O_2 \rightarrow 2\,OH^{\bullet} \qquad k_2 \quad \mathbf{(2)}$$

$$H_2O_2 \leftrightarrow HO_2^- + H^+ \qquad pK_a{=}11.8 \quad \mathbf{(3)}$$

$$HO_2^- + O_3 \rightarrow \rightarrow \rightarrow \quad OH^{\bullet} \qquad k_3 \quad \mathbf{(4)}$$

Based on the reactions above, an increase in pH is expected to increase radical formation, and thereby increase the nitrobenzene removal. However, undetectable ozone concentration in the solutions indicated that the ozone concentration is too low to enhance $OH^{\bullet}$ production pH increases. On the other hand, at higher pH values, carbonate equilibrium is shifted from HCO_3^- to $CO_3^=$ which is a more effective $OH^{\bullet}$ scavenger as indicated previously (the reaction rate constants of $OH^{\bullet}$ with HCO_3^- and $CO_3^=$ are $3.6 \cdot 10^7$ and $4.2 \cdot 10^8$ $M^{-1}sec^{-1}$, respectively). Hence, the removal of nitrobenzene was retarded by the consumption reactions of $CO_3^=$ that are more pronounced at high pH.

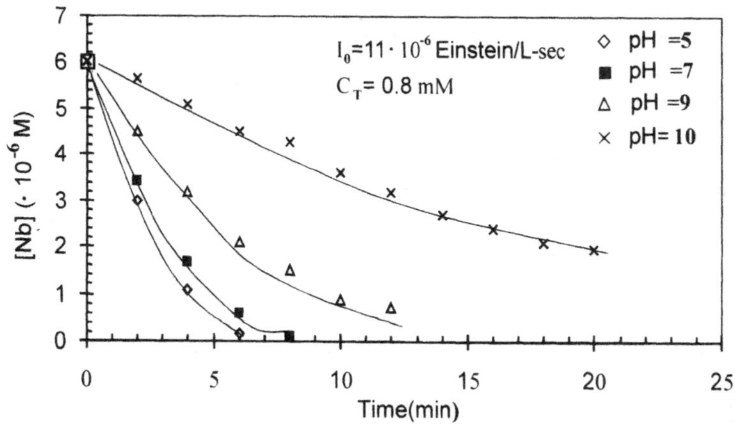

Fig. 4. The effect of pH on the removal of nitrobenzene by O_3/UV

Conclusions

The effectiveness of H_2O_2/UV, O_3/UV and ozonation alone for the removal of nitrobenzene in water was investigated. Laboratory experiments were conducted in a semi-batch reactor. The experimental results indicated that H_2O_2/UV was the least effective process for removing nitrobenzene under the operating conditions. O_3/UV was much more effective than ozonation for removing nitrobenzene, especially at higher C_T. However, the removal rates became comparable for both processes at low C_T, since nitrobenzene promotes radical reactions in the ozonation process. This promoting effect is not that important in O_3/U, so the increase in C_T did not decrease the nitrobenzene removal rate as much as it did in ozonation alone. An increase in pH in the range of 5 to 10 decreased the nitrobenzene removal in water as a result of the shift of the equilibrium from bicarbonate to carbonate, which was more effective in removing $OH^{\bullet}$.

References

1. Farhataziz, Ross, A.B.: Specific Rates of Reactions of Transients from Water in Aqueous Solution, Natural Bureau of Standards, NSRDS-NBS59 (1977) Washington D.C.
2. Garrison, R.L., Mauk, C.E., Prengle, Jr., H.W.: Advanced Oxidation System for Complexed Cyanides, In: 1st International Symposium on Ozone for Water and Wastewater Treatment, (Rice, R.G. and Browning, M.E., Eds.) International Ozone Institute, Syracuse, NY, (1975) p. 551
3. Ikemizu, I., Orita, M., Sagiike, Morooka S., Kato, Y.: Ozonation of Organic Refractory Compounds in Water in Combination with UV Radiation, J. Chem. Eng. Japan 20 (1987) 369
4. U.S. Environmental Protection Agency – Risk Reduction Engineering Laboratory, Site Program Demonstration of the Ultrox International Ultraviolet radiation/Oxidation Technology; Technical Evaluation Report, EPA/540/5-89/012, January, 1990
5. Jones, B. M., Langlois, G.W., Sakaji, R.H.: Effect of Ozonation and UV Radiation on Biorefractory Organic Solutes in Oil Shale Retort Water, Environ. Prog. 4 (1985) 252
6. Glaze, W.H., Peyton, G.R., Lin, S., Huang, F.Y., Burleson, J.L.: Destruction of Pollutants in Water with Ozone in Combination with Ultraviolet Radiation. 2. Natural Trihalomethane Precursors, Environ. Sci. and Technol. 16 (1983) 454
7. Akata, A., Gurol, M.D.: Photocatalytic Oxidation Processes in the Presence of Polymers, Ozone Sci. Eng. 14 (1992) 367
8. Omura, K., Matsuura, T.: Photo-induced Reactions – IX: The Hydroxylation of Phenols by the Photo-decomposition of Hydrogen Peroxide in Aqueous Media, Tetrahedron 24 (1968) 3478
9. Ogata, Y., Tomimaza, K., Takagi, K.: Photo-oxidation of Formic, Acetic and Propionic Acids with Aqueous Hydrogen Peroxide, Canadian Journal of Chemistry 59 (1981) 14
10. Sundstrom, D.W., Weir, B.A., Klei, H.G.: Destruction of Aromatic Pollutants by UV Catalyzed Oxidation with Hydrogen Peroxide, Environmental Progress 8 (1) (1989) 6
11. Mabey, W.R., Tse D., Baraze A., Mill T.: Photolysis of Nitroaromatics in Aquatic Systems. I. 2,4,6-trinitrotoulene. Chemosphere 12 (1983) 3

12. Bader, H., Hoigne, J.: Determination of Ozone in Water by the Indigo Method, Wat. Res. 15 (1981) 449
13. Bader, H., Sturzenegger,V., Hoigne, J.: Photometric Methods for the Determination of Low Concentrations of Hydrogen Peroxide by the Peroxidase Catalyzed Oxidation of N,N-Diethyl-p-phenylene-diamine (DPD), Wat. Res. 22 (9) (1988) 1109
14. Parker, C.A.: A New Sensitive Chemical Actinometer: II. Potassium ferrioxalate as a Standard Chemical Actinometer, Proc. R. Soc. London Ser. A. 235 (1956) 518
15. Hoigne, J., Bader, H.: Rate Constants of Reactions of Ozone with Organic and Inorganic Compounds in Water, II – Dissociating Organic Compounds, Wat. Res. 17 (1983) 181

Control of Chemical Treatment Processes

Flow Cytometry as an Operational Tool to Improve Particle Removal in Drinking Water Treatment

O. Bergstedt*, H. Rydberg** and L. Werner**

*Water Environment Transport, Chalmers University of Technology, Göteborg, Sweden and the Recycling Board, City of Göteborg, Sweden
olof.bergstedt@krets.goteborg.se
**Göteborg Water and Sewage Works, City of Göteborg, Sweden

Abstract.

Waterborne disease caused by parasitic protozoa, e.g. *Giardia* and *Cryptosporidium*, is one of the major challenges to those responsible for public water supply. These parasites are able to survive for long periods in cold water and can be present in the raw water in spite of absence of fecal indicators. Lake Vänern and river Göta älv have been pointed out as the Swedish water system most affected by parasites. Current Swedish regulations stipulate multiple microbiological barriers for surface waters and continuos turbidity measurements of the drinking water. The city of Göteborg has initiated a research and development program to evaluate if the safety concerning *Giardia* and *Cryptosporidium* is good enough and if not how to improve it in a sustainable way. This paper shows that turbidity measurements are not always sufficient to ensure a well functioning removal of parasite sized particles. Flow cytometry has been shown to give additional information about the particles in parasite size fast enough for operational use. Total log-reductions of parasite sized particles varied between 1-3. The largest variations occurred in the filtration step even though turbidity never exceeded 0.1 FNU. The results indicate that choice of coagulant, flocculation design, separation technique and filter operation could have potential for improving parasite removal.

Introduction

River Göta älv is the main water source for the city of Göteborg as well as several other Swedish communities. The river water sometimes contains relatively large numbers of fecal indicator organisms. Göteborg operate with an active choice of best available raw water and treatment with multiple microbiological barriers. No fecal indicators have been detected in the drinking water leaving the treatment plants of Göteborg during a long series of years with several samples per week.

Parasitic protozoa have caused a large number of waterborne outbreaks in the industrialized world. *Cryptosporidium* was the causative agent to the largest waterborne outbreak in U.S. history [1]. It has been estimated that 400,000 people fell ill and potentially 112 died.

Giardia and *Cryptosporidium* have been detected in natural waters throughout the world and Sweden is no exception. The number of reported waterborne outbreaks has increased in Sweden, but the causative agent is unknown in most of the cases [2]. Routines for isolating and identifying parasites are less developed in Sweden than in U.S., but parasites have been identified in several Swedish outbreaks. A survey of Swedish raw waters showed 26% positive samples for *Giardia* and 32% positive samples for *Cryptosporidium* [3]. Lake Vänern/river Göta älv was pointed out as the water system with particular prevalence of *Giardia* and/or *Cryptosporidium*. None of the parasites were detected in the drinking water of Göteborg. However, *Cryptosporidium* was detected in the drinking water of one of the communities using raw water from the same raw water system.

Ottosson [4] estimated that the densities in the source water of Göta älv could cause 5 *Cryptosporidium* and up to 75 *Giardia* infections per 10 000 inhabitants and year with an assumed 2.7 log reduction of (oo)cysts in a water treatment plant. *Giardiasis* cases have to be reported in accordance with the Communicable Disease Act, and the total number in Sweden in 1998 was 1475 cases [5]. Only 125 of these were from Göteborg county. Of the total number of cases reported in Sweden only 20% of the infections actually originated in Sweden. Applying this percentage to the Göteborg county, this would mean less than 0.5 reported infections per 10 000 inhabitants. The difference between the actual number reported and the prediction shows a great need to refine the risk assessments.

The city of Göteborg has initiated a research and development program to evaluate and if necessary decrease the risk of waterborne outbreaks caused by *Giardia* and *Cryptosporidium.*

The purpose of the work that this paper describes was to evaluate whether flow cytometry is a practical method to measure removal of parasite sized particles. A number of case studies have been performed to study whether flow cytometry can give additional information compared to turbidity.

The Lackarebäck water treatment plant

The Lackarebäck water treatment plant is one of the two plants supplying drinking water to half a million consumers in the Göteborg area. The main raw water source is the river Göta älv via the Delsjö lakes reservoir. The water is treated by chemical precipitation with aluminum sulfate, sedimentation and rapid filtration through granular activated carbon.

The Lackarebäck plant has seven parallel lines and six flocculation chambers in each line. Originally, there were 3 horizontal baffles in the two first tanks and two in the following (Figure 1).

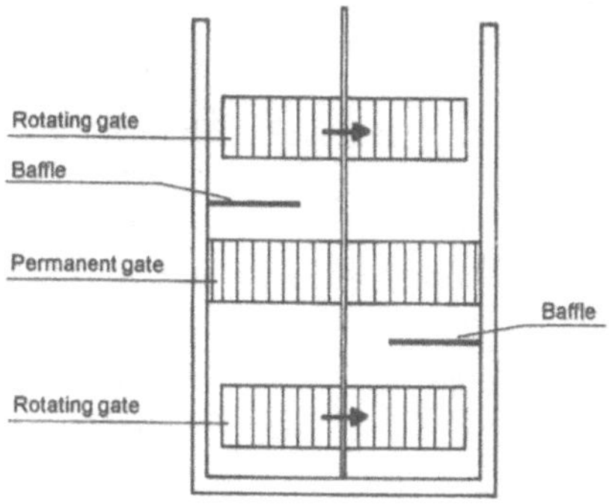

Fig. 1. Example of baffles in a flocculation chamber at the Lackarebäck plant

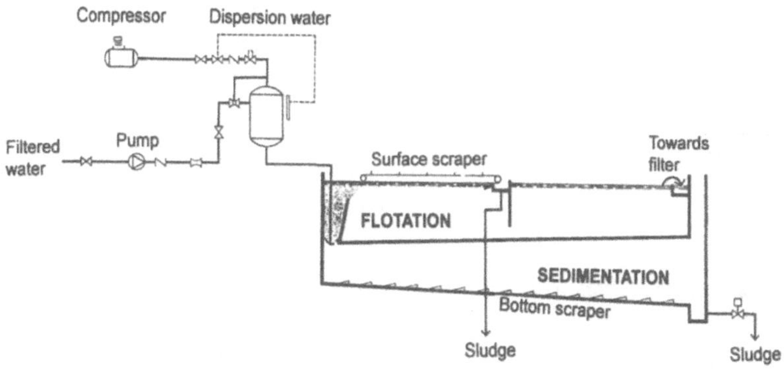

Fig. 2. Flotation in series with sedimentation at the Lackarebäck water treatment plant

Algae blooms in the Delsjö lake reservoir include species that give taste and odor to the water. A flotation step was included in one of the seven full-scale lines for algae removal, (Figure 2). Investigations showed algae mass being reduced by 2.6 log and the reduction of certain algae species known to cause taste and odor showed a log reduction of 4 or more [6].

Wasting first filtrate after flushing was not considered necessary when the Lackarebäck plant was designed in the 1960's. Now the plant will be changed to the first filtrate practice to make it possible to avoid that water with low quality is delivered.

River Göta älv is subjected to fecal contamination as indicated by the relatively high counts of *E. coli* and (oo)cysts (Table 1). The passage through the raw water reservoir lake Stora Delsjön brought both turbidity and *E. coli* counts down, but there was no significant reduction of (oo)cysts. The drinking water quality had excellent values, but neither of the differences between gives assurance of the treatments ability to prevent waterborne disease from parasites.

Table 1. Water quality values during 1999 and parasites between August 1996 and March 2000, (Annual report 1999 and analyses reports [7])

		River Göta älv	Raw water from Lake Stora Delsjön	Drinking water from the Lackarebäck plant
E. coli (L^{-1})	n	104	99	154
	minimum	100	< 10	< 1
	median	930	< 10	< 10
	maximum	7800	80	< 10
Turbidity (FNU)	n	156	157	159
	minimum	2.5	0.55	< 0.05
	median	7	0.74	< 0.05
	maximum	40	1.2	0.08
Giardia (L^{-1})	n	8	5	1
	minimum	n.d.	n.d	n.d
	median	n.d.	0.01	
	maximum	0.19	0.08	
Cryptosporidium (L^{-1})	n	8	5	1
	minimum	n.d.	n.d.	n.d.
	median	0.01	0.02	
	maximum	0.46	0.164	

n = number of analyses n.d. = non detectable

Materials and methods

For most samples a number of analyses were performed including aluminum content and bulk parameters of organic matter. The results presented in this paper were focused on particle related parameters such as turbidity and particle counting to describe particle reduction. All samples were analyzed the same day as the sampling.

Turbidity measurements

Measurements were made with a Hach Ratio Turbidimeter model 18900 according to the (accredited) method SS 028125-2. The range was 0.05 – 200 FNU with inaccuracy of ± 2%.

Particle counting with flow cytometer

Water samples were taken in 50 ml sterile centrifuge tubes (FALCON). To achieve low particle background (total counts <10/ml; autofluorescent <1/ml) all tubes were pre-rinsed twice with the water to be sampled. Before analyses tubes were vortexes and approximately 1ml sample transferred to 1.5 ml micro-centrifuge tubes. When performing jar tests, samples were taken directly in pre-rinsed micro-centrifuge tubes.

The particle counting was performed with a Microcyte flow cytometer (Optoflow AS). Using a 635 nm diode laser this instrument measures light scatter and particle associated fluorescence (650-900 nm). Light scattered at small angels, forward light scatter, depends mainly on particle size, but also on the surface structure and optical properties of the particle. The instrument is particularly suitable for quantifying particles, in any interval, between 0.4 and 20 µm. The instrument was calibrated to generate absolute counts by using 5.5 µm polystyrene standards (10^6/ml ± 5%, Optoflow AS).

The definition of particle size is dependant on the method of analysis. Using flow cytometry, cells scatter the light so that they appear to be about half their actual size. This means that bacteria being 1 µm in diameter scatters the light approximately as a 0.5 µm polystyrene particle would do. A *Cryptosporidium* oocyst being spherical with a diameter of 4 - 6 µm spreads the light approximately as a 2.5 µm polystyrene particle. A *Giardia* cyst being oval 8 – 9 x 13 – 16 µm spreads the light approximately as 5 µm polystyrene particle. An indicator parameter was defined by 1 µm and 15 µm polystyrene standards. The channels for maximum counts of 1 µm and 15 µm particles were recorded and the interval between the peaks was defined as 1-15 µm particles. Thus, the parameter included 50% of the 1µm particles and 50 % of the 15 µm particles generating the parameter particles 1-15 µm.

The choice of the indicator parameter size interval was checked by using formalinized *Giardia* and *Cryptosporidium* (oo)cyst preparations (Waterborne, Inc). Parasites were labeled using Cy-5 (Nycomed Amersham, Ltd) conjugated monoclonal antibodies (Giardi-a-glo/Crypt-a-glo, Waterborne) and verified by fluorescence microscopy. (Oo)cyst preparations were also stained with SYTO 62 (Molecular Probes, Inc). When analyzed on the Microcyte instrument (oo)cysts were all included in the 1-15 µm interval. Bacteria suspensions generally spread the light as 0.5 µm polystyrene particles and were never included in the 1-15 µm interval.

Results were displayed as light scatter histograms, corresponding to size distribution. In the 1-15 µm interval these results were presented as total counts and the share of fluorescent particles. Particles showing autofluorescence were mainly algae, probably due to 635 nm excitation of photosynthetic pigments such as chlorophyll a, chlorophyll b and allophycocyanine. This assumption was verified by fluorescence microscopy (Carl Zeiss filter set 26).

Jar testing

The jar tests were performed with a Flocculator 2000 (Kemira Kemi AB), on raw water from Lackarebäck. The equipment consisted of six 1L, cylinder shaped glass beakers with stirrers controlled by a micro-processor. Chemicals were added with a micro pipette. The pH was adjusted with sodium hydroxide. The coagulants were added at 400 rpm, G = 300 s^{-1} during 20 s. Flocculation was done in one step at 20 rpm, G = 3 s^{-1} during twenty minutes and the time for sedimentation was 20 minutes.

Full scale sampling

Results from an 11 week period of full-scale sampling with samples of raw water and water before filtration from all lines was evaluated to see whether flocculation design or flotation influenced the particle removal.

Results and discussion

Flow cytometry

The flow cytometer proved easy to operate and calibrations with polystyrene standards showed high stability. Samples of high particle counts were analyzed in 5 minutes. To get acceptable reproducibility, samples with lower particle counts (< 1000/ml) had to be processed up to 10 minutes.

There was a large variation in parasite sized particles, especially in low turbidity (< 0.1 FNU) filtrates. (Figure 3). This showed that filtrates with low turbidity was no guarantee of low (oo)cyst concentrations. Particle counts 1-15 µm yielded a range enabling estimates of log 3 reduction between raw water and filtrate.

There were some differences in the reduction of fluorescent and non-fluorescent particles in the parasite size range, as shown in Tables 2 and 3. The fluorescent component of the particles increased from about 5% in the winter to 25% in the spring (Figure 4).

Jar test comparison of coagulants

A comparison of six different coagulants at different dosages and pH indicates that changing coagulant could improve particle reduction. Turbidity removals were poor, possibly because of the particle content of the coagulants in combination with short flocculation and sedimentation times (data not shown).

Higher basicity is supposed to improve the particle removal [8]. The coagulant with the highest basicity tested PAX XL60 also showed the best log reduction of parasite sized particles (Figure 5). Another positive property of the PAX XL60 improving particle reduction might be that it is the only one with a silicate additive, about 1 % of the content. The results were inconclusive regarding the effect of higher dosage. The particle contents of the coagulants and sodium hydroxide might have decreased the resulting reduction at higher dosages. If there is a better reduction at higher dosages comparisons should be made of cost and environmental impact rather than equivalent dosages.

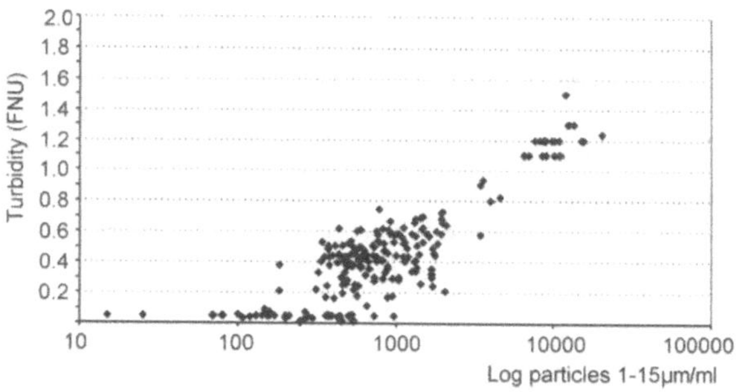

Fig. 3. Correlation between turbidity and flow cytometer particle count for all 220 samples during the 11 week test period

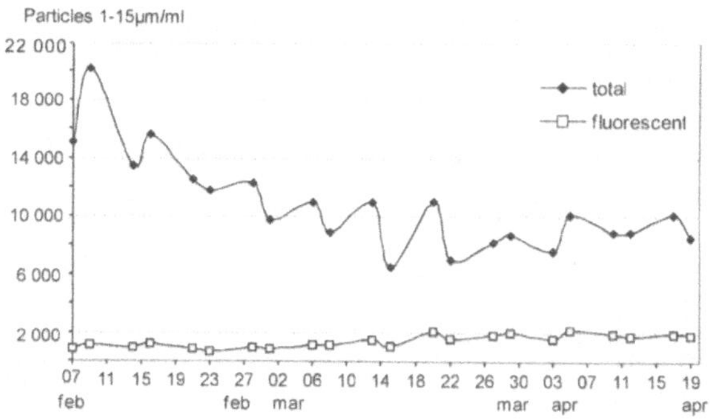

Fig. 4. Development of particle counts in raw water, winter and spring 2000

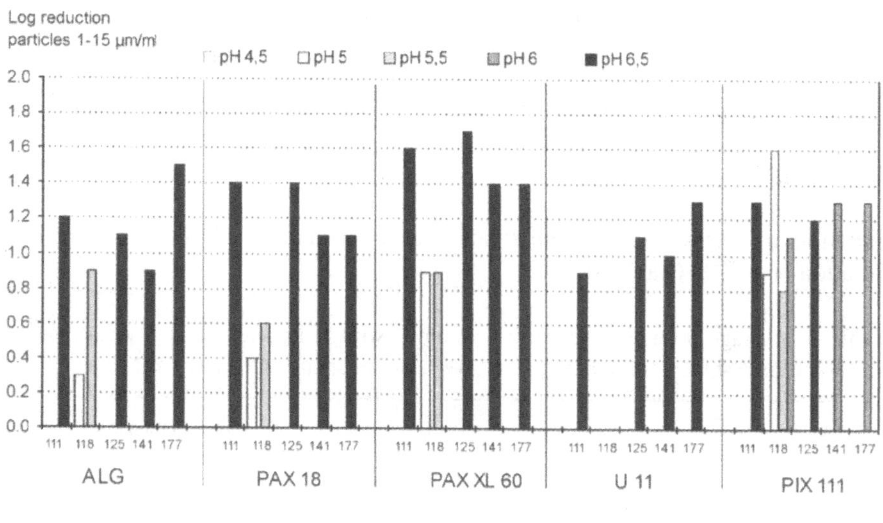

Fig. 5. Comparison of log-reduction of parasite sized particles using different coagulants

Flocculation design

Flocculation and sedimentation lines with baffles in the two first chambers and line with baffles in all chambers were compared to lines without baffles.

The turbidity reductions indicated better particle reduction without baffles while the parasite sized particles showed no effect of baffles or possibly a slight improvement (Table 2 and Figure 6). The individual differences between the lines without baffles were larger than the influence of baffles.

Table 2. Average turbidity and particle removal with different flocculation designs at the Lackarebäck plant, between of February and April 2000

	No baffles lines	Baffles in first two chambers	Baffles in all chambers
Turbidity log-reduction	0.41	0.33	0.34
Particle 1-15 μm log-reduction	1.03	1.04	1.03
Fluorescent particle 1-15 μm log –reduction	1.23	1.21	1.25

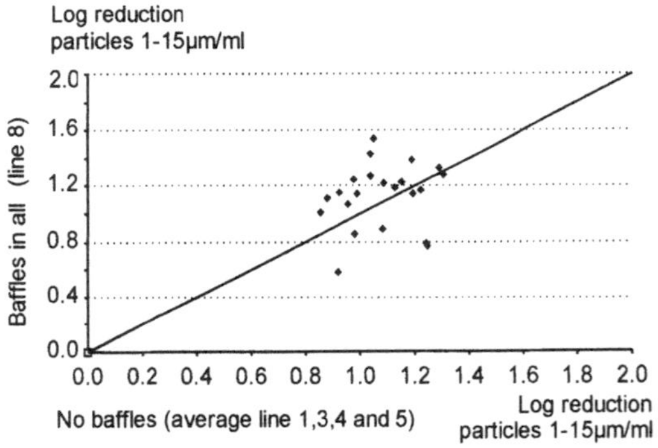

Fig. 6. Comparison of particle reduction with and without baffles in the flocculation

Sedimentation and flotation

Flotation in series with sedimentation was compared to all the lines without flotation or baffles (Table 3).

Table 3. Average turbidity and particle removal with sedimentation and sedimentation-flotation design at the Lackarebäck plant, between of February and April 2000

	Sedimentation	Sedimentation and flotation
Turbidity log-reduction	0.41	0.70
Particle 1-15 μm log-reduction	1.03	1.12
Fluorescent particle 15 μm log-reduction	1.25	1.33

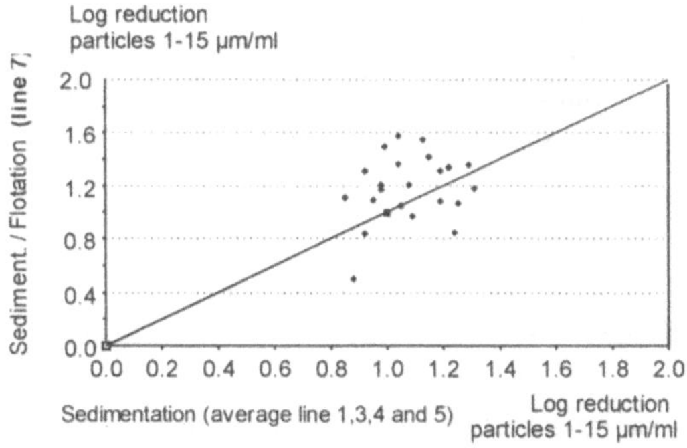

Fig.7. Comparison between sedimentation and sedimentation-flotation systems

The turbidity reduction was better with flotation than without. Particle reduction was generally better than the turbidity reduction. The difference between the two systems was less pronounced for particles 1-15µm than turbidity. The log-reductions with flotation showed a large variation between 0.5 and 1.6 (Figure 7). The intermittent operation of the scrapers (Figure 2) could have something to do with this.

Filter operation

Figure 8 shows an example of the turbidity and particle counts of filtered water at the Lackarebäck plant over the course of a 2 day filter cycle in November 1999. After backwashing the turbidity exceeded the official lower range limit 0.05 FNU only a few minutes.

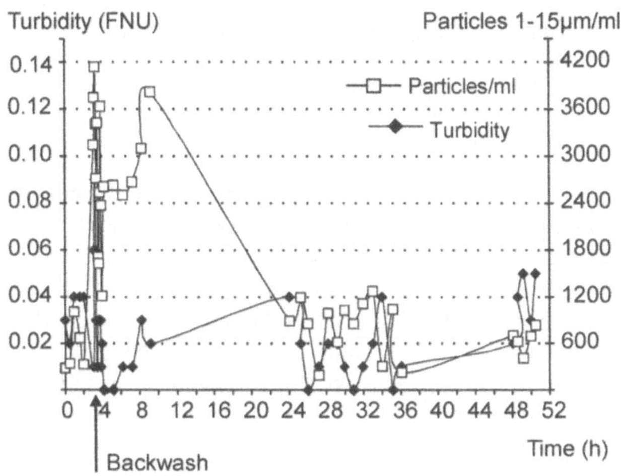

Fig. 8. Turbidity and particle counts after filtration with 6.8 m/h at the Lackarebäck plant November 22-23, 2000

The particle counts were several thousands per ml, similar to water before filtration at Lackarebäck, for at least six hours. The high particle counts suggest that a longer time for wasting the first filtered water may be necessary than turbidity measurements indicate. This indication also occurred at low filtration rate.

The graph in Figure 8 shows turbidity and particle counts during one whole filter cycle.

Conclusions and further work

Although turbidity can be a useful parameter to optimize particle removal, it could lead to wrong assumptions about removal of particles in the size range of protozoa and the extra information from particle counts should be useful in optimization.

It is useful to know as much as possible about different sorts of particles in the parasite size range to be able to develop accurate indicator parameters for parasite removal. Using an instrument with a 635 nm laser gives an easy opportunity to distinguish certain algae from the other particles. In water with high contents of these algae, this could be valuable when further developing the use of reduction of naturally occurring particles to assess parasite reduction.

Total log-reductions of parasite sized particles varied between 1-3. The largest variations occurred in the filtration step even though turbidity never exceeded 0.1 FNU.

Further research is needed to verify that the results reflect actual parasite removal. Stained or antibody labeled (oo)cyst can be one way and model organisms another way.

More data are needed to evaluate advantages of baffles in flocculation units. The reason why the flotation log-reduction of parasite sized particles varied so much should be investigated before further installations are made. The time needed to waste filtered water after flushing should be studied further.

Flow cytometry is a fast and simple analysis for particle reduction that can make it possible to measure log-removals also in low turbidity raw waters.

References

1. Miller, M.J.: "Point of View" Cryptosporidios: a waterborne disease, The microbiological quality of water, FBA/IWSA 1997. Reprinted from J. of AWWA 86 (12), (January 1994)
2. Stenström, T.A., Boisen, F., Georgson, F., Lahti, K., Lund, V. Andersson, Y., Ormerod, K.: Vattenburna infektioner i Norden (Waterborne infection in Nordic countries) TemaNord 1994: 585, Nordiska ministerrådet, (1994)
3. Hansen, A., Stenström, T.A.: Kartläggning av Giardia och Cryptosporidium i svenska ytvattentäkter (Giardia and Cryptosporidium i swedish surfacewaters, Smittskyddsinstitutet och Livsmedelsverket), (1998)
4. Ottoson, J.: Giardia and Cryptosporidium in Swedish sewage treatment plants, Master thesis, Swedish Institute for Infectious Disease Control, 1998
5. SMI, Swedish Institute for Infectious Disease Control: Smittsamma sjukdomar i Sverige 1998, 1999

6. Bergstedt, O.: Upgrading of water treatment for improved particle separation in Lovo sedimentation basins, IAWQ-conference, Upgrading of water and waste water systems, Kalmar, 1997
7. Göteborg Water and Waste Water Works: Årsberättelse 1999 (Annual report 1999 in Swedish)
8. Kemwater: Product chemistry and flocculation, 1999

Determining the Fate of Flocculants by Fluorescent Tagging

D.M. Bennett, B.A. Bolto, D.R. Dixon, R.J. Eldridge, N.P. Le and C.S. Rye

CSIRO Molecular Science and Cooperative Research Centre for Water Quality and Treatment, Bag 10, Clayton South, Victoria 3169, Australia.
David.dixon@molsci.csiro.au

Abstract

PolyDADMAC flocculants were synthesized containing 1-2% of an amine-functional monomer. These copolymers were tagged with a reactive fluorophore and used in jar tests on waters with different colour and turbidity. The performance of the copolymers was indistinguishable from that of DADMAC homopolymers of comparable MW. Residual flocculant could be detected fluorometrically at concentrations below 0.1 mg/L. Measurable residuals were always present in treated water, especially at doses above or below the optimum. Solid particles - either turbidity in the raw water or alum flocs - lowered both the flocculant demand and the residual concentration.

Introduction

Synthetic polymeric flocculants are widely used in potable water treatment, but there is no satisfactory way of quantifying any flocculant residual in the treated water. Their use is therefore regulated arbitrarily, with dose restrictions varying widely from place to place. Cationic polymers in particular are prohibited in some countries but routinely used in others. There is a clear need for better understanding of the fate of polymeric additives in water treatment to enable their regulation to be placed on a more scientific footing.

Water-soluble polymers can be detected at low concentrations by means of fluorescent tagging, but commonly used flocculants contain no point of attachment for a fluorescent tag. We synthesized analogues of commercial flocculants containing low levels of an amine-functional comonomer and tagged them with reactive fluorophores, either 5-dimethylaminonaphthalenesulfonyl (dansyl) chloride or fluorescein 5-isothiocyanate (FITC). The tagged flocculants can be used in jar tests in the usual way and quantified fluorometrically in the product water. Initial work was done with poly(diallyldimethylammonium chloride) (polyDADMAC) because this is by far the most commonly used flocculant in the polymer-only mode and because the higher doses used in this mode make the measurement of polymer concentrations easier. We report here studies of the effectiveness of these polymers as primary coagulants and as coagulant aids in the treatment of surface waters. For each water, residual flocculant was measured as a function of polymer dose.

Experimental

Water sources

Water was collected from Hope Valley Reservoir, Adelaide, South Australia, and Mount Zero Reservoir near Horsham, Victoria. Both these waters are highly coloured but low in turbidity. Natural organic matter (NOM) was isolated from Hope Valley Reservoir water by ion exchange as reported previously [1].

Jar tests

Jar test procedures were detailed previously [1]. Settled water was filtered through Whatman No 1 paper to simulate a sand filter. Samples were then buffered to pH 9 for fluorometric analysis. Fluorescence intensities were measured with a Perkin Elmer LS 50B luminescence spectrometer. Dansyl tags were excited at 215 nm and fluorescein tags at 493 nm. Emission peaked near 501 and 515 nm, respectively.

Tagged polyDADMACs

DADMAC was copolymerized at several monomer concentrations, yielding taggable polymers having weight-average molecular weights ranging from 100 000 to 500 000 (determined viscometrically according to Burkhardt et al. [2]). Copolymers containing 1 or 2 mol-% of taggable comonomer were tagged with FITC in pH 9 buffer at 75 °C or with dansyl chloride in water at room temperature. Residual taggant was removed by washing successively with sodium sulfate, sodium chloride and deionized water in an ultrafiltration cell until the permeate fluorescence was negligible. The purified polymers were freeze-dried and stored in vacuum desiccators. Standard solutions were made up as needed and stored in the dark.

Results

Sensitivity of the fluorometric analysis

Figure 1 shows fluorescence intensities (arbitrary units) for two tagged DADMAC polymers. In buffered high-purity water both yield linear calibration plots in the mg/L range. The slope for the FITC-tagged polymer is about 250, while that for the dansyl-tagged polymer is only 40. Since baseline readings under our conditions are typically a few units, the detection limit for the former polymer is ~ 10 µg/L. However, emission from the FITC tag is severely quenched in the presence of 1 mg/L (as total organic carbon) of Hope Valley NOM, complicating its use in real waters, while the dansyl tag is unaffected. Figure 1 also shows a finite emission from NOM in the absence of tagged polymer. Measurements on treated water samples require a correction for this effect. Centrifuging solutions containing FITC-tagged polyDADMAC and 1 mg/L NOM reduced the fluorescence intensity by a further 10-20 units at all polymer concentrations. This implies that the cationic polymer associates with anionic NOM to form aggregates that are on the margin of colloidal stability, and that these aggregates contribute measurably to the observed fluorescence. NOM from other sources also quenches emission from FITC-tagged polyDADMAC, as do transition metal ions such as Fe(III). Silica has little effect at concentrations of a few mg/L; higher levels reduce the measured intensity, apparently by a simple light scattering mechanism. Quenching in NOM-containing water is greatly reduced, but not completely eliminated, by prior treatment with alum.

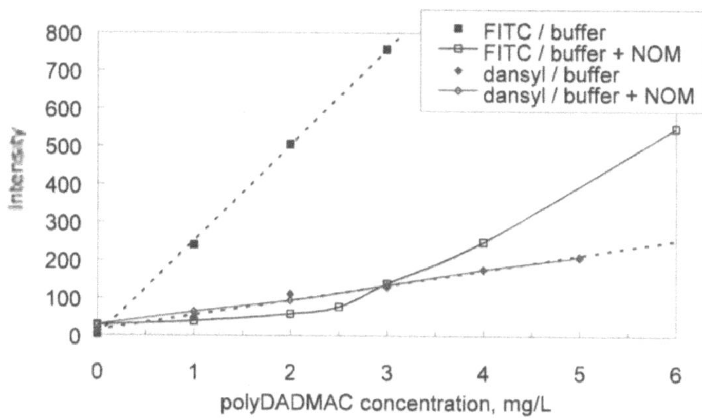

Fig. 1. Calibration plots for two tagged polyDADMACs in buffered high-purity water and in the presence of 1 mg/L Hope Valley NOM

Performance of the tagged flocculants

Tagged polyDADMACs were compared against a range of commercial polyDADMAC flocculants in jar tests with synthetic and real waters. Some synthetic waters made up from clay and NOM concentrates appeared to demand marginally higher doses of FITC-tagged polymer to achieve a given level of clarification, but no such effect was noted with real waters. Figures 2 and 3 show the results of jar tests on Mount Zero Reservoir water with a tagged polyDADMAC of MW ~ 250 000 and two commercial polyDADMACs, one comparable in MW and one of lower MW. In all cases there is little change in turbidity below about 4 mg/L, then a sharp fall in the range 5-8 mg/L. True colour decreases monotonically with increasing dose at least up to 7-8 mg/L, reaching values of 2-4 Hazen units (HU). The tagged polymer seems to differ from the other two by no more than they differ from each other.

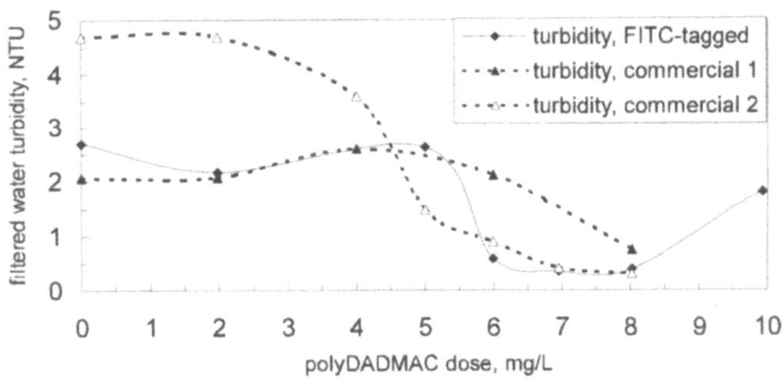

Fig. 2. Treatment of Mount Zero Reservoir water with tagged and commercial polyDADMACs – turbidity as a function of dose

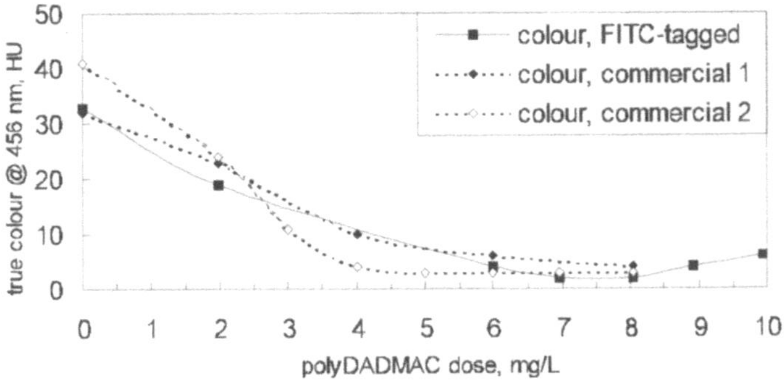

Fig. 3. Treatment of Mount Zero Reservoir water with tagged and commercial polyDADMACs – colour as a function of dose

Figures 4 and 5 compare two tagged polyDADMACs with a commercial polyDADMAC of comparable MW in treating Hope Valley Reservoir water. The turbidity increases from an initial very low level at low polymer doses, peaks at about 1.5 NTU for a dose of 2 mg/L, then decreases as the optimum dose is approached. True colour decreases steadily over the same dose range. The turbidity increase can be attributed to the formation from the remaining NOM of colloidal polymer-NOM complexes that are not removed by settling or filtration. Again the tagged polymers appear to be similar in effectiveness to the commercial flocculant.

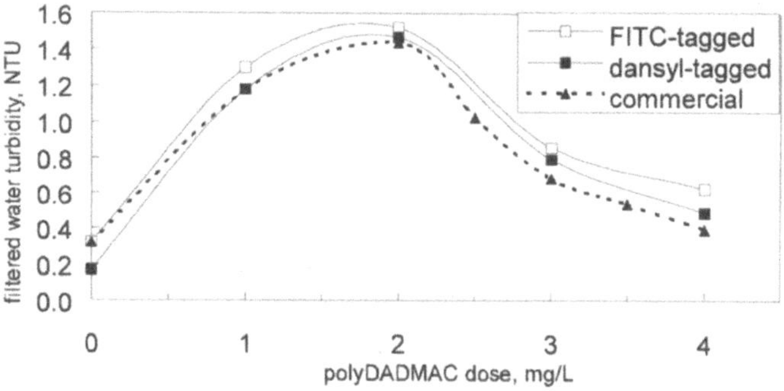

Fig. 4. Treatment of Hope Valley Reservoir water with tagged and commercial polyDADMACs – turbidity as a function of dose

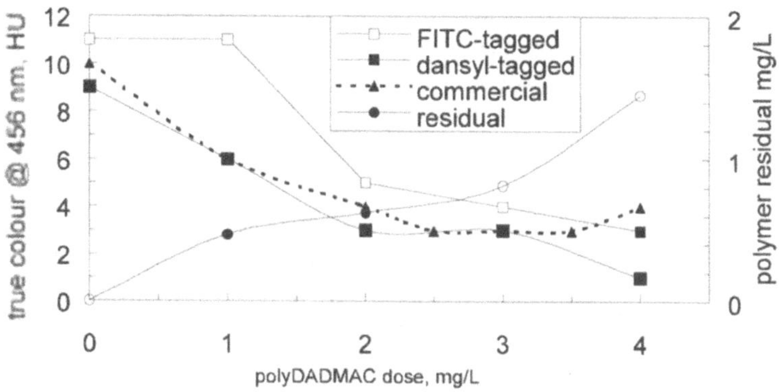

Fig. 5. Treatment of Hope Valley Reservoir water with tagged and commercial polyDADMACs – colour and polymer residual as a function of dose

Flocculant residuals

Luminescence measurements revealed the presence of residual flocculant in all waters treated with tagged polyDADMAC in the polymer-alone mode. Figure 5 shows that when Hope Valley reservoir water was dosed with 1 mg/L of polyDADMAC, well below optimal, 0.45 mg/L remained after settling and filtration. The residual increased slowly at moderate doses, falling in percentage terms to about 25-35% of the dose, then rose steeply at higher doses.

In Figure 6 measured flocculant residuals in Mount Zero Reservoir water are overlaid on turbidity and colour results. The residual increases with dose in the low dose range, where relatively high turbidity and colour values indicate that most of the original organic matter remains in the treated water in colloidal form. Then the residual falls to a minimum (0.4-0.7 mg/L, 7-8%) near the optimum dose, before rising again when excess polymer is added.

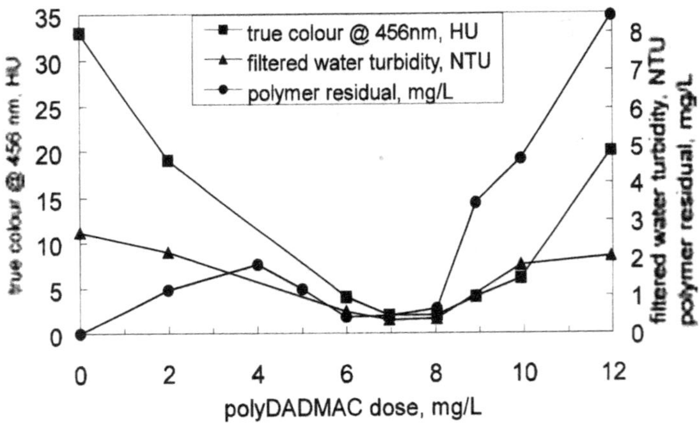

Fig. 6. Treatment of Mount Zero Reservoir water with tagged polyDADMAC. Residual polymer (RH axis) is measurable at all doses.

Effect of solid particles

We previously reported for several waters that removal of colour and UV-absorbing organics by polymer-only treatment is improved by the addition of clays or metal oxides [1, 3] and that a small dose of cationic polymer can be substituted for much of the alum demand [3]. When Mount Zero Reservoir water spiked with 50 mg/L of ball clay was treated with polyDADMAC, turbidity and colour after treatment were lower than for unspiked water (Figures 7 and 8). The optimum dose was significantly reduced. Polymer residuals were much lower, not exceeding 0.1 mg/L at doses up to 6 mg/L. The presence of alum flocs has a similar effect: Figures 9 and 10 compare the treatment of Mount Zero Reservoir water with polyDADMAC plus 24 mg/L alum (half the optimum alum-only dose) with polymer-only treatment. Colour removal is much greater at low polymer dose (and also greater than was achieved by alum-only treatment at any dose). Turbidity removal is also greatly improved. The optimum polymer dose is now less than 1 mg/L. Polymer residuals are much smaller except at the lowest dose, being roughly constant at 0.1 mg/L for doses up to 1 mg/L and rising to 0.2 mg/L at 2 mg/L polymer.

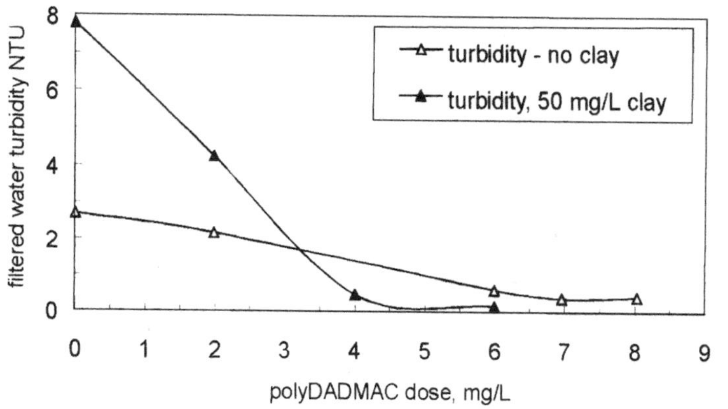

Fig. 7. Treatment of Mount Zero Reservoir water with and without added clay. Turbidity as a function of dose of tagged polyDADMAC.

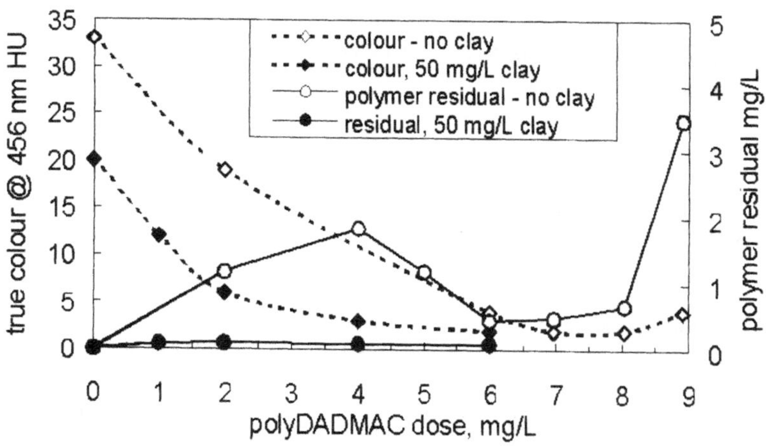

Fig. 8. Treatment of Mount Zero Reservoir water with and without added clay. Colour and polymer residual as a function of dose of tagged polyDADMAC.

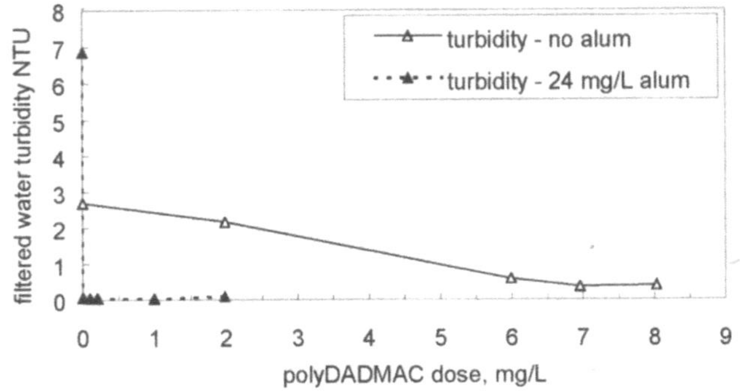

Fig. 9. Treatment of Mount Zero Reservoir water with alum plus polyDADMAC or with polyDADMAC alone

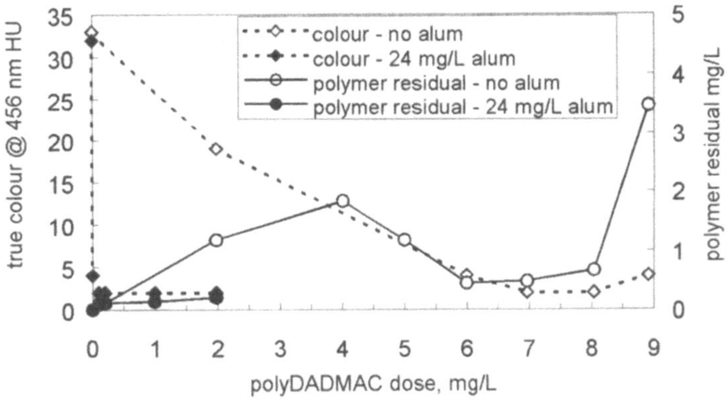

Fig. 10. Treatment of Mount Zero Reservoir water with alum plus polyDADMAC or with polyDADMAC alone

Discussion

Fluorometric analysis

As noted above, measured fluorescence intensities were corrected for the contribution arising from residual NOM. The correction was estimated in either of two ways. In the first method, we assume that the emission I_{NOM} from NOM is proportional to the UV absorbance, A, of the water, which was routinely measured at 456 nm. For any polymer dose d the correction is then given by

$$I^{d}_{NOM} = I^{0}_{NOM} \cdot A^{d} / A^{0} \quad (1)$$

where the superscript 0 refers to raw water. In the second method, fluorescence intensity I^{λ} and UV absorbance A^{λ} were measured for all samples at two wavelengths, that corresponding to the emission maximum from the tagged polymer (501 nm for dansyl tags) and a shorter wavelength where only NOM fluoresces, typically 475 nm. We then assume that the correction is given by

$$I^{501}_{NOM} = I^{475}_{NOM} \cdot A^{501} / A^{475} \qquad \textbf{(2)}$$

since the absorbance of the polymer is negligible at both wavelengths. The two methods gave the same result within experimental error.

Flocculant residuals

As part of a pioneering study of the role of NOM in flocculation, Narkis and Rebhun [4] carried out jar tests using a cationic polyelectrolyte labeled with ^{14}C. The radioactive tag enabled them to measure flocculant residuals ~ 0.1 mg/L. They found that humate and fulvate were effectively removed from solids-free solution only in a narrow flocculant dose range; at higher and lower doses, close to 100% of both NOM and flocculant remained in solution. Outside the range of effective flocculation the turbidity remained constant or increased as polymer-NOM complexes formed, but remained colloidally stable. Figures 2 and 4 of the present paper show similar behaviour. In the optimum dose range the residual concentration of radioactive flocculant (after centrifuging) was very low. For example, the optimum range for a 10 mg/L solution of Aldrich sodium humate was 12-14 mg/L, leaving a residual of 0.17 mg/L (1.2-1.4%). The residuals shown in Figure 5 are markedly higher. This may be partly because our treated waters were not centrifuged. In addition, since Hope Valley NOM is highly oxidized and much lower in MW than Aldrich humate, its complexes with cationic polymers are probably more soluble.

Narkis and Rebhun found clay (150 mg/L calcium montmorillonite) to be effectively flocculated at low dose (0.5 - 1 mg/L); no flocculant residual was detected in this range. NOM inhibits the flocculation of clay by preferentially consuming flocculant. Clay and NOM are removed together at a similar dose to that required for NOM alone, and again a small flocculant residual (~0.2 mg/L) was detected in this range. Above and below the optimum dose a substantial fraction of the flocculant remained in the treated water, yielding plots strikingly similar to the residual flocculant curve of Figure 6. The residuals were lower than in the absence of clay, suggesting that polymer-NOM complexes are removed to some extent by adsorption on clay particles. However, the effect was smaller than that shown in our Figure 8.

Conclusions and Significance

Fluorescent tagging enables flocculant residuals to be quantified at concentrations below 0.1 mg/L. Tagging has no significant effect on flocculation performance. Measurable amounts of tagged polyDADMAC were detected in all treated water samples. The residual in a low-turbidity surface water passed through a minimum of about 0.5 mg/L near the optimum dose for colour and turbidity removal. At higher and lower doses a high percentage of the polymer remained in the treated water. A water containing clay had a smaller polymer demand and residuals were only 0.1 mg/L at all doses up to the optimum. When polyDADMAC was used with alum, optimum polymer doses were much smaller, ~ 0.1-1 mg/L. The polymer residual was only 0.1-0.2 mg/L at doses up to 2 mg/L. Alum-polyelectrolyte combinations are an attractive treatment option, giving good clarification performance with low sludge volumes and minimal flocculant residual.

At present polymeric flocculants are regulated without any real understanding of the amount likely to reach the finished water. The results of this work suggest that water treated with polyelectrolytes, alone or with inorganic coagulants, is likely to contain traces of polymer, but that the residual can be minimized by careful control of dose. There is a clear need for water plant operators to monitor these residuals and, where possible, adjust treatment conditions to minimize polymer levels in finished water, consistent with satisfactory clarification. Our results also suggest that polymers should be more strictly regulated when they are used as primary coagulants than when they are used together with an inorganic coagulant.

The high cost of fluorescently tagged flocculants prohibits their routine use, but if produced in volume at an acceptable cost they would be suitable for occasional use in water plants as a diagnostic tool.

Experiments with a bench-scale continuous flocculation train followed by a sand filter are in progress to check the jar test findings. Tagged polyacrylamides have been prepared for jar tests in which a low polyacrylamide dose is used to strengthen alum flocs.

References

1. Bolto, B.A., Dixon, D.R., Eldridge, R.J., King, S.J.: The Use of Cationic Polymers as Primary Coagulants in Water Treatment. In: Chemical Water and Wastewater Treatment V, H.H. Hahn, E. Hoffmann and H. Ødegaard (Eds.). Springer, Berlin 1998, p. 173
2. Burkhardt, C.W., McCarthy, K.J., Parazak, D.P.: Solution Properties of Poly(dimethyldiallylammonium Chloride). J. Polym. Sci. C25 (1987) 209
3. Bolto, B.A., Abbt-Braun, G., Dixon, D.R., Eldridge, R.J., Frimmel, F., Hesse, S., King, S.J., Toifl, M.: Experimental Evaluation of Cationic Polyelectrolytes for Removing Natural Organic Matter from Water. Water Science and Technology 40 (9) (1999) 70
4. Narkis, N., Rebhun, M.: The Mechanism of Flocculation Processes in the Presence of Humic Substances. J. Amer. Water Works Assoc. 67 (1975) 101

Drinking Water Treatment

Coagulation-Microfiltration Processes for NOM Removal from Drinking Water

T. Carroll, D. Vogel, A. Rodig, K. Simbeck and N. Booker

Cooperative Research Centre for Water Quality and Treatment, CSIRO Molecular Science, Bag 10, Clayton South, VIC 3169, Australia.
t.carroll@molsci.csiro.au

Abstract

A variety of coagulants were investigated for potential as a microfiltration pretreatment to improve NOM removal from drinking water. The impact of pretreatment upon subsequent membrane performance was investigated in both laboratory and pilot-scale trials. Membrane flux decline rates were reduced by pretreatment, but improvements were similar for all coagulants tested, including alum and prepolymerised aluminium and iron-based coagulants. These findings suggest that physical property differences between the flocs of different coagulants do not significantly impact membrane performance when formed and removed under typical microfiltration plant operating conditions.

Introduction

Microfiltration is particularly effective for the removal of particulate contaminants, such as clay, bacteria, algae and protozoa from drinking water. However, microfiltration is ineffective for the removal of dissolved contaminants such as natural organic matter (NOM). Residual NOM in drinking water is a cause of colour, disinfection by-products, and microbial regrowth [1,2,3]. Chemical coagulation is routinely used in Australia as a microfiltration pretreatment for enhanced NOM removal to meet colour targets. NOM associated with precipitated flocs in pretreatment is retained by the membrane, improving permeate quality.

Membrane filtration processes are subject to fouling by retained material. In the case of microfiltration, particles or flocs form a cake on the membrane surface, which contributes a hydraulic resistance to filtration [4]. As the cake grows, filtration pressure is increased to maintain permeate flowrate. Eventually, the cake is removed by periodic backwashing and chemical cleaning. Recovery of permeate flowrate in this way occurs at the expense of productivity and yield. Energy costs and water losses are related to the rate of fouling. In Australia, minimising washwater volume losses is particularly important, as water resources are scarce and washwater disposal often requires further processing or transport off-site.

The rate of membrane fouling is an important selection criterion for coagulants in microfiltration pretreatment. Microfiltration processes have a substantially shorter time available for flocculation than a conventional sedimentation process, and substantially higher shear forces and pressures at the filter surface. The kinetics of flocculation and the physical properties of the resulting flocs may therefore influence the resistance of the microfiltration cake. In this paper, a variety of commercial coagulants are investigated for potential use in microfiltration pretreatment. The membrane fouling rates and permeate water quality are investigated at laboratory scale under conditions of low shear and long flocculation time, and at pilot scale under more physically stressful, though realistic, conditions.

Experimental

The raw water sources used in the microfiltration trials were the Moorabool River near Anakie, and the Maroondah Aqueduct near Yerring Gorge, both in Victoria, Australia. The Moorabool water had a relatively high dissolved organic carbon content (8-12 mg/L) and relatively low turbidity (3.9 NTU). The Maroondah water had a moderate dissolved organic carbon (2.5-3.5 mg/L), and a low turbidity (0.6-2.6 NTU).

NOM removal from both waters was investigated by treatment in standard jar tests with several different aluminium and iron-based coagulants. The coagulants used were alum, polyaluminium chloride (PAC), polyaluminium chlorohydrate (PAC-AC), ferric chloride, polyferric sulphate (PFS), and a 50/50 alum/polyferric sulphate mixture (alum-PFS). Moorabool water was treated as two lots collected at different times of year. The aluminium-based coagulants were tested on lot #1, and the iron-based coagulants were tested on lot #2. Both lots were treated with alum to provide a basis for comparison across all coagulants. In the jar test procedure the pH of 1.00 kg of raw water was adjusted from 7.5 to 6.0 with sulphuric acid. The

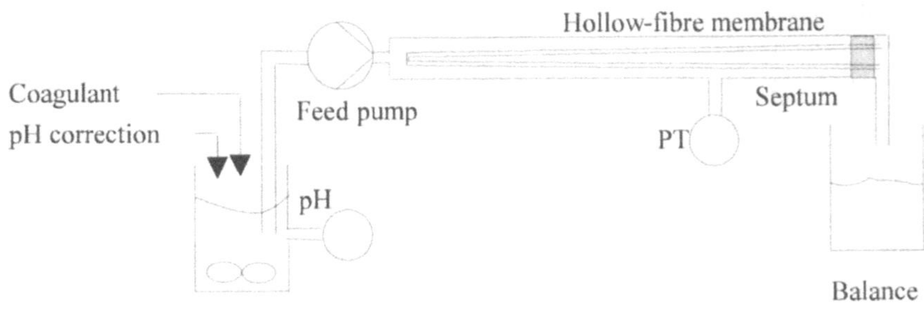

Fig. 1. Single-fibre microfiltration rig

water was then dosed with coagulant as required and stirred for 60 s at 250 RPM followed by 90 s at 100 RPM. The pH was maintained at 6.0 throughout with sodium hydroxide. Dissolved organic carbon (DOC) and UV absorbance (254 nm) were measured on both feed and treated water after 0.45 μm prefiltration.

Fouling of a microfiltration membrane after coagulation pretreatment was investigated at both laboratory and pilot scale. Laboratory-scale microfiltration runs were carried out on Moorabool water filtered through single 0.2 μm polypropylene hollow-fibres using the apparatus shown in Figure 1. Each fibre had a nominal surface area of 700 mm^2, and a typical initial permeate flux of 860 L/m^2h at 80 kPa. A virgin membrane was used in each run, and the permeate flux declined as fouling proceeded. Trans-membrane pressure and permeate throughput were measured by a pressure transducer (PT) and balance, respectively, and logged with data acquisition software. The pressure followed the pump curve uncontrolled, and the permeate flux was calculated from the differential throughput. The water was pretreated using the jar test procedure described above (although the 100 RPM stirring was omitted), and pumped onto the hollow-fibre membrane. The feed suspension was stirred at 250 RPM and the pH was controlled at 6.0 throughout the run. Coagulant doses were fixed for both lots of water.

Fig. 2. Coagulation-microfiltration pilot plant

Pilot-scale microfiltration runs were performed on the plant shown in Figure 2. The plant was housed within a 20 ft. container located below the Maroondah Aqueduct at Yerring Gorge. The feed water was gravity-fed from the aqueduct through a flow control valve at 20 L/min. The water was then treated in a series of four gravity-fed 10 L tanks. Coagulant was dosed into the first tank, and caustic and acid were dosed into the second tank to correct the pH to 6.0. Flocculation occurred in the third and fourth tanks, under low-shear stirring, with a mean residence time of 1 min. The treated water was then fed to a 3 m^2 Memcor microfiltration unit with automatic air backwash. The membrane consisted of 3 x 1 m^2 modules of 0.2 μm polypropylene hollow fibres. The unit was run at a constant permeate flux of 360 L/m^2h, and an initial trans-membrane pressure (TMP) of 40 kPa. The pressure increased as fouling proceeded, and the membrane was backwashed to a waste tank at 60 min intervals with 600 kPa air. At a pressure of 180 kPa the membrane was chemically cleaned with a proprietary cleaning solution (Memclean) mixed with hydrogen peroxide. The membrane was then flushed with pure water (reverse osmosis) until the initial pressure was achieved. Turbidity and UV absorbance (254 nm) were measured on both feed and permeate.

Results and Discussion

NOM Removals After Coagulation Treatment

The NOM removals from Moorabool water by the aluminium and iron-based coagulants are shown in Figure 3. The aluminium-based coagulants produced a minimum DOC of approximately 4 mg/L from lot #1, at a dose of 3.2 mg/L as Al^{3+}. The DOC removals for alum, PAC and PAC-AC were similar. The iron-based coagulants produced a minimum DOC of approximately 5.5 mg/L from lot #2, at a dose of 4.0 mg/L as Al^{3+} equivalent (i.e. molar basis). The DOC removals for $FeCl_3$, PFS and alum-PFS were similar. The DOC removal for alum on lot #2 was similar to removals for the iron-based coagulants, although higher than for alum on lot #1. The differences in DOC removal between the two lots can therefore be attributed to differences in water quality rather than differences in coagulant effectiveness.

The NOM removals from Maroondah water by the aluminium and iron-based coagulants are shown in Figure 4. Both iron and aluminium-based coagulants produced a minimum UV absorbance at a dose of approximately 1.4 mg/L as Al^{3+}. The UV absorbance after PFS treatment was slightly higher than after treatment by the other coagulants at the same dose.

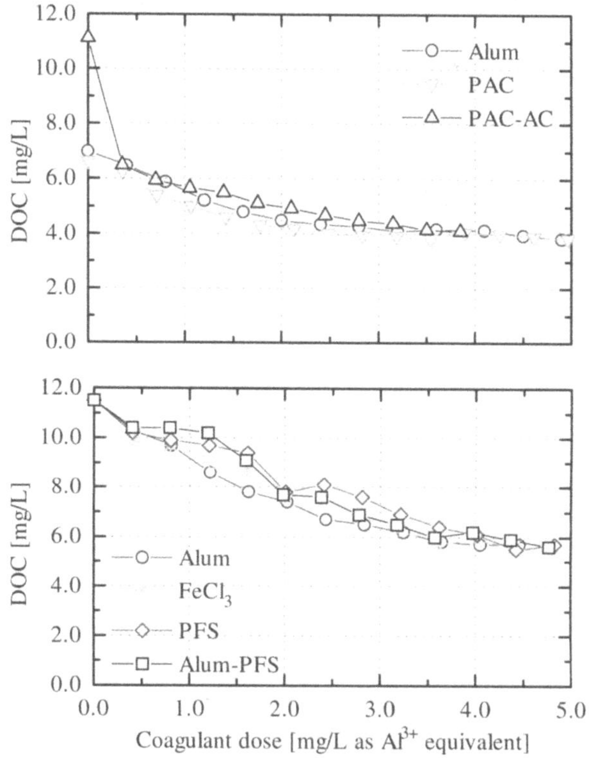

Fig. 3. Dissolved organic carbon (DOC) concentration from Moorabool water after treatment with alum, PAC, PAC-AC, $FeCl_3$, PFS and alum-PFS at pH = 6.0

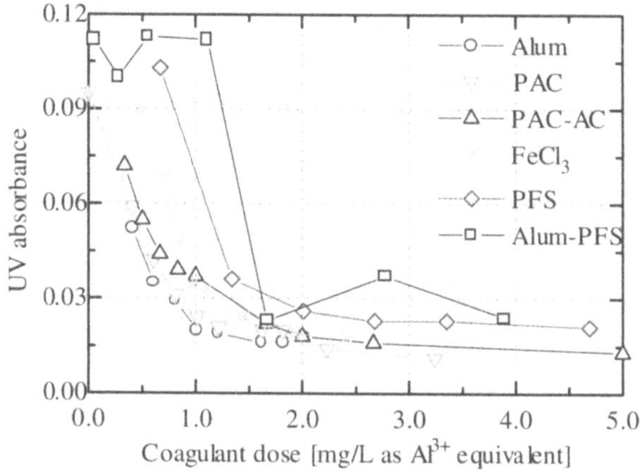

Fig. 4. UV absorbance (254 nm) of Maroondah water after treatment with alum, PAC, PAC-AC, $FeCl_3$, PFS and alum-PFS at pH = 6.0

Iron-based and aluminium-based coagulants typically remove hydrophobic and charged hydrophilic NOM, leaving a residual NOM of predominantly small hydrophilic neutral NOM [5,6,7,8,9]. Therefore, the residual NOM concentration after coagulation should be independent of the type of coagulant used, for a particular water lot.

Microfiltration Fouling After Coagulation Pretreatment

Single Hollow-fibre Membrane Laboratory Runs

Fouling of the single hollow-fibre membrane by Moorabool water after coagulation pretreatment is shown in Figure 5. The permeate flux is plotted as a function of permeate throughput after pretreatment at pH = 6.0 using the minimum coagulant dose determined from the jar tests (3.2 mg/L as Al^{3+} for Moorabool lot #1 and 4.0 mg/L as Al^{3+} equivalent for Moorabool lot #2). In all cases, the rate of fouling after pretreatment was substantially lower than the rate for untreated water. Coagulation may either prevent sub-micron particles and NOM from entering membrane pores when associated with larger flocs, or the cake resistance may be reduced as particles are aggregated into larger and more open structures [4]. The fouling rates were similar for all of the aluminium-based coagulants used on lot #1.

In each case, the flux dropped to approximately 500 L/m^{2}h after 500 mL was filtered. The rate of fouling for alum and alum-PFS on lot #2 was also similar. After 500 mL was filtered, the flux was 400 L/m^{2}h in both cases. However, for both $FeCl_3$ and PFS, the rate of fouling was slightly higher than for alum; the flux declined to 300 L/m^{2}h after 500 mL was filtered.

The similarity between the rates of fouling achieved with traditional coagulants such as alum or $FeCl_3$ and the respective prepolymerised coagulants (PAC, PAC-AC and PFS) may be caused by a number of factors. The faster kinetics of prepolymerised coagulants, and the improved filterability in conventional filtration are well-established [10,11]. However, the filtration resistance of all laboratory-filtered flocs may be similar, either intrinsically, or because kinetic and filterability-related differences are not manifest under laboratory conditions. The single-fibre runs are done in batch, so flocs are typically contacted for hours rather than seconds. Furthermore, single-fibre fluxes are too low to produce appreciable hydrodynamic shear at the membrane surface, although flocs were contacted under high mechanical shear before filtration. Secondly, as the smallest flocs dominate the contribution to filtration resistance, it is possible that the fouling rate is determined by size rather than structure. Alternatively, the filtration resistance of the flocs, though coagulant-dependent, may be lower than those from other rate-determining sources of fouling such as residual NOM [12,13]. In either case laboratory-scale filtration is limited as a predictor of ultimate process performance, and pilot-scale trials are necessary.

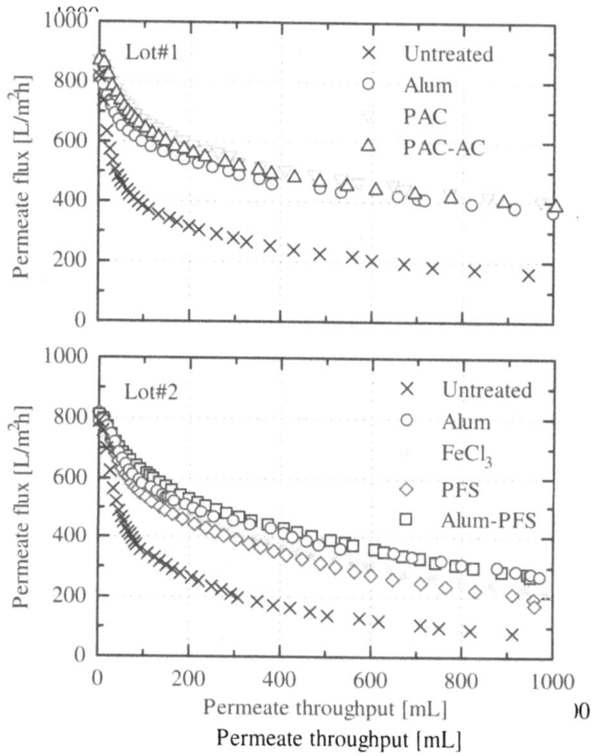

Fig. 5. Rates of membrane fouling in single hollow-fibre laboratory runs by Moorabool water after treatment with alum, PAC, PAC-AC, $FeCl_3$, PFS and alum-PFS at pH = 6.0

Memcor Pilot Plant Runs

Pilot plant membrane fouling runs after coagulation pretreatment are shown in Figure 6. The coagulants were each dosed at 1.4 mg/L as Al^{3+} equivalent as determined from the Maroondah jar tests. The trans-membrane pressure is plotted as a function of permeate throughput at constant flux. The rates of fouling for different coagulants were determined from the TMP rise rates before the first backwash (at approximately 1000 L). These fouling rates are given in Table 1, with the feed and permeate water quality. As for the single-fibre runs, the rate of fouling after pretreatment was substantially lower than the rate for untreated water. The throughput after alum treatment was approximately 10 times higher than without treatment. All coagulants fouled the membrane at similar rates initially (0.02-0.03 Pa/L), except for $FeCl_3$ which fouled substantially faster (0.07 Pa/L). The rate of fouling increased with throughput in all cases. The permeate quality was similar for all coagulants except for PAC-AC, which treated the feed with highest UV absorbance. Permeate UV-absorbance was substantially reduced by coagulation pretreatment in all cases. The rate of fouling was relatively insensitive to variations in the feed UV-absorbance (NOM content).

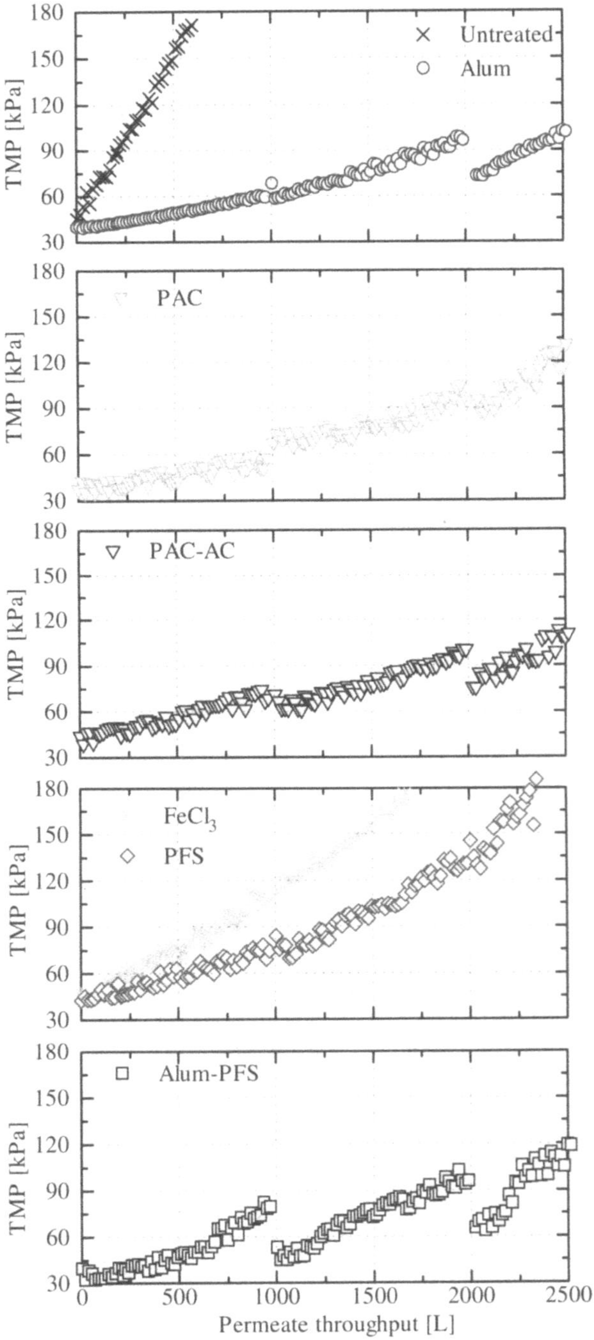

Fig. 6. Rates of membrane fouling in Memcor pilot plant runs of Maroondah water after treatment with alum, PAC, PAC-AC, $FeCl_3$, PFS and alum-PFS at pH = 6.0

Table 1. Initial fouling rates and water quality in Memcor pilot plant runs of Maroondah water after treatment with alum, PAC, PAC-AC, $FeCl_3$, PFS and alum-PFS at pH = 6.0

Coagulant	Initial Fouling Rate[Pa/L]	Turbidity [NTU]		UV_{254}	
		Feed	Permeate	Feed	Permeate
Untreated	0.213	0.6	0.03	0.083	0.065
Alum	0.022	2.3	0.01	0.132	0.022
PAC	0.018	2.2	0.00	0.113	0.017
PAC-AC	0.031	2.2	0.02	0.202	0.038
$FeCl_3$	0.069	2.1	0.10	0.160	0.022
PFS	0.034	2.6	0.03	0.121	0.021
Alum-PFS	0.030	2.5	0.03	0.182	0.025

The pilot-plant runs showed the same coagulant-independent trend in fouling rate as the laboratory runs. On this basis, it seems likely that the fouling of a 0.2 µm polypropylene microfiltration membrane in this study was insensitive to structural differences in the flocs formed from the different coagulants tested. Structural features such as floc permeability and fracture tendency under hydrodynamic shear at the membrane surface may be of secondary importance compared to the size and quantity of the smallest flocs. If this is the case, fouling may be reduced by using a polyelectrolyte with the coagulant. However, alum and a high molecular-weight, non-ionic polyacrylamide (LT-20) had the same rate of fouling as alum alone in laboratory runs, albeit under kinetically-ideal conditions. Pilot plant runs are needed to assess polyelectrolyte benefits under kinetically-limited conditions.

Previous laboratory experiments on fractionated Moorabool water indicated that hydrophilic neutral NOM played an important role in the fouling of polypropylene microfiltration membranes [13]. However this NOM fraction was not removed by coagulation/flocculation [14], and may determine the rate of fouling after coagulation pretreatment of this water source. Further work is needed before the same conclusions can be drawn for the Maroondah water used in pilot plant runs, as it has a significantly lower concentration of hydrophilic neutral NOM than Moorabool water [15].

Conclusions

A variety of traditional and prepolymerised aluminium and iron-based coagulants were tested for suitability as a pretreatment for drinking-water microfiltration. All coagulants reduced the rate of fouling, although differences in membrane performance between coagulants was minimal. Floc structure, and kinetic effects appear to play a secondary role compared to floc size or residual NOM in the fouling of polypropylene microfiltration membranes by the surface waters studied.

Acknowledgements

The funding for this work was provided by the Cooperative Research Centre for Water Quality and Treatment. The assistance of Melbourne Water in providing the pilot plant location is also appreciated.

References

1. MacCormick, A.B.: The application of microfiltration in water and wastewater treatment. In Modern Techniques in Water and Wastewater Treatment, L.O. Kolarik and A.J. Priestley (Eds.), CSIRO Publishing, Australia, (1995), pp. 45-51
2. Rook, J.J.: Formation of haloforms during chlorination of natural waters. Water Treatment Exam. 23 (1974) 234
3. van der Kooij, D.: Assimilable organic carbon as an indicator of bacterial regrowth. J. AWWA 84 (1992) 57
4. Wiesner, M.R. and Laine, J.M.: Coagulation and membrane separation. In Water Treatment Membrane Processes, J. Mallevialle, P.E. Odenaal and M.R. Wiesner (Eds.), McGraw-Hill, U.S.A. (1996) pp. 4.1-4.40
5. Bose, P., Reckhow D., Bezbura B.: Isolation and characterisation of aquatic natural organic matter: Implications for removal during water treatment. Proc. AWWA Annual Conf., San Antonio, Texas (1993) pp. 417-429
6. Dryfuse, M., Miltner, R., Summers, R.: The removal of molecular size and humic/non-humic fractions of DBP precursors by optimised coagulation. Proc. AWWA Annual Conf., Anaheim, California (1995) pp. 217-241
7. Singer, P. and Harrington, G.: Coagulation of DBP precursors: Theoretical and practical implications. Proc. AWWA WQTC Conf., Miami, Florida (1993) pp. 1-19.
8. Korshin, G.V., Benjamin, M., Sletten, R.: Adsorption of natural organic matter (NOM) on iron oxide: Effect of NOM composition and formation of organo-halide compounds during chlorination. Water Res. 31 (1997) 1643
9. White, M., Thompson, J., Harrington G., Singer, P.: Evaluating criteria for enhanced coagulation compliance. J. AWWA 89 (1997) 64
10. Wiesner, M.R., Lahoussine-Turcaud, V., Fiessinger, F.: Organic removal and particulate formation using a partially neutralised $AlCl_3$. Proc. AWWA Annual Conf., Denver, Colorado (1986) pp. 1685-1704
11. Matsui, Y., Yuasa, A., Furuya, Y., Kamei, T.: Dynamic analysis of coagulation with alum and PACl. J. AWWA 90 (1998) 96
12. Lahoussine-Turcaud, V., Wiesner, M., Bottero, J.Y., Mallevialle, J.: Coagulation pretreatment for ultrafiltration of a surface water. J. AWWA 81 (1990) 76
13. Carroll, T., King, S., Gray, S.R., Bolto, B.A., Booker, N.A.: The fouling of microfiltration membranes by NOM after coagulation treatment. Water Res. in press.
14. Bolto, B., Dixon, D., Eldridge, R., King, S., Toifl, M.: The use of cationic polymers as primary coagulants in water treatment. In Chemical Water and Wastewater Treatment V, H.H. Hahn, E. Hoffmann and H. Ødegaard (Eds.), Springer, Berlin (1998) pp. 171-185
15. Fan, L.-H., RMIT University, Melbourne, Australia, unpublished results

Potential of Ferric and Polyaluminium Coagulants for Nanofiltration Pretreatment

J. Yli-Kuivila*, I.T. Miettinen** and R. Laukkanen*

*Helsinki Univ. of Tech., Lab. of Env. Eng. P.O. Box 5300, FIN–02015 HUT, Finland
jukka.yli-kuivila@hut.fi
**National Public Health Institute, Div. of Env. Health, P.O. Box 95, FIN-70701 Kuopio

Abstract

The conventional chemical treatment of soft, cold and humic surface water was studied on a pilot scale, and fouling indexes were used to evaluate the suitability of the pretreated water for nanofiltration. The recommended values for fouling indexes were easily achieved. The best modified fouling index results were about 4 s/l^2 with two ferric salts and about 5 s/l^2 with polyaluminium chloride. Optimal flocculation pH and coagulant dosages for the modified fouling index were also suitable for minimizing organic matter and turbidity at the same time. In addition to removing organic carbon, nanofiltration can be applied to decrease ionic substances such as coagulant residues. Chemical treatment reduces organic matter of higher molar mass efficiently, but organic residue together with residual iron or aluminium may induce fouling. In nanofiltration pilot-scale studies at a waterworks using polyaluminium chloride, residual aluminium was observed to increase the fouling of membranes. The original water flux through NF membranes was, however, recovered by chemical cleaning, indicating the fouling was reversible.

Introduction

Surface waters in northern regions generally contain rather high amounts of natural organic matter (NOM). Efficient removal of NOM can usually be expected in a conventional chemical treatment process. However, residual total organic carbon (TOC) typically varies between 2 and 3 mg/l in drinking waters, causing a risk of by-product formation during disinfection, microbial growth in the distribution network, and taste and odour in potable water. The high coagulant dosages required for efficient NOM removal may cause significant increases of residual chloride, sulphate, aluminium, iron or manganese.

Nanofiltration (NF) improves efficiently the organic quality of water and rejects especially multivalent ions. NF membranes are so tight that macromolecules with a molar mass greater than about 300 g/mol are rejected, including most NOM. Accordingly, TOC in the nanofiltration permeate comes close to the detection limit [4]. The fraction of microbially available organic matter describes the bacterial regrowth potential better than TOC. Low assimilable organic carbon (AOC) concentrations of 5 to 35 μg C/l have been reported in earlier NF studies in Finnish surface water plants [4]. Removal of organic matter decreases significantly the formation of disinfection by-products.

Although the quality of the water produced by NF has been excellent, pretreatment should be optimised to minimise the fouling of nanofiltration membranes. It is recommended that the Silt Density Index (SDI) in NF feed water should be below 5 or 3 and the modified fouling index (MFI) below 10 s/l^2 [2,8]. Since iron and aluminium may accumulate on the surface of membranes, often in combination with organic matter, and reduce the water flux, their concentration in feed water should be limited [3,7,10].

In pilot-scale studies, using a dissolved-air flotation (DAF) – sand filtration process, optimal coagulant dosages and flocculation pH were determined to minimise MFI and SDI. The results obtained for organic matter, turbidity and coagulant residuals were compared to fouling index results to determine the necessary changes in process optimization. Three coagulants, namely, ferric chloride sulphate (PIX-110), low manganese ferric sulphate (PIX-322), and polyaluminium chloride (PAX-14), were studied.

Materials and Methods

DAF – Sand Filtration Pilot-Scale Studies

The studies with a DAF pilot unit, which was followed by a sand filter, were completed in January to March 1999. The configuration is illustrated in Figure 1.

Raw water was pumped into the DAF pilot from the intake pipe of Espoo City Waterworks (ECW). The feed flow was controlled to 6.5 m^3/h with a rotameter.

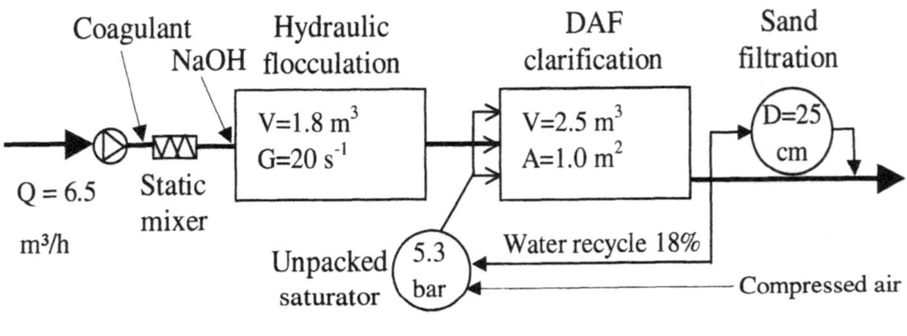

Fig. 1. Process schematic of the DAF – sand filtration pilot unit

The surface load of the flotation basin was 7.7 m/h, since the recycle ratio was 18%. Water with dissolved air was dispersed into the flotation zone via three needle valves. Clarified water was collected by a horizontal pipe, which was in a depth of 1.5 m. Water level in the flotation basin was raised at a few hours' interval to remove the formed sludge blanket. The pressure filter was filled with quartz sand (0.7–1.2 mm). The filter load was adjusted to approximately 5 m/h.

Three coagulants were tested. Constituent product specifications, as given by the manufacturer (Kemira Chemicals Ltd), are shown in Table 1, and the flocculation pH values and coagulant dosage ranges used are summarized in Table 2. In each pilot test run, the pH was adjusted to the desired value after the target coagulant dose was stabilised. If the process was steady after an hour, the sludge blanket was removed. After 30 min, the sand filter was backwashed for 10 min and then fast-rinsed for 10 min. Samples of the clarified and filtered waters were taken an hour later, and MFI and SDI indexes of the filtered water were determined.

Table 1. Constituent product specifications of the studied coagulants

		PIX-110	PIX-322	PAX-14
Fe^{3+}	%	12.0 ± 0.5	11.7 ± 0.3	< 0.02
Al^{3+}	%	*	*	7.2 ± 0.3
Me^{3+}	mol/kg	2.15	2.10	2.67
Cl^-	%	7.3 ± 0.5	*	22.0 ± 0.2
SO_4^{2-}	%	18 **	30.0±1.0	*
Specific weight		1.5 kg/l	1.54 kg/l	1.32 kg/l
Mn	mg/kg	740**	< 50.0	

*low, not specified by the manufacturer **calculated on the basis of water analyses in pilot-plant studies

Table 2. Coagulant doses and flocculation pH ranges

Coagulant	Dosage range g/m^3	Al/Fe dosage mmol/m^3	pH range
PAX-14	35 – 120	95 – 320	5.2 – 6.4
PIX-110	66 – 183	140 – 390	4.0 – 5.0
PIX-322	77 – 165	160 – 350	4.0 – 4.7

Raw Water Quality

Raw water for the treatment was taken from a small lake in southern Finland. During the DAF pilot studies, the temperature of the raw water remained below 1°C, which is most unfavorable for the treatment. TOC concentration in raw water remained slightly below 11 mg/l. The minimum turbidity was 5 FNU (formazine nephelometric units) and maximum 14 FNU. Hardness was below 1°dH.

Analytical Methods

MFI and SDI fouling indexes were measured simultaneously and in duplicate using individually packed ME 25/21 ST membrane filters of Schleicher & Schuell. All membrane filters were from the same production batch. For calculation of SDI, the times required to filter 250 ml or 500 ml water through a 0.45 µm membrane at a pressure of 207 kPa are measured [1]. The initial time is compared to filtering times after 5, 10 or 15 minutes. To measure MFI simultaneously with SDI, total filtered water volume was weighed every 30 seconds. Time/volume-values were plotted against volume values and MFI was equal to the slope of the linear part of this filtering curve [6].

TOC, $KMnO_4$ consumption (COD), pH, turbidity and iron were determined according to relevant national standards. TOC was measured using an Astro 2001 TOC analyzer. Aluminium was analysed with a spectrophotometer according to a national standard proposal. Cations were analysed using inductively coupled plasma atomic emission spectrometry, or by mass spectrometry. Samples for those analyses were pressed through a 0.45 µm filter before acidification, whereas acid was added straight to the water samples as a pretreatment procedure for aluminium and iron analyses. UV_{254}-absorption was measured according to a Standard Methods modification with the sample pre-treatment including a 10 min centrifugation (4000 rpm) instead of filtration. Chloride and sulphate were analysed by an accredited ion chromatography technique. In the heterotrophic growth response (HGR) assay, the growth of indigenous heterotrophic bacteria in the water sample was followed using a spread plate counting technique [5]. High-performance size exclusion chromatography (HPSEC) with the macroporous silicic-particle TSK columns was used for the fractionation of organic matter [9].

Results

MFI and SDI Values Reached with Three Coagulants

A summary of the best MFI and SDI results and conditions, where they were achieved, are shown in Table 3. The same flocculation pH and coagulant dosages were optimal for MFI and SDI. The MFI results with different coagulant dosages are illustrated in Figure 2. The lowest fouling indexes were achieved with ferric coagulants. With conventional coagulant dosages (Me^{3+} 170 mmol/m^3), coagulation with PAX-14 resulted in better values than coagulation with ferric salts. The results with ferric chloride sulphate and low manganese ferric sulphate were rather similar to each other.

When dosage of PIX-110 was below 200 mmol/m^3, MFI values lower than 10 s/l^2 could only be achieved at pH values of approximately 4.2. With increasing dosage, however, the fouling index results improved and a slightly higher pH was beneficial. Using PIX-322, almost all measured MFI values were below 10 s/l^2. No upper limit for ferric coagulant addition was found (maximum usage 390 mmol/m^3). Optimal pH for flocculation using PAX-14 was 5.7.

Table 3. pH values and coagulant dosages resulting in good fouling indexes

Coagulant	pH after flocculation	Dosage	Best MFI	Best SDI
PAX-14	5.4 – 5.8	65 g/m^3 (Al^{3+} 150-200 mmol/ m^3)	5.5	3.9
PIX-110	4.1– 4.8	≥ 140 g/m^3 (Fe^{3+} 300 mmol/ m^3)	4.1	3.4
PIX-322	4.0 – 4.5	≥ 120 g/m^3 (Fe^{3+} 250 mmol/ m^3)	4.4	3.4

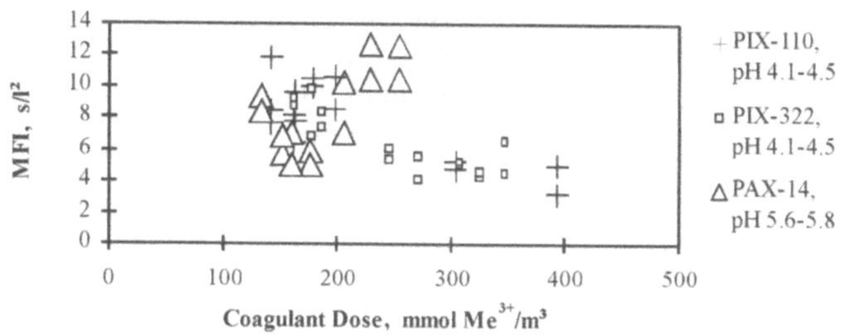

Fig. 2. MFI results in pilot-scale DAF – sand filtration process

Quality parameters for conventional process optimisation

To evaluate the pilot performance in terms of conventional process optimisation, organic matter, turbidity and residual metal were measured. Raw water quality and representative results are shown in Table 4.

Sand filtration removed organic matter more efficiently when either of the ferric coagulants was used instead of PAX-14. TOC removal in sand filtration was improved only slightly with dosages above 200 mmol/m^3. The same dosages and pH values as described in the previous section were optimal for TOC removal, though TOC was less sensitive to changes in pH at optimal coagulant dosages.

Table 4. Raw water quality and results in DAF – sand filtration pilot studies

C, clarified F, filtered	TOC mg/l	$KMnO_4$ consumption	UV_{254}-abs. cm^{-1}	Al / Fe mg/l	Turbidity FNU
Raw water	10.8	52	0.48	0.7 / 1.0	7
PAX-14, C	3.5	13	0.050	0.71 / -	1.8
PIX-110, C	2.6	9.1	0.035	- / 1.8	1.3
PIX-322, C	2.8	10.5	0.044	- / 1.8	1.7
PAX-14, F	2.7	8.5	0.047	0.10 / -	0.14
PIX-110, F	2.2	7.2	0.041	- / 0.14	0.13
PIX-322, F	2.3	7.5	0.048	- / 0.21	0.17

The turbidity results indicate that PAX-14 was better than the ferric salts at dosages of 130 to 200 mmol/m^3. Increasing the dosage of ferric coagulants decreased the turbidity in treated water, while increased PAX-14 dosage increased the turbidity. The best turbidity results were obtained at pH 5.5-6.1 for PAX-14, and at a pH of approximately 4.5 for ferric salts.

Iron and aluminium residuals were fairly high in the DAF pilot studies. Using PAX-14 instead of ferric salts, the residues after flotation were considerably lower at all conditions studied. After sand filtration, however, the residuals were approximately equal in molar terms. The results of aluminium analyses with PAX-14 were similar regardless of 0.45 µm prefiltration (3 samples). Using the ferric coagulants, the iron concentration was much lower in the prefiltered sample. As shown in Figure 3, a major proportion of the residual iron in the clarified water was insoluble, but this portion was efficiently removed in sand filtration. In four cases pH rose significantly in sand filtration.

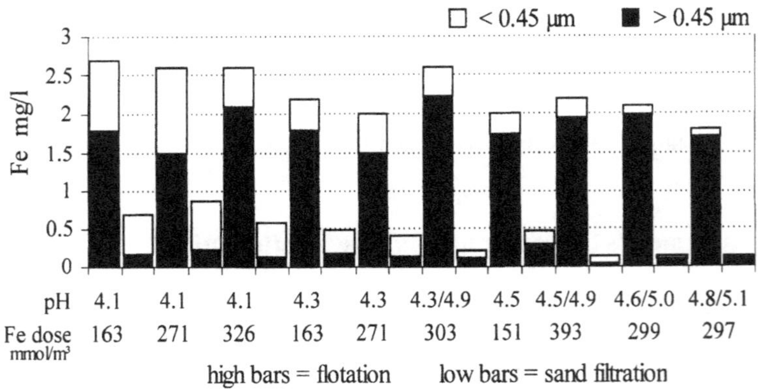

Fig. 3. Iron residuals in DAF pilot study divided into two size groups

Quality of Organic Carbon Residual

To predict the potential of microbial growth on NF membranes, HGR values were analysed from nine filtered water samples having been treated with different coagulants. The maximum HGR growth levels were $9{\cdot}10^3$ CFU/ml, $12{\cdot}10^3$ CFU/ml and $56{\cdot}10^3$ CFU/ml for PIX-322, PIX-110 and PAX-14, respectively. Using PAX-14, 8 to 11 days' incubation time was required before the growth actually started. The delay was slightly shorter with ferric coagulants. The maximum microbial growth was attained within approximately 10 days' incubation in the case of ferric salts, but required 21 days with PAX-14.

To study the changes in molecular size distribution of organic matter some samples were fractionated using HPSEC [9]. Results are illustrated in Figure 4. As shown in Figure 4 a), three heaviest fractions absorbing UV_{254} light were retained almost totally in the DAF – sand filtration process and the fourth fraction of treated water contained approximately 85% less organic matter than raw water. The lightest fraction was not affected by the treatment. The size distribution of fractions was nearly similar for all coagulants in pilot-scale studies as well as in the full-scale process using PAX-14 (Fig. 4 b).

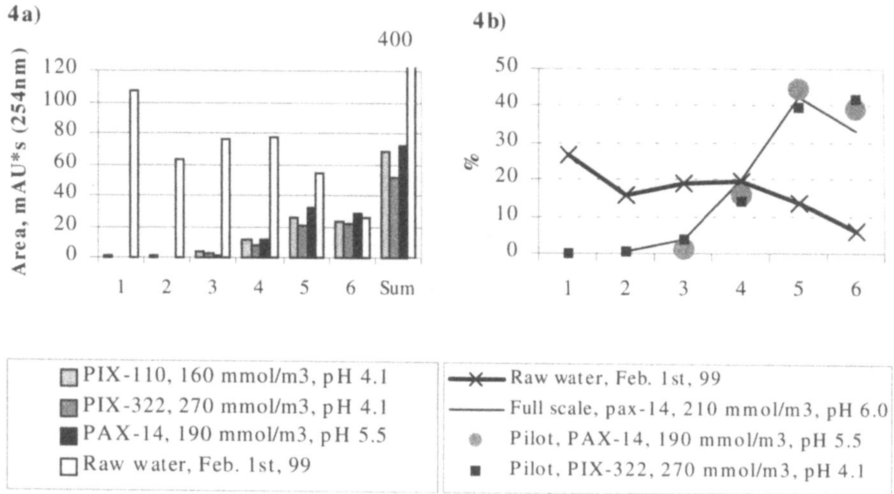

Fig. 4. Molecular size distribution of organic matter in raw water and after DAF – sand filtration treatment

Chloride, Sulphate and Other Coagulant Residuals

The concentration of chloride after DAF pilot plant treatment was equal to the sum of the chloride in the raw water and that contained in the coagulant, about 36 mg/l at the highest. Using PIX-322 in the DAF pilot, about 94% of the added sulphate remained in the water phase and the highest concentration was approximately 50 mg/l.

There was considerable manganese impurity in some coagulants. The concentrations of other heavy metals in the coagulants were negligible. With a dosage of PIX-110 of 300 mmol/m^3, the manganese concentration after the process was 160 μg/l, which was about 100 μg/l more than in raw water.

Discussion

Comparison of Conventional Process Optimisation and Fouling Indexes

Since MFI and SDI recommendations [2,8] were met, the DAF – sand filtration process, with the proper use of any of the studied coagulants, could be followed by nanofiltration. The lowest MFI values were achieved when both turbidity and organic matter were low. However, the relationship between MFI and turbidity, organic matter, or both was not statistically significant ($R^2 < 0.5$ in linear estimation). The importance of iron and aluminium residual to fouling indexes is questionable, since higher pH would favor the removal but retard MFI. Besides, equal results were obtained with aluminium concentrations of 1.9 and 0.2 mg/l. A comparison of the conditions required for good MFI and SDI results and for efficient removal of TOC, turbidity and coagulation metal is presented in Table 5. Optimal conditions for MFI and SDI were shown in Table 3.

Table 5. Optimal conditions for the removal of TOC, turbidity and metal residuals as compared to MFI

Coagulant	TOC		Turbidity		Al or Fe residual	
	pH	Dosage	pH	Dosage	pH	Dosage
PAX-14	=	+	+	=	+	=
PIX-110	=	=	=	=	+	=
PIX-322	=	=	=	=	+	=

+ increasing pH or dosage from optimal level for MFI improves removal = same optimal range

Organic Carbon Residuals after conventional Treatment

Organic matter removal, according to TOC and COD analyses, was more efficient when using PIX-110 or PIX-322 in equimolar dosages instead of PAX-14. However, the absorption of UV_{254}-light was lowest with PAX-14, and there were no differences in absorptivity between clarified and sand-filtered samples. This can be explained by the centrifugation pretreatment for the UV_{254}-absorbance analysis: with PAX-14, organic matter was bound to a larger extent to colloids too small to be separated in sand filtration, but large enough to be removed by centrifugation. Since small particles have a strong tendency to deposit on membrane surfaces [11], PIX-110 and PIX-322 might suit better than PAX-14 for NF pretreatment.

Although the removal of TOC was reasonably high, plenty of microbially available organic matter remained in the water. This resulted in rather high heterotrophic growth response values. HGR analyses indicated a significant microbial growth potential in the treated water. The stronger potency of aluminium over iron to inhibit bacterial growth could be an explanation for the longer delay preceding the actual growth in the case of PAX-14. The differences in growth might not be significant, since even the lowest microbial inocula can lead to rapid organic growth in suitable conditions.

Effects of Aluminium and Iron Residuals on Nanofiltration

Although the concentration of aluminium or iron did not seem to affect the MFI, their residuals have been reported to accumulate on the surface of the membranes and to reduce the water flux of the membrane [3,7]. The limits of 50 to 100 µg/l aluminium and iron have been suggested for membrane feed water [7,10]. Since the solubility product constants of aluminium hydroxide (K_{SP} $2 \cdot 10^{-32}$) and iron(III) hydroxide (K_{SP} $4 \cdot 10^{-38}$) are very low, even much lower concentrations may precipitate. The solubility minimum is close to pH 7.

The average aluminium reduction with the Filmtec NF255 nanofiltration membrane was 92% in Espoo. A proof of the fouling feature was the release of high concentrations of aluminium into the weak acid-based cleaning solution. Since the water flux through NF membranes was retained with chemical cleanings, the fouling was reversible.

Nanofiltration Decreases Sulphate, Chloride and Manganese Residuals

Sulphate and chloride increase the corrosive characteristics of water. Since the waters are soft in Finland, tight limits are recommended: 25 mg/l for chloride and 50 mg/l for sulphate. The concentration of chloride and sulphate in raw water was approximately 10 mg/l and 8 mg/l, respectively. The recommended limit for chloride is often exceeded when using PAX-14. If high dosages of PIX-322 were used, the recommended sulphate limit could be exceeded.

Negatively charged NF membranes retain sulphate as a divalent ion more effectively than chloride, which is monovalent. Sulphate concentrations in the NF permeate remained below the detection limit of 0.1 mg/l in most of the samples in the NF studies in Espoo. Chloride retention rate in Espoo was over 90% with Filmtec NF70 membrane, but varied between 2% and 44% with NF255, the average level being 27%.

Without the additional manganese load of a flocculation chemical, manganese removal might not be required at all to keep the manganese concentration in potable water below the national limit of 50 µg/l. The conditions in our studies were not favourable for oxidation and precipitation of manganese, and almost all added manganese remained in the water phase. Above pH 8, the oxidation of Mn^{2+} with molecular oxygen can take place, and the removal of manganese precipitates should take place before nanofiltration. The retention of the divalent ion Mn^{2+} in nanofiltration is high and $Mn(OH)_2$ precipitation is not expected.

Conclusions

Based on the studies, the following conclusions are made:

- With conventional coagulant dosages (Me^{3+}170 mmol/m^3, flocculation pH Al: 5.7, Fe: 4.5), better MFI results were achieved with polyaluminium chloride than with ferric salts. The best results, however, were obtained at higher ferric salt dosages (300 mmol Fe/m^3).
- Since the optimal pH values and coagulant dosages for MFI minimisation were in good accordance with simultaneous decrease in organic matter and turbidity, no major changes in conventional process optimisation will be needed when nanofiltration is introduced to improve the water quality.
- Aluminium has proved to be a significant foulant of NF membranes at ECW. To be able to keep MFI values low, and, at the same time, maintain efficient aluminium and iron removal, pH should be 5.7 - 6.0 or 4.5 - 4.8 for PAX-14 and ferric salts, respectively.

- Since the sand filtration more efficiently retained aluminium than iron, and a more significant portion of the residual organic carbon was found to be present in colloidal form by using PAX-14, the coagulants PIX-110 and PIX-322 might be more suitable for the NF pretreatment.
- If low desalting NF membranes such as Filmtec NF255 are used, sulphate-based coagulants should be preferred to keep the concentrations of sulphate and chloride below recommended limits in potable water. The other alternative is to use high desalting membranes such as Filmtec NF70.

Acknowledgments.

The main sponsors of this research were the National Technology Agency of Finland, Soil and Water Ltd. and Kemira Chemicals Ltd. Espoo City Waterworks is specially acknowledged for technical assistance and laboratory analyses, and YIT Corporation for lending us the DAF pilot plant.

References

1. ASTM D 4189-95 Standard Test Method for Silt Density Index (SDI) of Water. In: Annual book of American Society for Testing and Methods standards. 11.01 (1997) 402-404
2. DOW Chemicals Ltd. FILMTEC Membrane Elements, Technical Manual (1995)
3. Dudley, L. Y., Darton, E. G.: Pretreatment Procedures to Control Biogrowth and Scale Formation in Membrane Systems. Desalination, 110 (1997) 11-20
4. Härmä, V.: Nanofiltration for Water Quality Improvement in Finnish Surface Waterworks. Helsinki Univ. of Technol., Lab. of Environ. Eng. TKK – VHT – 22, Licentiate thesis in Finnish (1999) pp. 133
5. Noble, P. A., Clark, D. L., Olson, B. H.: Biological stability of groundwater. J. Am. Water Works Assn. 88 (5) (1996) 87-96
6. Schippers, J. C., Verdouw, J.: The Modified Fouling Index, a Method of Determining the Fouling Characteristics of Water. Desalination, 32 (1980) 137-148
7. Taniguchi, Y.: An Overview of Pretreatment Technology for Reverse Osmosis Desalination Plants in Japan. Desalination, 110 (1997) 21-36
8. Taylor, J. S., Jacobs, E. P.: Reverse Osmosis and Nanofiltration, In: Water Treatment Membrane Processes (Mallevialle J., Odendaal P. E. & Wiesner M. R., eds.) New York, USA, McGraw-Hill (1996) pp. 9.1-9.70
9. Vartiainen, T., Liimatainen, A., Kauranen, P.: The use of TSK size exclusion columns in determination of the quality and quantity of humus in raw waters and drinking waters. The Science of the Total Environment. 62 (1987) 75-84
10. Ventresque, C., Turner, G., Bablon, G.: Nanofiltration: from prototype to full scale. J. Am. Water Works Assn. 89(10)(1997) 65-76
11. Wiesner, M. R., Chellam, S.. Mass Transport Considerations for Pressure-Driven Membrane Processes. J. Am. Water Works Assn. 84(1) (1992) 88-95

The Importance of Coagulation for the Removal of Cryptosporidium and Surrogates by Filtration

P.M. Huck*, B.M. Coffey**, M.B. Emelko* and C.R. O'Melia***

*University of Waterloo,Department of Civil Engineering, 200 University Avenue; Waterloo, Ontario, Canada
pm2huck@uwaterloo.ca
**Metropolitan Water District of Southern California, Water Quality Monitoring Section, 700 Moreno Avenue; La Verne, California, USA
***The Johns Hopkins University, Department of Geography and Environmental Engineering, 3400 N. Charles St., Baltimore, Maryland, USA

Abstract

This investigation examined the removal of *Cryptosporidium* and several surrogates by granular medium filtration. The study focused on coagulation impacts on removals. Inactivated *C. parvum* oocysts and non-inactivated *Bacillus subtilis* were seeded in two pilot plants. The pilot plant locations were chosen to represent different coagulation regimes. Under optimized coagulation conditions (filter effluent turbidity < 0.1 NTU), approximately 5 $\log_{10}$ removal of *Crypto.* was obtained in one location, and approximately 3 $\log_{10}$ in the other. This difference is likely related at least in part to the difference in coagulation regimes. Suboptimal coagulation produced a substantial deterioration of *Crypto.* removal in both locations. The trends seen for *Bacillus* were generally similar to those observed for *Crypto.*. Although turbidity may be of value for assessing coagulation impacts on *Crypto.* removal, particle counts may be a more sensitive parameter in this regard.

Introduction

Physical removal of protozoan pathogens is receiving increased attention because of the difficulty of chemically inactivating these organisms, particularly *Cryptosporidium parvum*. Although numerous investigations have examined the removal of these and other pathogens through filtration processes, those studies have generally been conducted under steady-state conditions with optimized pre-treatment. Adequate chemical pretreatment during coagulation and flocculation is critical for maintaining good particle removal during filtration [1,2]. Furthermore, several studies have indicated the importance of coagulation processes for improving filter removal efficiencies of *Cryptosporidium* and *Giardia* [3,4,5,6,7].

In this paper, *Cryptosporidium* passage through filters during periods of suboptimal coagulation is compared to removals under optimal coagulation conditions. The removal of *Cryptosporidium* was compared to the removal of particles, turbidity, and *Bacillus subtilis* spores.

Methods and materials

Pilot plants at two locations were used in this investigation, and, except when suboptimal coagulation was deliberately induced, were operated to mimic the conditions of their respective full-scale plants. One pilot plant was located at the Britannia Water Treatment Plant in Ottawa, Canada, while the other was located at the F.E. Weymouth Filtration Plant of the Metropolitan Water District (MWD) in La Verne, California. The Ottawa pilot plant used a high coagulant dose (~40 mg/L alum and 2 mg/L activated silica) to achieve both total organic carbon (TOC) and particle removal. The Metropolitan pilot plant used a low coagulant dose (5 mg/L alum and 1.5 mg/L cationic polymer) for particulate removal only. Chlorine (~2 mg/L) was added at rapid mix as a pre-oxidant at both pilot plants. The optimized coagulation conditions were selected to meet the 0.1 NTU filtered water turbidity goal of the Partnership for Safe Water, a voluntary treatment optimization program sponsored by the U.S. Environmental Protection Agency and the American Water Works Association. The suboptimal conditions were selected to mimic treatment plant operations which - though still meeting the U.S. Interim Enhanced Surface Water Treatment Rule (IESWTR) requirements of 0.3 NTU - targeted turbidities at the upper range of compliance.

The benchmark system chosen was conventional treatment with dual-media filtration. Both filters contained media depths and sizes typical of many existing North American treatment plants. At Metropolitan, the benchmark filter design contained 508 mm (20 in.) of anthracite over 203 mm (8 in.) of sand. At Ottawa, the filter design contained 457 mm (18 in.) of anthracite over 279 mm (11 in.) of sand. At Metropolitan, the backwashing regime consisted of chlorinated water with surface wash. At Ottawa, chlorinated water and air scour was used.

In each location a pilot-scale, dual-media filter was seeded with jar-coagulated suspensions of $\sim 10^8$ formalin-inactivated *Cryptosporidium parvum* oocysts and $\sim 10^9$ (non-inactivated) *Bacillus subtilis* spores. (A previous bench-scale study showed that formalin inactivation did not significantly change the removal of *Cryptosporidium* oocysts through filters after either optimized coagulation or no coagulation [7]).

Experimental Design

The pilot plant locations were selected to represent two different coagulation regimes: a relatively low dose optimized for particle removal (Metropolitan, treating primarily Colorado River water) and a relatively high dose for combined TOC and particle removal (Ottawa River water). Three basic experiments were performed in each location. They were (1) no coagulation; (2) optimized coagulation; and (3) suboptimal coagulation.

No coagulation

This experiment served as a control to determine the losses of oocysts and endospores through the pilot-plant facilities and also as a worst-case condition of total coagulant failure. Previous seeding of *Cryptosporidium* oocysts at Metropolitan's pilot plant indicated a loss of approximately 0.3-log (50 percent) of oocysts when no chemicals were added to the water [8,9]. In the tests reported here, each seeding experiment consisted of four replicate filter influent samples and four replicate filter effluent samples taken at 15-min intervals over one hour.

Optimized coagulation

This experiment served as another control and was used to determine the maximum removal of oocysts and endospores through the pilot-plant filters. It is this optimized condition that has been the focus of much previous work [2,8,9,10]. In this experiment, seeding continued for one hour and four replicate samples were taken from the filter influent and filter effluent at 15-min intervals. Seeding began 2-4 hours into the filter run, which was after the filter ripening period. Previous seeding of *Cryptosporidium* oocysts at Metropolitan's pilot-plant showed removals of >3 $\log_{10}$ during stable filter operation [8,9].

Suboptimal coagulation

Because the residence time for water in treatment plants is relatively short, in locations where influent water quality can change rapidly coagulant dosage optimization can be difficult to attain. Even though the treatment plant may be capable of producing low-turbidity water when high-turbidity water consistently enters the plant, the speed of the raw water quality changes relative to the coagulation response may lead to increased particle passage. The slow response of the coagulation process to changing raw water conditions was implicated in the Milwaukee cryptosporidiosis outbreak [11]. In the present research, the sub-optimal coagulation experiments determined the effects of changing coagulation conditions (without a change in raw water quality) on pathogen passage. These experiments - where the coagulant dose was changed 40 to 60 percent from optimum - were then compared to the no-coagulant and optimized coagulation tests above. The targeted suboptimal turbidity was 0.2 - 0.3 NTU. In some tests, however, the target effluent turbidity was exceeded.

Seeding Protocols

The seed suspension of oocysts or spores was diluted to 1.5 L with pre-oxidized influent water and jar-coagulated under coagulant and mixing conditions that mimicked pilot-scale treatment. The jar-coagulated organisms were then seeded directly into the influent of a single filter by a peristaltic pump for 60 min. This seed location was selected to minimize significant losses of microorganisms in upstream unit processes and to better characterize their removal during filtration. During seeding, the seed suspension was constantly agitated with a magnetic stirrer to ensure steady distribution of the organisms during the seeding procedure.

The targeted seeded influent concentration for each organism was ~10^5 oocysts or spores per liter. Samples were collected in sterile bottles containing sodium thiosulfate. Filter effluent samples were collected in 1 L Wheaton bottles from the continuously running effluent line. Filter influent samples were collected in 250 mL amber bottles from the water column directly above the filter media using a continuous recirculation peristaltic pump.

Analytical Methods

Bacillus subtilis

The analysis for *Bacillus subtilis* spores was performed according to a previously described method [12] Sodium thiosulfate (100 mg/L final concentration) was added to all containers to remove any residual chlorine. Typically, duplicate 0.1 L samples were collected to enumerate filter influent samples, and duplicate 1.0 L samples to enumerate filter effluent samples. For the endospore seeding, *B. subtilis* (ATCC 6051) were obtained from a commercial laboratory.

Cryptosporidium parvum

The oocysts were obtained from a commercial laboratory (Waterborne, Inc., New Orleans, La.). For each test, ~10^8 oocysts were obtained—already inactivated with 5 percent formalin. Prior to seeding, a small portion of the stock suspension was removed to perform a hemacytometer count.

Filter influent samples were analyzed in sample volumes of 10 mL and filter effluent samples were analyzed in volumes of 500 mL (or less if the filter effluent turbidity was elevated). Oocysts were collected by direct vacuum filtration of the sample through 27-mm diameter, 0.45-μm pore size polycarbonate membranes. Standard immunofluorescent assay (IFA) techniques were followed to stain the samples. The slides were then either analyzed at the University of Waterloo (for samples from Ottawa) or shipped (from Metropolitan) to a commercial laboratory (CH Diagnostic & Consulting Services, Inc., Loveland, Colorado), for presumptive microscopic analysis. Recovery experiments were performed at both locations using both filter influent and effluent water matrices, as a procedural check. The measured *Cryptosporidium* levels reported in this paper were not adjusted by the recovery.

Results and Discussion

Table 1 shows major water quality characteristics and operating conditions of the two pilot plants. The coagulant dosage in Ottawa was nearly eight times greater than at Metropolitan and the coagulation pH was lower. Both of the pilot plants received raw water which was low in turbidity (averages of 1.6 to 2.5 NTU at Ottawa and 0.6 to 1.1 NTU at Metropolitan) and particles (averages of 4,400 to 7,300 particles/mL at Ottawa and 3,000 to 4,000 particles/mL at Metropolitan). Major differences between the raw waters included temperature (Ottawa's coldest water was 1°C, whereas Metropolitan's coldest water was 13°C) and alkalinity (Ottawa's alkalinity was ~20 mg/L as $CaCO_3$, whereas Metropolitan's alkalinity was 110-130 mg/L as $CaCO_3$).

Table 2 summarizes the removal of seeded *Cryptosporidium* and *Bacillus* at both locations. The box-and-whisker plots in Figure 1 represent the minimum, 25th percentile, median, 75th percentile, and maximum values for the removals of *Cryptosporidium.* Generally similar trends were seen for *Bacillus.* Each removal is expressed as the $\log_{10}$ difference between paired sets of data taken at the filter influent and filter effluent. The statistical comparisons (single-factor analysis of variance) for *Cryptosporidium* were based on 4 to 35 pairs of data.

Table 1. Water quality characteristics and operating conditions at Ottawa and Metropolitan

Parameter	Type/Units	Ottawa	Metropolitan
Coagulant dose	alum (mg/L)	38	5
	SiO_2 (mg/L)	2	--
	cationic polymer (mg/L)	--	1.5
Coagulation/ filtration pH	--	5.9-6.1	7.7-8.0
Filtration rate	gpm/ft^2	2.6	4.0
Media depth	anthracite (in.)	18	20
	sand (in.)	11	8
Media size	anthracite (mm)	1.07	1.0-1.1
	sand (mm)	0.52	0.43-0.50
Media uniformity coefficient	Anthracite	1.35	<1.65
	sand	1.32	<1.65
Raw water quality	alkalinity (mg/L as $CaCO_3$)	19-23	107-134
	pH	7.1-7.4	7.7-8.4
	temperature (°C)	1-24	13-25
	TOC (mg/L)	approx. 5	2.6-2.9
	turbidity (NTU)	1.0-2.7	0.4-2.4

Table 2. Filter performance at Ottawa and Metropolitan*

Location	**Ottawa**			**Metropolitan**		
Operational Period	Opt.	Sub-Opt.	No Coag.	Opt.	Sub-Opt.	No Coag.
Turbidity (NTU)						
Plant influent	1.60	2.40	2.50	0.92	0.60	1.10
Filter effluent	0.03	0.69	2.30	0.05	0.16	0.56
$\log_{10}$ removal	1.7	0.7	<0.1	1.2	0.6	0.2
Particles $\geq$ 2 μm (#/mL)						
Plant influent	$5.1 \cdot 10^3$	$4.4 \cdot 10^3$	$7.3 \cdot 10^3$	$3.0 \cdot 10^3$	$4.0 \cdot 10^3$	$4.0 \cdot 10^3$
Filter effluent	2	$2.3\ 10^2$**	$5.2 \cdot 10^3$	14	$3.2 \cdot 10^2$	$2.5 \cdot 10^3$
$\log_{10}$ removal	3.7	1.7	0.3	2.5	1.1	0.2
Cryptosporidium (oocysts/L)						
Filter influent	$1.0 \cdot 10^6$	$1.2 \cdot 10^6$	$4.9 \cdot 10^5$	$2.1 \cdot 10^5$	$1.6 \cdot 10^5$	$2.6 \cdot 10^5$
Filter effluent	4	$7.1 \cdot 10^2$	$2.5 \cdot 10^5$	$7.0 \cdot 10^2$	$2.0 \cdot 10^4$	$1.8 \cdot 10^5$
$\log_{10}$ removal	5.4	3.4	0.3	3.1	1.1	0.2
Bacillus spores (cfu/L)						
Filter influent	$8.5 \cdot 10^4$	$4.0 \cdot 10^4$	$7.9 \cdot 10^4$	$6.1 \cdot 10^4$	$3.5 \cdot 10^5$	$2.0 \cdot 10^6$
Filter effluent	$5.1 \cdot 10^2$	$3.3 \cdot 10^3$	$4.0 \cdot 10^4$	$6.0 \cdot 10^2$	$1.1 \cdot 10^5$	$1.9 \cdot 10^6$
$\log_{10}$ removal	2.6	1.2	0.3	2.2	0.5	0.0

Opt. = optimized, Sub-Opt.= suboptimal, No Coag. = no coagulants,
*Results shown are arithmetic averages of 4-5 samples within each of several replicate runs. The $\log_{10}$ removals are also expressed as the average of individual data pairs, so the removal may not be exactly computed from the columns above.**High variability

At both Metropolitan and Ottawa greater $\log_{10}$ removals were obtained during optimized coagulation (i.e., 2-4 hours into the filter cycle when effluent turbidity was 0.05 NTU or less) than during suboptimal coagulation or coagulant failure. These differences were all statistically significant at the one percent level (α = 0.01). Removals of *Cryptosporidium* were higher than those of *Bacillus* (Table 2), though the differences in *Bacillus* removals were also statistically significant at the one percent level (α = 0.01). The essentially zero removals of both organisms for the no-coagulant condition confirmed that seeded organisms were not being lost in the pilot plants.

For *Cryptosporidium*, removals at Ottawa were higher than at Metropolitan. In addition, no *Cryptosporidium* oocysts were detected in a number of filter effluent samples at Ottawa during optimized coagulation; thus, the calculated removals of *Cryptosporidium* may have been limited by the influent concentration. The removals of *Cryptosporidium* during optimized coagulation at Ottawa and Metropolitan were 5.4 ± 0.4 and 3.1 ± 0.7, respectively. The reasons for this difference are not fully understood at this time but may relate to the substantially different coagulation conditions at the two locations. However, the removals of *Bacillus* at Ottawa during optimized coagulation ($2.6 \pm 1.1\ \log_{10}$) were not substantially greater than at Metropolitan ($2.2 \pm 0.5\ \log_{10}$).

Suboptimal coagulation showed a substantial deterioration in *Cryptosporidium* and *Bacillus* removal. At both locations, removals were reduced by 2 $\log_{10}$. The suboptimal coagulation conditions were achieved by reducing the optimum coagulant dosage (alum and polymer at Metropolitan or alum and activated silica at Ottawa) by 40 to 60 percent. At Metropolitan, the coagulant reduction resulted in an average effluent turbidity of 0.16 NTU (see Table 2), which is well below the IESWTR specified level of 0.3 NTU. At Ottawa, the coagulant reduction resulted in an average effluent turbidity of 0.56 NTU. The suboptimal coagulation experiments at Ottawa varied more substantially than at Metropolitan, in terms of both *Cryptosporidium* (Figure 1) and turbidity.

The relationship between seeded *Cryptosporidium* or *Bacillus* and particle removal was examined under all of the coagulation conditions (results not shown). At Metropolitan, *Cryptosporidium* and *Bacillus* removal were highly correlated to particle removal (r^2 values of 0.87 and 0.82, respectively). At Ottawa, the strength of the correlation was not as high (0.60 and 0.25 for *Cryptosporidium* and *Bacillus*, respectively).

Figures 2 and 3 show the effect of coagulation condition on filter effluent turbidity and *Cryptosporidium* removal. The *Cryptosporidium* removals are shown for individual influent-effluent data pairs. In Figure 2 (optimal conditions) turbidity was always less than 0.1 NTU. Although Ottawa *Cryptosporidium* removals were almost always greater than MWD's, in each location removals calculated from individual influent-effluent sample pairs varied considerably. This shows the need for replication in this type of work.

Figure 3 (suboptimal coagulation conditions) shows that in all MWD experiments of this type the filter effluent turbidity was in the range of 0.1 to 0.2 NTU. Some Ottawa data are available for this range, but in other Ottawa experiments the effluent turbidity was closer to 1 NTU. When turbidity was 0.1 to 0.2 NTU, the Ottawa *Cryptosporidium* removal did not appear to decrease, whereas MWD's did. This would suggest that the sensitivity of turbidity for monitoring coagulation impacts on *Cryptosporidium* removal may be site specific and perhaps dependent on the coagulation regime used. It is possible that particle counts may be a more sensitive indicator of poor coagulation performance.

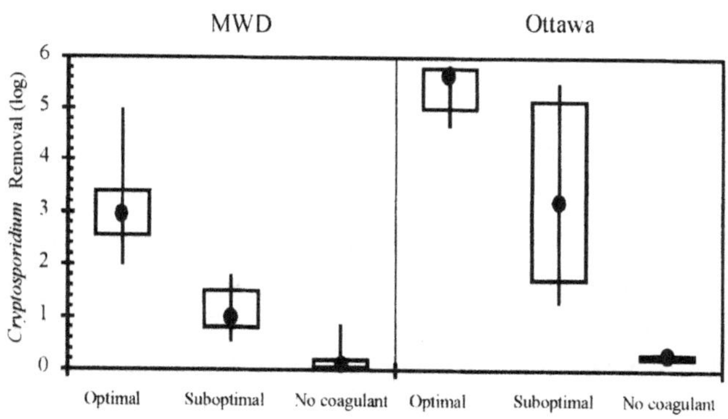

Fig. 1. Effect of coagulation condition on removal of *Cryptosporidium* oocysts. Graph shows minimum, 25th percentile, median, 75th percentile, and maximum values.

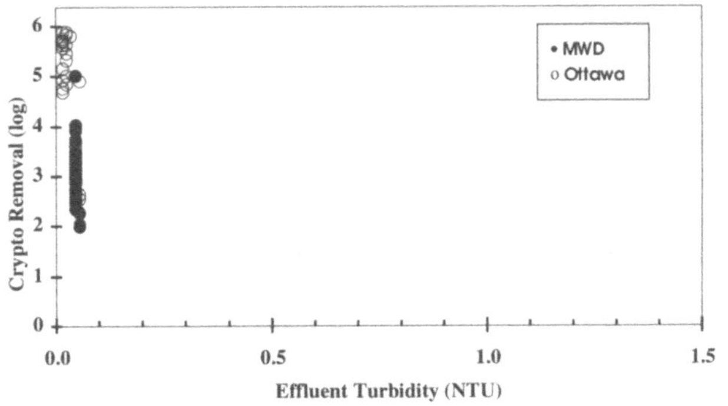

Fig. 2. *Cryptosporidium* removal as a function of filter effluent turbidity (optimal coagulation conditions)

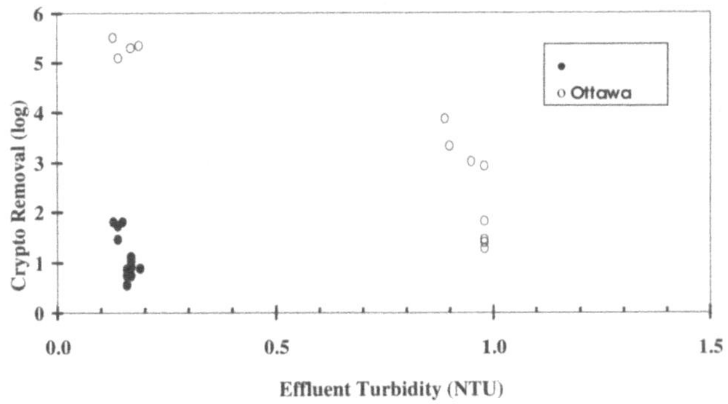

Fig. 3. *Cryptosporidium* removal as a function of filter effluent turbidity (suboptimal coagulation conditions)

Conclusions

This pilot study at two locations compared removals of *Cryptosporidium, Bacillus,* turbidity and particles during optimized coagulation, suboptimal coagulation, and coagulant failure. The results from this work indicated that:

- Under optimized coagulation conditions with similar effluent turbidities (<0.1 NTU), a 2 $\log_{10}$ difference in *Cryptosporidium* removals was observed at the two locations. This difference is likely related at least in part to the different coagulation regimes.
- Even at filter effluent turbidity levels less than 0.3 NTU, substantial deterioration of filtration performance may result if coagulation is not optimized.

- *Cryptosporidium* removals were more sensitive to coagulation conditions than was turbidity removal. The sensitivity of turbidity for measuring coagulation impacts on *Cryptosporidium* removal may depend on the coagulation regime.
- Filter effluent particle monitoring may provide a sensitive measure of coagulation performance and *Cryptosporidium* removal.

Acknowledgements

Funding for this research was provided in part by the American Water Works Association Research Foundation. The authors express their appreciation to members of the project teams at Ottawa and Metropolitan. In particular, the operation of the pilot facilities and the substantial analytical support is greatly appreciated. Support for the work at the University of Waterloo is also provided by the partners of the Natural Sciences and Engineering Research Council of Canada Industrial Research Chair in Water Treatment.

References

1. Tobiason, J.E., O'Melia, C.R.: Physicochemical Aspects of Particle Removal in Depth Filtration. Jour. AWWA, 80(12) (1988) 54
2. Patania, N.L., Jacangelo, J.G., Cummings, L., Wilczak, A., Rilet, K., Oppenheimer, J.: Optimization of Filtration for Cyst Removal. AWWA Research Foundation and AWWA. Denver, Colorado (1995)
3. Charles, G., Morgan, J.M., MacPhee, M., Kim, M., Frederickson, D.W.: Bench-Scale Parasite Spiking – An Alternative to Pilot-Scale Giardia and Cryptosporidium Spiking Investigations. Proc. AWWA Water Qual. Tech. Conf., New Orleans, Louisiana (1995)
4. Plummer, J.D., Edzwald, J.K., Kelley, M.B.: Removing Cryptosporidium by Dissolved-Air Flotation. Jour. AWWA 87 (9) (1995) 85
5. Ongerth, J.E., Pecoraro, J.P.: Removing Cryptosporidium Using Multimedia Filters. Jour. AWWA, 87(12) (1995) 83
6. Huck, P.M., Emelko, M.B., Douglas, I.M., Coffey, B.M.: Using Spores and Particles to Assess the Robustness of Filters for Cryptosporidium Removal. Proc. Particle Measurement and Characterization in Drinking Water Treatment Symposium, Nashville, Tennessee (1999)
7. Emelko, M.B., Huck, P.M., Slawson, R.M.: Optimizing Filter Removal of Viable and Inactivated Cryptosporidium During Challenge Periods. Proc. International Symposium on Waterborne Pathogens, Milwaukee, Wisconsin (1999)
8. Yates, R.S., Green, J.F., Liang, S., Merlo, R.P., De Leon, R.: Optimizing Coagulation / Filtration Processes for Cryptosporidium Removal. Proc. International Symposium on Waterborne Cryptosporidium, Newport Beach, California (1997)
9. Yates, R.S., Green, J.F., Liang, S., Merlo, R.P., De Leon, R.: Optimizing Direct Filtration Processes for Cryptosporidium Removal. Proc. AWWA Water Qual. Tech. Conf., Denver, Colorado (1997)
10. Nieminski, E.C., Ongerth, J.E.: Removing Giardia and Cryptosporidium by Conventional Treatment and Direct Filtration. Jour. AWWA, 87 (1995) 9

11. Fox, K.R., Lytle, D.A.: Milwaukee's Crypto Outbreak: Investigation and Recommendations. Jour. AWWA, 88 (9) (1996) 87
12. Rice, E.W., Fox, K.R., Miltner, R.J., Lytle, D.A., Johnson, C.H.: Evaluating Plant Performance with Endospores. Jour. AWWA, 88 (9) (1996) 122

Biological Pre-Treatment for Improved Removal of Manganese in Chemical Drinking Water Treatment

G. Heinicke*, T. Hedberg*, F. Persson** and M. Hermansson**

*Water Environment Transport, Chalmers University of Technology, Gothenburg, Sweden
gerald.heinicke@wet.chalmers.se
**Department of Cell and Molecular Biology, Microbiology. Göteborg University, Gothenburg, Sweden
frank.persson@gmm.gu.se

Abstract

Biological pre-treatment was found to enhance the removal of manganese from colored surface waters at two Swedish waterworks, where conventional chemical treatment at times was insufficient. The pilot plant studies also indicated the potential of the bioreactors for removal of odor.

Introduction

High concentrations of iron and manganese are common problems in ground water, as well as certain types of surface waters. The conventional treatment process is aeration and pH adjustment, thereby precipitating the metals. If aeration alone is not sufficient, strong oxidants such as potassium permanganate, chlorine or ozone may be used. Alternatively, an array of biological treatment processes has successfully been applied, taking advantage of the oxidizing capabilities of iron and manganese bacteria. Examples are *in-situ* (VYR) treatment, trickling-, sand- and contact filters [1-5].

Particularly in the Nordic countries, surface waters may contain high concentrations of natural organic matter (NOM). Here, the predominant soil conditions and relatively high precipitation in combination with low temperatures, cause humic substances to be washed into rivers and lakes, rendering many of them distinctly colored. The presence of humic substances can make it difficult or even impossible to effectively remove iron and manganese by chemical oxidation processes [5]. Some waterworks frequently fail to produce drinking water that complies with quality standards, concerning its contents of organic matter (color, COD) as well as iron and manganese. If less difficult raw water is not at hand, the problem has to be addressed by the waterworks.

Recent advances in membrane technology offer another viable treatment alternative for colored raw waters. Particularly in Norway, nanofiltration has become a widespread method for removing humic substances [6].

In this study, biological processes have been tested as a pre-treatment step for chemical coagulation and filtration of colored surface water. The purpose was to find out if a bioreactor harboring biofilm on plastic carriers could enhance the effect of the subsequent chemical treatment, so that an acceptable tap water quality, particularly concerning manganese, can be maintained consistently.

Sotenäs and Karlskrona Waterworks

The municipal waterworks at Sotenäs and Karlskrona initiated pilot plant studies for biological pre-treatment. Raw water characteristics at the two waterworks are summarized in Table 1.

At Sotenäs, raw water is from a shallow eutrophic lake, Lilla Dalevattnet. During summer stagnation, oxygen deficiency and high water temperatures may occur, resulting in peak loads of dissolved iron and manganese. At times it has also been impossible to comply with the Swedish manganese standard (0.05 mg L^{-1}) in drinking water. During autumn, odor and taste of the treated water have been reported, probably related to excessive growth of algae in the lake. Consumers, in particular the local cannery, have complained about the quality of the tap water. Average production at Sotenäs is around 4000 m^3 d^{-1}.

Table 1. Raw water quality during the study.

	Waterworks					
	Sotenäs			Karlskrona		
Parameter	Mean	Min	Max	Mean	Min	Max
pH	7.0	6.7	8.1	6.9	6.3	7.2
Turbidity (NTU)	7.0	2.1	13.0	2.4	1.4	4.2
Dissolved O_2 (mg L^{-1})	9.3	4.5	13.5	11.8	7.6	16.4
Alkalinity (mg L^{-1} HCO_3^-)	31	27	39	13	4	21
Calcium (mg L^{-1})	11	9	13	8	6	13
Fe, total (mg L^{-1})	0.49	0.02	1.3	1.65	0.99	2.71
Fe, dissolved (mg L^{-1})*	0.25	0.06	0.41	1.13	0.66	2.00
Mn, total (mg L^{-1})	0.065	0.017	0.253	0.130	0.030	0.539
Mn, dissolved (mg L^{-1})*	0.034	0.003	0.175	0.086	0.006	0.443
Color (mg L^{-1} Pt) **	84	55	159	144	95	222
DOC (mg L^{-1})	9.6	7.6	12.0	16.6	12.5	21.7
AOC (µg L^{-1})***	51	32	68	87	30	125

*Filtered through a 0.45 µm membrane **apparent color (unfiltered sample) ***Assimilable Organic Carbon, 4 sampling occasions

Karlskrona waterworks takes water from River Lyckebyån, one of the most difficult to treat raw water sources exploited in Sweden. Lyckebyån's water is distinctly colored, and also iron and manganese often exceed the values required by Swedish raw water quality recommendations. Even the delivered tap water does at times not comply with the standards, concerning COD^{*}_{Mn}, high temperature, iron, manganese, aluminum and nitrite. Average production is around 11000 $m^3 d^{-1}$.

Both waterworks apply conventional chemical treatment, comprising coagulation with aluminum sulfate carried out in continuous upflow filters (Dynasand), filtration through activated carbon and finally chlorination.

Materials and Methods

Bioreactors

As a reaction to the problems described above, the waterworks set up several experimental bioreactors, and co-operation was established with Chalmers and Göteborg Universities. Different reactor types and carrier materials were tested. The pilot plants described in this paper consisted of tanks filled with cross-flow plastic carrier material followed by coagulation in continuous upflow filters. The bioreactor at Sotenäs was backwashed at intervals of 2-3 weeks, in contrast to the Karlskrona reactor that was not. Details about the pilot plants are found in Table 2.

At Karlskrona waterworks one of the existing full-scale treatment trains was used in combination with the bioreactor. At Sotenäs waterworks a pilot plant Dynasand filter was used, operating at the same conditions (pH, Al-sulfate dose) as the full-scale plant.

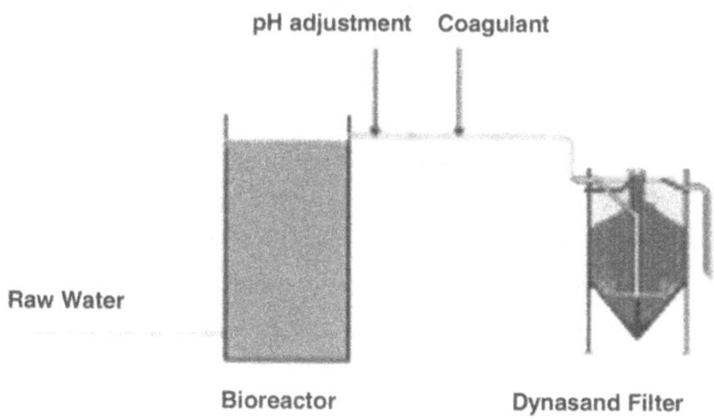

Fig. 1. Experimental set-up at Sotenäs and Karlskrona waterworks

* COD_{Mn}: Chemical Oxygen Demand using potassium permanganate. The dichromate method is to be banned in Sweden for its use of mercury. Current limit: 4 mg L^{-1}

Table 2. Description of the pilot-plant process combinations

	Sotenäs	**Karlskrona**
Bioreactor		
Height (m)	4.5	3 (approximately)
Cross-section (m)	0.6 ∅	2.6 x 3.4
Volume (m^3)	1.27	27
Flow (m^3h^{-1})	1.2 – 2	14 – 20
Reactor type	Plug-flow	Completely mixed
Carrier material* (m^3)	1.13 (BioFlow 30)	12.7 (KMT)
Retention time (min)	39 – 63	81 – 116
Aeration	No	Yes, by perforated tube
Chemical dosage		
pH adjustment	NaOH to pH 6.1	Lime water & CO_2 to pH 6
Al-Sulfate (mg L^{-1})	45 - 55	70 - 90
Dynasand filter		
Filter bed area (m^2)	0.3	5

* BioFlow 30 (Watergroup-Rauschert a/s): PP/PE Regranulate, 320 m^2 m^{-3}, 1.05 g cm^{-3}
KMT (Kaldnes Miljøteknologi AS): PE, 500 m^2m^{-3}, density less than 1 g cm^{-3}

Analysis

At Sotenäs, preceding the construction of the pilot plant described above, a conventional jar test procedure was applied to determine the optimal pH and Al-sulfate dose for biologically treated water and raw water. Glass beakers with a volume of 2 L were used. Chemical water analysis was performed at the waterworks' laboratory.

At Karlskrona all analysis except TOC and DOC were done at the waterworks laboratory. A commercial laboratory supplied the latter ones. Jar tests to optimize chemical dosage were done on biologically pre-treated water and raw water. After chemical treatment, the odor was investigated at Karlskrona waterworks. The test was done according to method number 1 described by the Swedish National Food Administration [7]. The odor of a drinking water sample was investigated both at room and elevated temperature (20 and 40-60 °C respectively). Special glass flasks with a wide opening were used. When the covered sample reached the specified temperature, the cover was removed and the type and strength of the odor determined. The odor scale applied here was: *no odor – insignificant – weak – distinct – strong*. At each occasion, 5 or 6 experienced persons took part in the test. The analysis was done as a blind test.

Biofilm carriers (BioFlow 30 and KMT respectively) were collected from the bioreactors at Sotenäs and Karlskrona and placed in 20 mM Tris-buffer* (pH 7,75). To remove the biofilm from the plastic carriers, the samples were sonicated (Branson 1510) for 5 min in three consecutive runs. The samples were cooled on

* Tris(hydroxymethyl)-aminomethanehydrochloride

ice between the runs. Pieces of plastic were also cut directly from the biofilm carriers for microscopy. These pieces were immobilized in agarose (low melting point), stained with Acridine Orange (AO) for 7 min, washed 3x3 min. in citrate-phosphate buffer (pH 6.6) and finally fixed in 2% formaldehyde. The samples were stored at 4°C for subsequent confocal microscopy.

Dry weight measurements: Samples of sonicated biofilm were filtered onto pre-weighed Whatman GF/F glass fiber filters. The filters were dried for 7 min in a microwave oven at full effect, allowed to cool and were weighed. The amount of volatile suspended solids (organic matter) was measured after combustion of the biofilm on filters for 1 h at 550 °C. At one occasion the amount of Mn in the biofilm was measured spectrophotometrically according to Swedish standard (SIS 028130), using a UNICAM Helios alfa spectrophotometer, after boiling duplicate samples of combusted biofilm in 4M H_2SO_4 for 30 min. in an autoclave (120 °C).

Total microbial counts: Sonicated samples of biofilm were filtered on Nuclepore filters (0.2 μm) pre-stained with Sudan Black. The filtered samples were stained with AO for 7 min. If the samples contained precipitations, 0.2% of oxalic acid was added before filtering. The microorganisms were counted in an epifluorescence microscope (Olympus BH-2, 100x lens).

Isolation of Mn-oxidizers: Samples of sonicated biofilm were spread on solid media for heterotrophic manganese-oxidizing bacteria (1.0 g malt extract, 2.0 g peptone, 0.5 g yeast extract, 0.2g $MnSO_4 \cdot H_2O$, 1 mg $FeSO_4 \cdot 7H_2O$, 15 g agar and 1 L distilled H_2O) and were incubated for 1 month in the dark at 20°C. Mn-oxidizers were detected visually as brownish colonies surrounded by a brownish area.

Confocal laser scanning microscopy (CLSM): The immobilized pieces of biofilm were visualized using CLSM (Nikon Diaphot 300 with a Molecular Dynamics 2100 and Image Space software) for measurements of thickness and qualitative characteristics of the biofilm.

Results and discussion

The practical studies presented in this paper were initiated by the involved waterworks in order to find solutions to the acute problems they faced. Where available, existing basins were used and the bioreactors thus had a number of different forms and volumes. Different approaches were taken. At Karlskrona, an old lime water tank was converted into a completely mixed bioreactor, with carrier material floating around. From this set-up, no physical filter-effect was expected. The reactor was supposed to accomplish oxidation of dissolved manganese only, converting it to the precipitated form (MnO_2). The Sotenäs reactor was designed as a plug-flow column, which was backwashed at intervals. Although the used carrier material was coarse, some retention of particles in the bioreactor was assumed.

The jar test at Sotenäs indicated that using the same aluminum sulfate dose (53 mg L^{-1}), the remaining Mn concentration was lower in the biologically pre-treated samples than in those with only chemical treatment (data not shown).

The manganese concentrations left in the water after Dynasand filtration with and without biological pre-treatment are shown in Figure 2. Despite the different bioreactor design at the two sites, the critical peaks of manganese that occurred in the ordinary production did not exist in the bio-line. During December 1998 and one week in September 1999, the coagulation at the Karlskrona bio-line was operated with iron(III)chloride, at an average pH of 5.3. During this period the removal of manganese was less efficient than in the ordinary part of the plant, while the removal of humic material was improved (compare also Figure 3). Apart from these incidents the chemically treated water in the bio-line complied with the standard at all times.

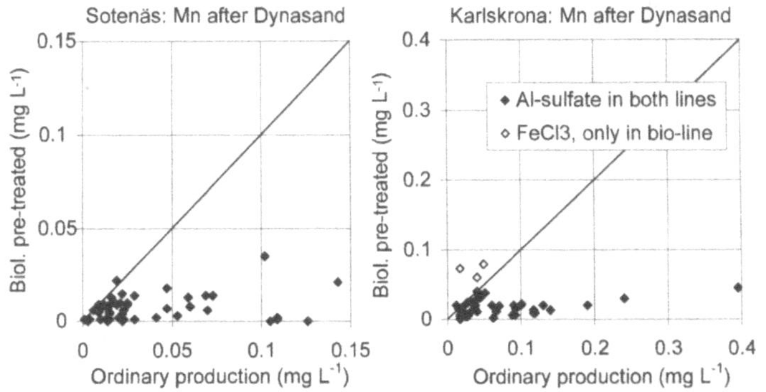

Fig. 2. Total Mn concentrations remaining after Dynasand filtration. Each point represents the same sampling occasion for the pilot plant treatment train and the ordinary production of the waterworks.

Figure 3. shows in which treatment step total manganese is removed. As could be expected from the bioreactor design, total manganese is not significantly removed in the Karlskrona bioreactor. However, the subsequent Dynasand filter removed manganese efficiently. At Sotenäs, a higher proportion of the influent total manganese was removed already in the bioreactor. There the incoming peaks of total manganese were removed in the bioreactor, while the Dynasand process took the rest. After starting up the bioreactor, there were some weeks of poor manganese removal. This is in accordance with results from other studies [3,4,8]. Note that in July 1999, the intake depth for raw water at Sotenäs was changed from 6 to 2 m, as high manganese concentrations had been measured in the raw water (up to 0.17 mg L^{-1}). Concentrations of particle-bound manganese were assumed to be the difference between the unfiltered and filtered water samples (Figure 4.). At Karlskrona, only the dissolved fraction is consistently removed in the bioreactor. On several sampling occasions, aggregates of particle bound Mn were washed out from the Karlskrona reactor, causing measured outflow concentrations of up to 14 times those in the raw water. The following Dynasand process easily retained these aggregates. At Sotenäs, removal of the two fractions was almost equal. Backwash water at Sotenäs was intensely brown colored, seemingly removing large amounts of accumulated sludge.

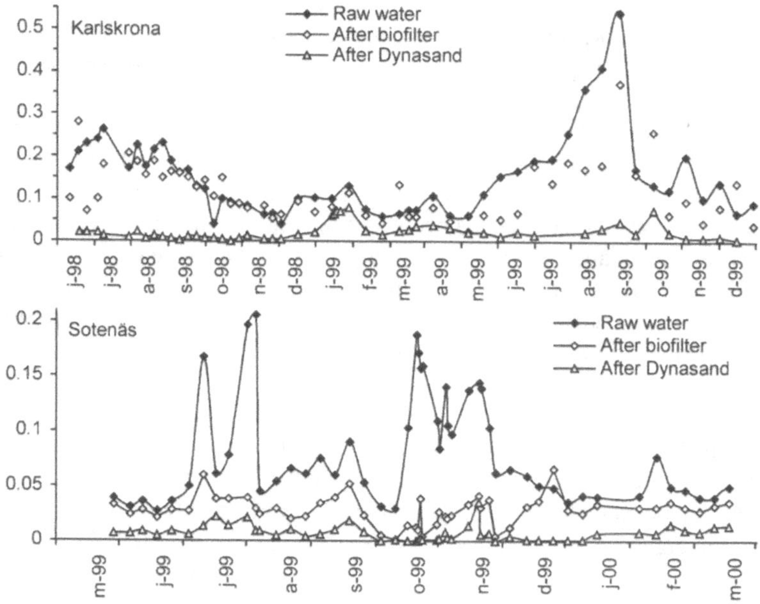

Fig. 3. Total manganese concentrations in the pilot plant (mg L^{-1}, unfiltered samples)

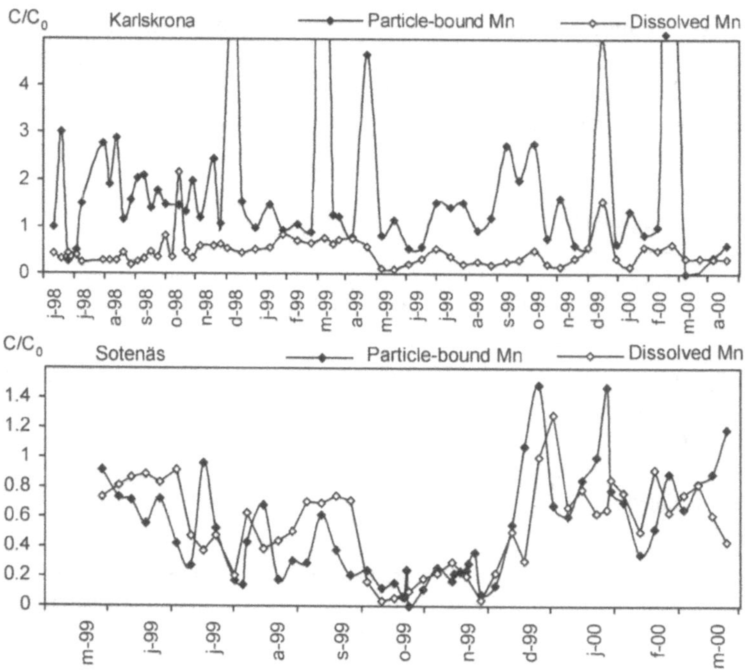

Fig. 4. Removal of dissolved and particle-bound manganese in the bioreactor

The biological pre-treatment described in this paper had no measurable effect on bulk parameters of organic matter, either directly after the reactor or after subsequent chemical treatment. It must be pointed out that these parameters are very rough measures and give little indication about possible interactions of iron and manganese with organic ligands. Concentrations of dissolved iron were not significantly decreased by the two bioreactors (data not shown). The ordinary coagulation process was however sufficient to consistently remove iron.

The biofilm analyses at Sotenäs were carried out on the predecessor of the bioreactor described above. This earlier version of the bioreactor was run in parallel with the bioreactor at Karlskrona during the time period for biofilm characterization. The carrier material was the same in these bioreactors at Sotenäs and Karlskrona. The results are included as a general indication of biofilm character.

The general characterization of the biofilms from Sotenäs and Karlskrona revealed considerable differences, with thicker biofilms in Karlskrona containing a higher proportion of organic matter and a higher abundance of bacteria than at Sotenäs. A single analysis of the Mn content within the biofilm indicated less precipitated manganese in the biofilm from Karlskrona (4.2 % ± 0.3 %) than in the sample from Sotenäs (38 % ± 4 %). From the biological data and pilot-plant performance observed in this study, it is impossible to determine the mechanism for the removal of manganese. Measurements from Karlskrona indicate a redox level around +300 mV. Compared to Eh/pH diagrams as found for example in [4], this is not high enough for chemical Mn oxidation. Biological processes suggested by other authors include intracellular oxidation, oxidation by extracellular enzymes and adsorption on already precipitated MnO_2 for subsequent chemical or biological oxidation [4,9]. The fact that the manganese removal was increased by the bioreactors at both sites, despite that the biofilms at the two sites were so different in character, may suggest that several processes are involved.

Table 3. Microbiological characterization of the biofilms

Period: Dec 98-Nov 99	Sotenäs (n=4)			Karlskrona (n=5)		
	Mean	Min	Max	Mean	Min	Max
Thickness (µm)	121	96	140	202	130	260
Microbial counts (10^6 cm^{-2})*	3.4	2.7	4.6	309	190	468
Organic matter (mass-%)	40	17	67	68	28	98
Mn-bacteria isolated	Yes			Yes		
Confocal Scanning Laser Microscopy	A homogeneous biofilm with some patchiness. Few or no filamentous organisms.			Considerable spatial heterogeneity and large morphological diversity. Many filamentous microorganisms.		

*Acridine Orange Direct Count. The microbial counts are lowest estimates of abundance since the sonication was not always sufficient to detach all microorganisms in the biofilm.

The analysis of odor at Karlskrona generally indicates less odor of the water receiving biological pre-treatment, compared to water from the ordinary production of the waterworks (Figure 5).

Most participants of the test panel described the odor as soil-like. The biologically pre-treated water was found to have less odor than water from the ordinary production, with one exception at very low odor level (November 2, 1998). This was valid for both cold and warm water samples. The components responsible for odor and taste are expected to be small organic molecules. Therefore a plausible explanation could be either adsorption of these molecules to the biofilm or utilization of the molecules by bacteria within the biofilm.

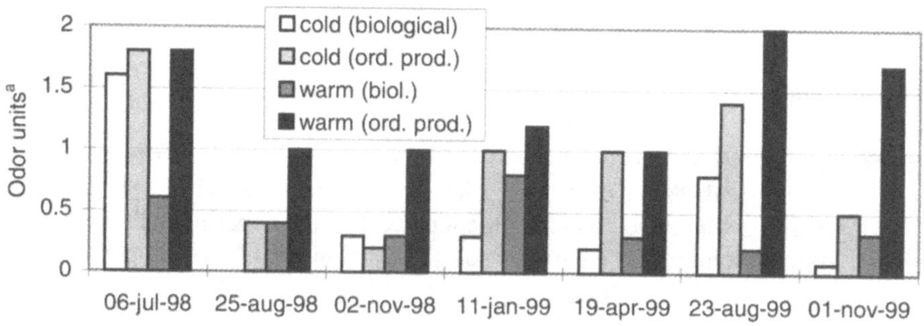

a) 0 = no odor; 0.5 = insignificant, 1 = weak, 2 = distinct, 3 = strong (Values are averages of the participants' judgment)

Fig. 5. Karlskrona waterworks – comparison of odor after chemical treatment with and without biological pre-treatment

Conclusions and further research

The pilot plant studies described in this paper show that biological pre-treatment could ensure a consistent water quality for colored raw waters with largely varying manganese concentrations. Encouraged by the results, Sotenäs waterworks has recently constructed a large pilot plant facility. Expanded clay granules were chosen as biofilm carrier material since tentative experiments indicated that bioreactors filled with this material achieved better manganese removal than the tested plastic carriers. Similar ideas for large-scale experiments exist at Karlskrona waterworks.

Further research will be necessary in order to fully understand the biochemical processes, and to optimize technical application of biological treatment prior to conventional coagulation in drinking water production.

This includes:

- Determination of which iron and manganese species occur in the raw water, and what complexes exist.
- What are the mechanisms contributing to the observed manganese oxidation? This would require more continuous biological monitoring and also biological experiments on a laboratory scale.
- Influence of parameters such as water temperature, pH, and redox on the oxidation process given the particular raw water characteristics.
- Optimization of bioreactor design, concerning biofilm carrier materials and flow regime, e.g. completely stirred tank or plug flow column.

Acknowledgements

The project has been financed by VA-FORSK, the research department of the Swedish Water and Wastewater Association. The authors wish to thank the following persons who have done the practical work at the experimental sites:
Tobias Salomonsson and Östen Andersson (VA-ingenjörerna AB, Stockholm and Göteborg), Anna-Kerstin Thell and Susanne Ekendahl (Göteborg University), Anders Claesson (Sotenäs waterworks) and Mats Strand (Karlskrona waterworks).

References

1 Seppänen, H.T.: Experiences of biological iron and manganese removal in Finland. J. IWEM 6 June (1992) 333-341

2 Michalakos, G.D., Nieva, J.M., Vayenas, D.V., Lyberatos, G.: Removal of iron from potable water using a trickling filter. Water Research 31 (5) (1997) 991-996

3 Hässelbarth, U., Lüdemann, D.: Die biologische Enteisenung und Entmanganung. Vom Wasser 38 (1971) 233 (in German)

4 Mouchet, P.: From Conventional to Biological Removal of Iron and Manganese in France. Journal AWWA 84 (4) (1992) 158-167

5 Hedberg, T., Wahlberg, T.A.: Upgrading of waterworks with a new biooxidation process for removal of manganese and iron. Wat. Sci. Tech. 37 (9) (1998) 121-126

6 Thorsen, T.: Membrane filtration of humic substances - state of the art. Wat. Sci. Tech. 40 (9) (1999) 105-112

7 Swedish National Food Administration / Livsmedelsverket Analysmetoder för bottensats, grumlighet, lukt, smak och temperatur vid undersökning av dricksvatten. Kungörelse SLV FS (1989) 30 (in Swedish)

8 Heinicke, G.: Biological Drinking Water Treatment - Evaluation of a Pilot Plant Study at Varberg Waterworks, Sweden. MSc-Thesis, Department of Sanitary Engineering, Chalmers University of Technology (1999) Gothenburg, Sweden

9 Peitchev, T., Semov, V.: Biotechnology for manganese removal from groundwaters. Wat. Sci. Tech. 20 (3) (1988) 173-178

Effects of Coagulant Type and Coagulation Conditions on NOM Removal from Drinking Water

B. Eikebrokk

SINTEF Civil and Environmental Engineering, Dept. Water and Wastewater,
N-7465 Trondheim, Norway.
bjornar.eikebrokk@civil.sintef.no

Abstract

Removal of natural organic matter (NOM) from soft and low alkalinity raw waters was studied in a coagulation-dual media anthracite-sand filter pilot plant. Three raw waters with colour levels of 15, 30 and 50 mg Pt/L, and five different coagulants were tested: alum (ALG), two poly-aluminium chlorides with different Ca-contents (PAX14 and Ca-PAX), iron chloride sulphate (JKL), and chitosan (Chi). The tests were run at water temperatures of 6 - 11 °C. Coagulation performance at low water temperatures (1-2 °C) using ALG and 3 different on-line static mixers was studied in a specific set of experiments. In terms of head loss and turbulence level, the 3 tested mixers were considered to represent the upper and lower extremes, as well as a more practically applicable unit. Optimum coagulant dose and pH levels were identified for the different combinations of coagulants and raw waters. The experiments demonstrated that coagulant dose requirements when using metal-based coagulants were controlled by the metal residues in treated water rather than colour or organic carbon. Thus, common problems related to excess metal content in treated water is normally caused by sub-optimum coagulant doses and improper pH control. On the other hand, coagulant overdosing adversively affects filter run length and head loss development. The results showed that the type of mixer did not affect ALG dose requirements and overall process performance to any great extent. Low water temperatures increased the pH-sensitivity of the process, especially with respect to residual aluminium. The natural cationic biopolymer chitosan was able to remove colour efficiently. With respect to organic carbon removal however, chitosan was less effective than metal-based coagulants, even at relatively high doses.

Introduction

Removal of turbidity and other components, rather than colour or organic carbon, have traditionally been used as the basis for optimisation of coagulation-filtration processes for drinking water treatment around the world. In recent years, however, more efforts have been put into optimising the removal of natural organic matter (NOM) from drinking water. The main motivation for this is normally to reduce the formation of potentially harmful, known and unknown disinfectant by-products (DBPs). Optimum NOM removal normally requires the use of elevated coagulant doses and a more strict pH control ("enhanced coagulation"). TOC removal requirements are normally expressed as a fraction, i.e. 0-50 % depending on raw water TOC and alkalinity levels [4], or by a maximum as well as a recommended concentration level. In the Norwegian water quality standard (adapted to the EU drinking water directive), the maximum TOC and colour levels are 5 mg TOC/L, and 20 mg Pt/L, with recommended values of 3 and 1 respectively. The maximum level of Al and Fe are 0.2 mg/L. When these metals are used for water treatment however, the maximum levels are 0.1 mg/L.

In order to obtain the required water quality, pH and coagulation dose optimisation and control is needed. In waters with high alkalinity levels, coagulant overdosing is normally used to depress the pH to optimum values. In soft and low alkalinity waters a base is normally required to prevent sub-optimum pH levels. As part of the Norwegian corrosion control strategy for such raw waters, the levels of calcium, alkalinity and pH are increased to 15-25 mg/L, 0.6-1.0 mmol/L, and 7.5-8.5, respectively [3]. NOM removal by metal coagulants can be characterised by high dose requirements and early breakthroughs. Sub-optimum doses may lead to excessive residual metal concentrations in finished water. On the other hand, it is important to minimise coagulant doses and sludge production, and thereby maximise filter run lengths. Coagulant overdosing also reduces pH which must be compensated with additional base. Coagulant dose optimisation and control is therefore cost-effective in most cases. From a hygienic point-of-view it is still imperative to maintain low turbidity levels in treated water while focusing on NOM removal. Chlorine resistant parasites like Crypto and Giardia are normally well controlled when treated turbidity is low. However, increased turbidity may be observed in some cases when minimising the coagulant doses and optimising the process with respect to NOM removal.

Materials and Methods

Coagulation-direct filtration pilot plant

A flow sheet of the pilot plant is presented elsewhere [1,2,3]. The plant includes units for raw water preparation, pH- and corrosion control, coagulant addition, filtration as well as on-line water quality control and data acquisition systems. The internal diameter of the 3.5 m high filter column was 0.15 m, and the dual media filter bed composition was 0.65 m of 0.8-1.6 mm hydroanthracite above 0.35 m of

0.4-0.8 mm silica sand. Below the sand there were 3 support layers, each 0.11 m deep: 1-2 mm, 2-5 mm and 5-10 mm. Pressure transmitters were located along the filter column at 5-15 cm intervals. In order to maintain a high precision in coagulant dosage, a relatively large water flow was coagulated (15 L/min). Flows in the range of 1.3-3.3 L/min were subsequently extracted from the coagulated water and fed to the filter column, corresponding to filtration rates in the range of 5-12.5 m/hr. Most coagulation optimisation experiments were however run at 7.5 or 10 m/hr.

Raw waters, coagulants and mixers

Raw water with colour levels of 15, 30 and 50 mg Pt/L were used in the experiments. The base water (RW15) was tap water from Lake Jonsvatnet, with levels of pH, turbidity, alkalinity and calcium normally in the range of 6.8-7.1, 0.1-0.3 NTU, 0.2-0.3 mmol/L, and 5-6 mg Ca/L, respectively. For the purpose of corrosion control, the levels of pH, alkalinity and calcium was raised to 7.5 - 8.5, 0.6 - 1 mmol/L, and 15-25 mg Ca/L, respectively. Important aspects of the combined NOM removal and corrosion control process is described elsewhere [3]. The raw waters RW30 and RW50 were prepared by adding small amounts (0.26 and 0.64 mL/L) of a neutralised regenerant solution to the base water. The regenerant was collected from a near-by treatment plant using anion exchange for NOM removal. The typical levels of total organic carbon in the 3 raw waters were 2.3, 3.5 and 5 mg/L, respectively. The water temperature during the tests was normally in the range of 6-11 °C.

Five different coagulants were tested: alum (ALG), two types of polyaluminium chlorides with different Ca-contents (PAX14 and Ca-PAX), iron chloride sulphate (JKL), and chitosan (Chi). Ca-PAX was tailor-made for combined NOM removal and corrosion control purposes, with a calcium to aluminium ratio of about 10 : 1. The experiments were run in such a way that several pH values were tested for every coagulant dose level. In this way possible interrelationships between optimum pH and the applied dose could be more easily detected.

In a specific set of experiments, the effect of water temperature was tested at two levels: LT (Low Temperature, 1-2 °C), and NT (Normal Temperature, 6-11 °C). A reversed heat pump was used to obtain the low temperature level. At the low temperature level, 3 static mixers were tested: 1) a high-intensity on-line mixer with an extreme head loss (OLM), 2) a very simple "empty pipe" mixer with practically no head loss (EPM), and 3) an intermediate mixer with a medium head loss (OLM+). The mixers OLM and EPM were considered to represent the upper and lower extremes, and the OLM+ a more practically applicable alternative in terms of mixing efficiency, turbulence and head loss levels. Some characteristics of the mixers are presented in Figure 1. The OLM+ was similar to the OLM mixer except for the fact that the minimum diameter of the centre cone was increased from 5 to 14 mm.

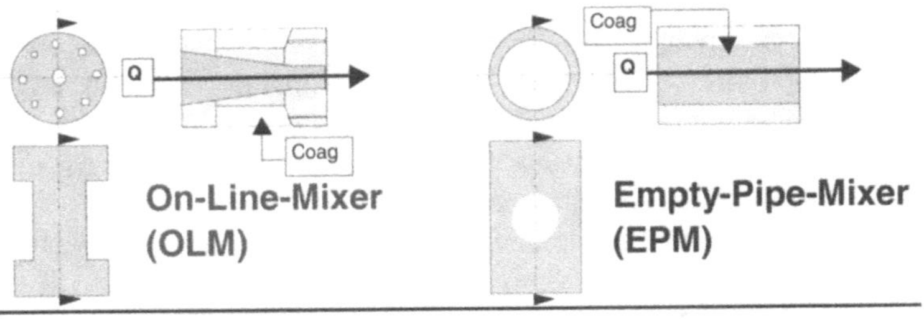

Mixer	Orifice Ø (mm)	Velocity (m/s)	Head loss (mH_2O)
OLM	5	12.7	6.05
OLM+	14	1.6	0.09
EPM	22	0.7	0.03

Fig. 1. Characteristics of the different mixers

Water sampling and analysis

Raw water, coagulated water and filtered water samples were taken 1-3 times during a filter run and analysed in accordance with the Norwegian Standards for Water and Wastewater Analysis. The parameters analysed included turbidity, pH, colour, total organic carbon NPOC (Dohrman DC-190), and residual Al or Fe (AA or ICP-MS) in raw and filtered water samples. Analysis of molecular weight distribution was also included in some samples.

Results and Discussion

Optimum coagulation – metal-based coagulants

Typical examples on the effects of coagulant dose and pH on process performance are presented in Figures 2 and 3. Raw waters RW15 (Figure 2) and RW50 (Figure 3) were coagulated with the aluminium based coagulants at doses of 0.8, 1.6 and 2.5 mg/L as Al. Regarding practical implications, the importance of pH control is obvious, and attention should also be paid to the interrelationship between the dose applied and the optimum pH. An increase in dose level widens the optimum pH range and tends to shift the optimum pH to a higher value. The results demonstrate that residual aluminium controls the dose requirements. NOM are effectively removed at dose levels required to comply with the water quality standard regarding residual aluminium (0.1 mg/L), and represent no problem with respect to the water quality standard regarding colour and organic carbon. Typically, the obtained colour and organic carbon removals are 80-95 %, and 50-70 %, respectively. The minimum coagulant dose requirements obtained for the different raw waters and coagulants are summarised in Table 1.

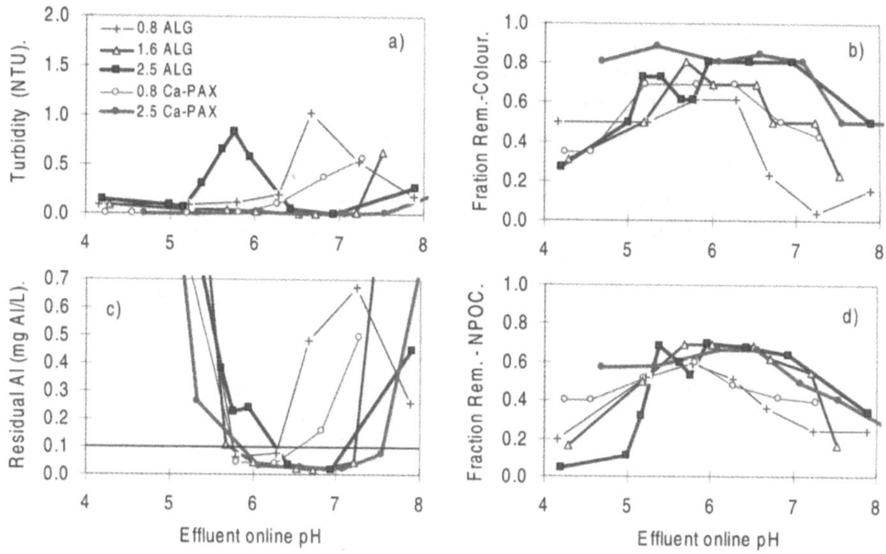

Fig. 2. Effects of dose and pH on filter effluent water quality. Coagulation of RW15 with ALG or Ca-PAX at dose levels of 0.8-2.5 mgAl/L.

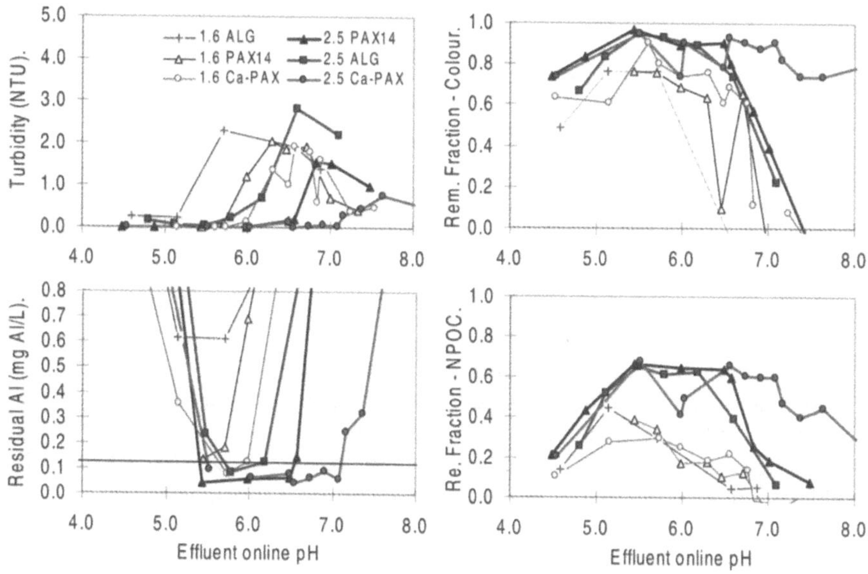

Fig. 3. Effects of dose and pH on filter effluent water quality. Coagulation of RW50 with ALG, PAX14 or Ca-PAX at dose levels of 0.8-2.5 mgAl/L.

It should be noted, however, that the optimum pH range is very narrow when using the minimum doses given in Table 1. In practice, it may be advantageous to increase the dose levels somewhat to simplify pH and process control. A 25 % increase is considered relevant for this purpose. Figure 4 illustrates the relationship between the molar dose requirements for the different coagulants and the raw water colour and organic carbon content. The difference in molar dose requirement between ALG and JKL was rather small. The poly aluminium coagulants required 10-30 % less dose compared to ALG. In terms of dose requirement, the advantage of using poly aluminium coagulants increases with increasing NOM content in the raw water. JKL is effective over a wider pH range (4.0-5.5) compared to ALG (5.5-6.5). However, the lower pH level required by JKL is disadvantageous in terms of coagulation and corrosion control chemical costs.

Table 1. Minimum dose requirements (mgMe/L) for the different coagulants

	ALG	PAX14	Ca-PAX	JKL
RW15	0.8	0.7	0.7	1.6
RW30	1.4	1.2	1.0	-
RW50	2.5	2.0	1.8	6.5

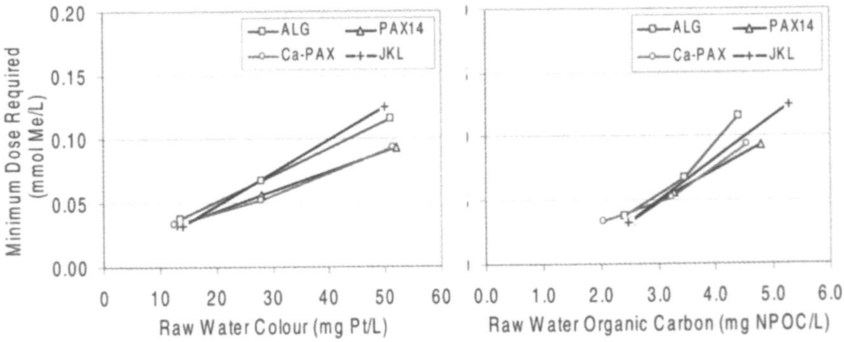

Fig. 4. Minimum dose requirements for the coagulants and raw waters used (mmol Me/L)

Although residual metal controls the minimum coagulant dose requirements, it is important to keep in mind that coagulant overdosing not only increases chemical costs and sludge production, but also increases the rate of filter head loss development and reduces filter run lengths. Figure 5 illustrates the effect on filter head loss and breakthrough of a stepwise increase in coagulant dose.

Effects of temperature and mixing

Temperature. The effects of low water temperatures on filtered water quality and process performance were tested on RW50 with ALG coagulation at doses of 2.5, 3.1 and 4.0 mgAl/L. Two temperature levels were investigated, low temperatures

1-2 °C (LT), and normal temperatures 6-11 °C (NT). For every dose level, 4 or 5 pH values in the range of pH 5.1-6.3 were used. The pH-values were practically identical for the NT and LT experiments. OLM+ was used for mixing.

The results show that NOM was effectively removed at all dose levels tested, relatively unaffected by the water temperature level. However, the sensitivity of residual aluminium to pH was greatly affected by water temperature, and the low water temperatures tend to narrow the range of optimum pH (Figure 6).

Although the average residual Al measured at pH 5.1-6.3 increased considerably at the low temperature level, the results at optimum pH (best run) were relatively unaffected by temperature (Table 2).

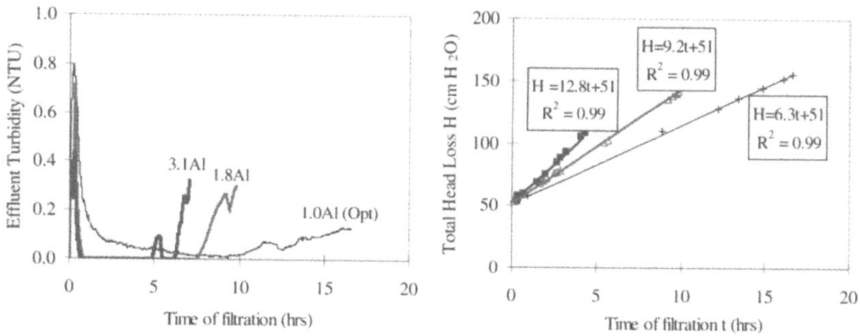

Fig. 5. Effects of coagulant dose level on filter run length and rate of head loss development (RW15, ALG, 10 m/hr)

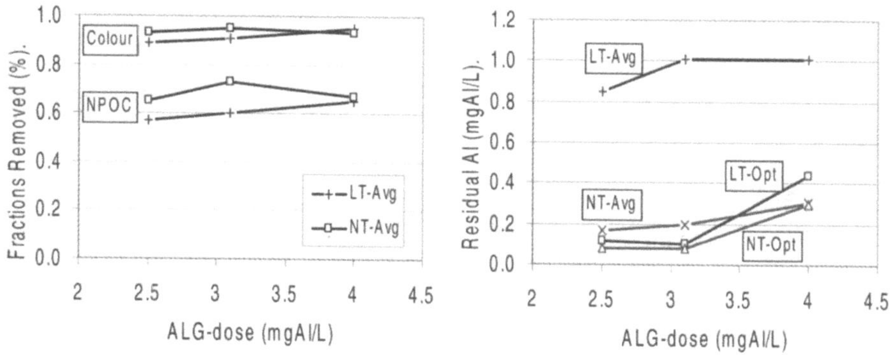

Fig. 6. Effects of water temperature levels (NT and LT) on coagulation. Average colour and organic carbon removal (left), and average and optimum residual aluminium concentrations (right). (RW50, OLM+, ALG, pH 5.3-6.3)

Table 2. Effects of water temperature on process performance – Optimum pH/Best runs (Mixer: OLM+; Coagulant: ALG in doses of 2.5-4 mg Al/L)

	D=2.5 mg Al/L		D= 3.1 mg Al/L		D= 4.0 mg Al/L	
	LT	NT	LT	NT	LT	NT
Turbidity (NTU)	0.11	0.12	0.11	0.28	0.13	0.83
Colour (%Rem)	94	99	95	98	96	98
NPOC (% Rem)	65	73	68	82	71	77
Res-Al (mg/L)	0.12	0.08	0.10	0.08	0.44	0.30

Mixing. Mixing was considered more important at low temperature levels, and specific tests using different mixers were run at the low temperature (LT) level of 1-2 °C. Dose levels applied were 2.5, 3.1 and 4.0 mgAl/L as ALG, and 4 different pH values were used at every dose level. The pH values used were practically identical for the different mixers. Figure 7 shows how produced water quality parameters were affected by the different mixers at the different levels of coagulant doses and pH, and Table 3 summarises the results obtained at optimum pH conditions. The results indicate only minor differences in performance between the 3 mixers, and no systematic differences were detected regarding NOM removal, i.e.colour and organic carbon.

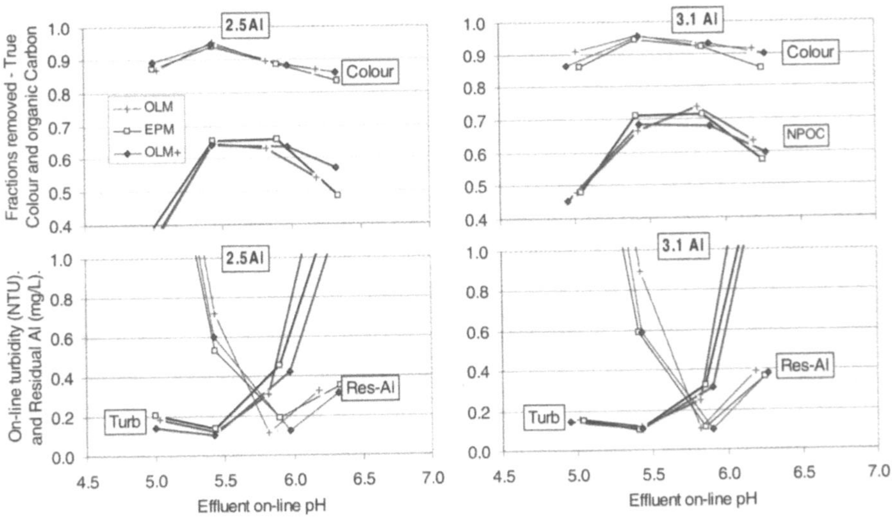

Fig. 7. Effect of mixer type on filter effluent quality (RW50 coagulated at 1-2 °C with 2.5 and 3.1 mgAl/L as ALG)

Table 3. Effects of type of mixer on effluent water quality and process performance – Best runs/optimum values (Low temperatures, ALG).

	D= 2.5 mgAl/L			D= 3.1 mg Al/L			D= 4.0
	OLM	OLM+	EPM	OLM	OLM+	EPM	OLM+
Turbidity (NTU)	0.12	0.11	0.14	0.12	0.11	0.10	0.13
Colour (%Rem)	95	94	94	96	95	94	96
NPOC (% Rem)	64	65	65	73	68	77	71
Res-Al (mg/L)	0.12	0.12	0.19	0.11	0.10	0.11	0.34

Coagulation by Chitosan

Due to the biodegradability and the absence of residual metal problems, the natural cationic polymer chitosan was tested for NOM removal as an alternative to metal-based coagulants (Figure 8). The results presented in Figure 8 show that colour was efficiently removed with chitosan. With respect to organic carbon removal however, chitosan was less effective than metal-based coagulants, even at relatively high doses.

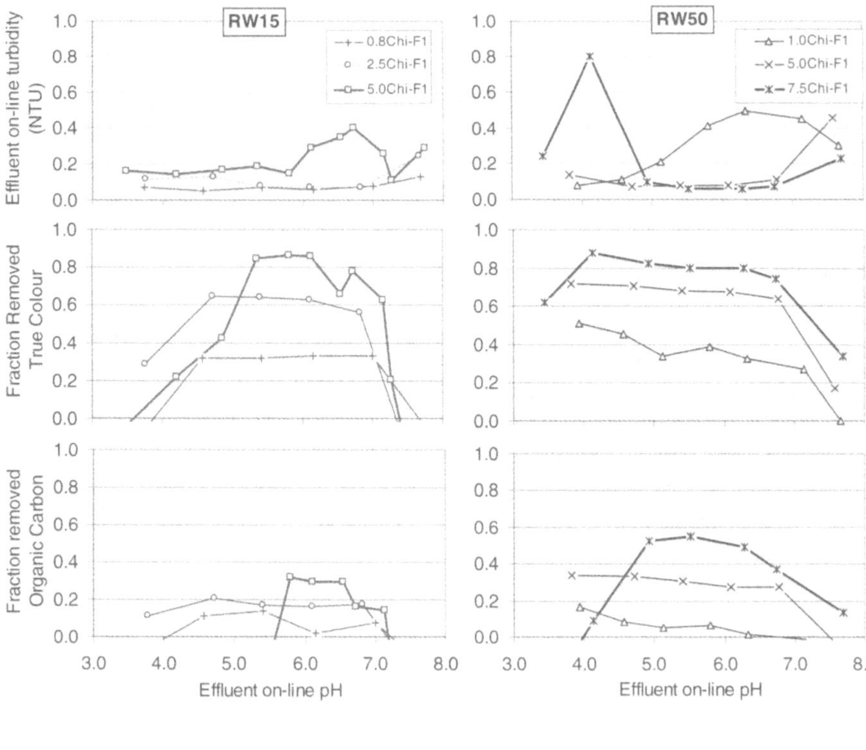

Fig. 8. Effects of dose and pH on filter effluent water quality. Coagulation of RW15 (left) and RW50 (right) with chitosan at dose levels of 0.8-7.5 mg/L (NT).

NOM fractions removed – ALG versus Chitosan

Figure 9 shows the molecular weight distribution with respect to TOC in untreated and treated waters RW15 and RW50. The coagulants used were ALG and Chitosan. Applied doses were 1.0 and 2.9 mg Al/L for RW15, and 1.6 and 7.3 mg/L of Chitosan for RW50. At the given doses, both coagulants seem to remove NOM primarily in the weight fraction of 5000-20000 Dalton , and ALG is more effective than chitosan in terms of overall TOC removal.

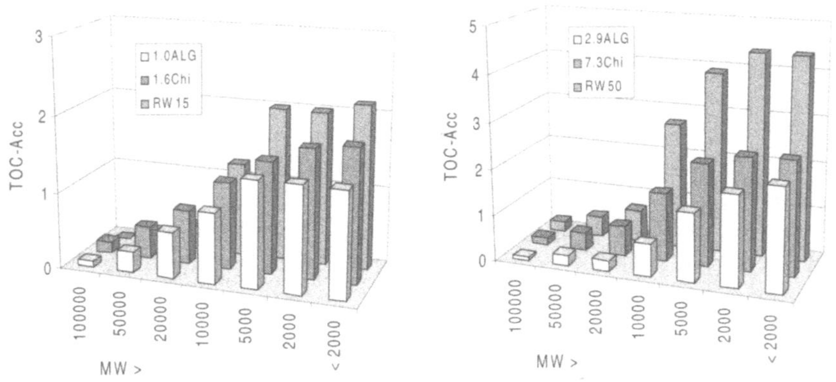

Fig. 9. Molecular weight distribution (as TOC) in raw and treated waters RW15 (left) and RW50 (right). ALG or chitosan was used for coagulation.

References

1 Eikebrokk, B.: Coagulation-direct filtration of soft, low alkalinity waters. Water Science and Technology 40 (9) (1999), pp. 55-62

2 Eikebrokk, B.: Removal of humic substances by coagulation. In: Hahn, H.H., Hoffmann, E. and Ødegaard, H. (Eds): Chemical Water and Wastewater Treatment V, Springer-Verlag, Berlin, Heidelberg (1996) pp 173-187

3 Østerhus, S.W. and Eikebrokk, B.: Coagulation and Corrosion Control for Soft and Coloured Drinking Water. In: Klute, R. and Hahn, H.H., (Eds): Chemical Water and Wastewater Treatment III, Springer-Verlag, Berlin, Heidelberg (1994) pp 137-153

4 USEPA Disinfectants and Disinfectants Byproducts. Final Rule. Fed. Reg., 63, 241, 69478 (1998)

Municipal Wastewater Treatment

The Role of Organic Polyelectrolytes in High Rate Alternatives to Primary Separation

N.S.C. Becker*, N.A. Booker*, A.Davey**, S.R. Gray*, R. Jago** and C. Ritchie*

* CSIRO Molecular Science, Private Bag 10, Clayton South, Victoria, Australia, 3169
stephen.gray@molsci.csiro.au,
**CDS Technologies, Nepean Hwy, Mornington, Australia

Abstract

Treatment of sewage with a CDS (continuously deflective separator) unit has produced a treated effluent quality of 5 NTU and 20 mg/l TSS at hydraulic loading rates of 57 $m^3m^{-2}h^{-1}$. This performance required the production of strong flocs formed using high doses (10 mg/l) of a low charge density (CD), high molecular weight (MW), cationic polyelectrolyte. Laboratory experiments suggested that this dose could be reduced to 2.5 mg/l if a medium CD, very high MW, cationic polyelectrolyte was used. The laboratory tests also indicated that the concentration of residual polyelectrolyte in the treated effluent would be reduced by a factor of ten if very high MW polyelectrolytes were used.

Introduction

The use of sedimentation processes for the separation of suspended solids from water and wastewater streams is limited by the settling velocity of the particles. Attempts to accelerate these settling processes through the use of inorganic coagulants and/or polymeric flocculants, have led to some improvements. However, improvement in separation rates is limited by the low apparent density of the floc and high drag coefficients. Past research at CSIRO has focussed on increasing the density of the floc through the attachment of heavy carrier particles such as magnetite.

The Sirofloc™ for sewage process rapidly flocculates (< 2 min) sewage around fine magnetite particles (< 20 μm) [1] using inorganic coagulant at typical doses of 10-20 mg/l Al^{3+}, following which magnetic flocculation of the magnetite produces large flocs that settle rapidly. Rise rates of 10-20 $m^3 \cdot m^{-2} \cdot h^{-1}$ are typically used in the settler. Treated effluent quality is normally 18 mg/l total suspended solids (TSS), 55 mg/l BOD and 128 mg/l COD [2].

Actiflo® for sewage is a similar process to Sirofloc™ except that the Actiflo® process uses sand instead of magnetite to weight the flocs. The sand particles are unable to magnetically flocculate and hence larger particles (130 μm) [3] are required to achieve the high settling rates required. The sewage is flocculated with inorganic coagulant, the sand added followed by flocculation with organic polymer (0.5 – 1 mg/l polymer [3]). The flocculation time is usually 5 minutes and typical effluent quality is 50 mg/l TSS [4].

While weighted particle systems are effective and have high hydraulic loading rates compared to primary sedimentation and enhanced sedimentation, the recycling systems make them complex and higher hydraulic loading rates may be possible using different separation processes.

Recent work at CSIRO has aimed at separating chemical floc using techniques other than sedimentation, and has been based on either high rate filtration or screening.

High rate filtration of sewage with high porosity, plastic media filter beds has been performed in pilot plant tirals, with TSS removal efficiencies of 70-90% [5,6] achieved at loading rates of 20–60 $m^3m^{-2}h^{-1}$. Typical chemical doses of 10 mg/l Al^{3+} and 5-6 mg/l organic polymer were required to achieve these levels of performance. High concentrations of organic polymer were necessary to produce flocs of sufficient strength to withstand the high levels of shear within the filter [5]. While the rates and removal efficiencies were impressive, the run times were short and frequent backwashing was required. The run time is limited by the volume of solids that the filter can retain before the shear rate in the filter increases to a level where the flocs break and filter breakthrough occurs. Increases in the filter porosity increase the solids loads that can be retained in the filter but simultaneously reduce the capture efficiency of the filter.

Current work seeks to advance the high rate filtration concept by moving to a continuous screening system. The screening system of choice was the CDS (continuously deflective separator) screen, which is currently marketed for gross solids removal from stormwater and sewage. A schematic diagram of the CDS system is shown in Figure 1. The inlet directs flow tangentially across an expanded mesh screen with an opening of 0.6-1.2 mm. Filtered sewage flows through the screen and the flow reverses direction before being discharged. The screen retains the solids, and the floatables are removed via a downcomer through the base of the unit along with any solids that sink. Discharge of the solids may be continuous or semi-continuous and discharge is regulated to increase the solids content of the sludge.

It was anticipated that running the device in a continuous operating mode would overcome the need for backwashing, the major limitation of high rate filtration for sewage treatment. Floc strength was expected to be of major importance for screening of sewage on the CDS unit, as high shear rates in the CDS unit will act to break up flocs and reduce solids capture.

Floc Strength

The strength of inorganic flocs can be greatly increased (up to 48 times) by the addition of polymeric flocculants [7], and floc strength is largely determined by the organic flocculant concentration [5]. Maximum floc strength does not occur at the

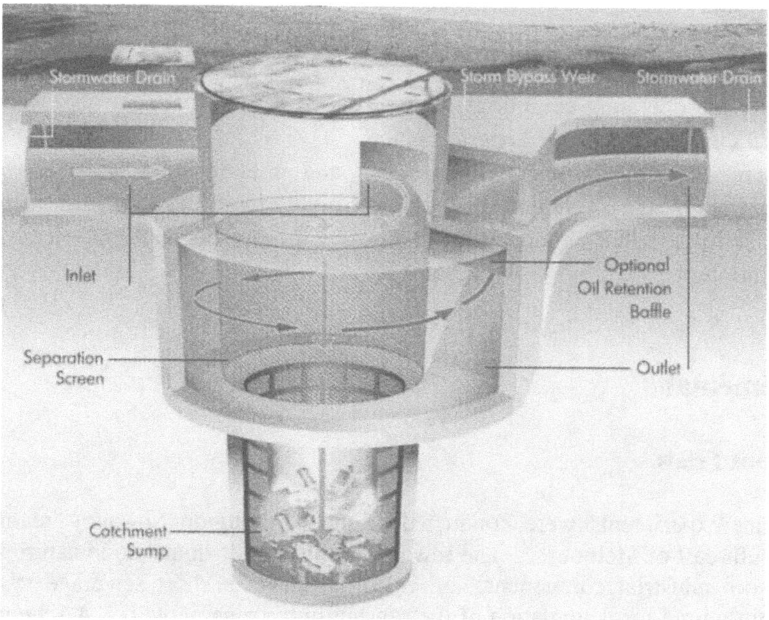

Fig. 1. CDS (continuously deflective separator) unit for gross solids separation in stormwater

optimum chemical conditions for flocculation [5,8,9], but typically occurs at higher doses.

Polymer type may also affect floc strength, with high molecular weight, low charge density polyelectrolytes producing sewage flocs of greater shear resistance than high molecular weight, high charge density cross linked polyectrolytes [5]. The strength of glass bead flocs has also been observed to increase 2.5 times [10] when flocculated with hydrolysed (30%) polyacrylamide (HPAM) of molecular weight 6 x 10^6 compared to flocculation with HPAM of molecular weight of 2.5 x 10^6. However, floc strength does not always vary when polyelectrolyte type is changed, as the strength of clay flocs were unaltered [11] when non-ionic polymer of 15 x 10^6 molecular weight was substituted for non-ionic polymer of 3.6 x 10^6 molecular weight.

High polyelectrolyte doses do not only increase floc strength, but may also increase the amount of residual polymer in solution. Gehr and Henry [12] have shown that polyelectrolyte concentrations dramatically increase above a critical value, usually corresponding to that required for charge neutralisation. Below the critical concentration the amount of residual polyelectrolyte is low. The polyelectrolyte concentration required for high rate filtration of sewage exceeds that required for charge neutralisation [5], and thus the possibility of significant residual polyelectrolyte concentrations exist when high floc strength is required. This is of concern, as cationic polyelectrolytes are known to be toxic to fish at high doses [13].

Objectives

The objectives of this research were two fold:

1. Identify the operating conditions required for effective treatment of sewage in the CDS unit and measure its performance.
2. Measure the floc strength of sewage and the residual polyelectrolyte concentration as a function of polyelectrolyte dose and type, to identify polyelectrolytes that produce strong flocs with low residual polyelectrolyte concentrations.

Experimental

Pilot Plant Trials

Pilot plant experiments were conducted at the Mornington Sewage Treatment Plant, south-east of Melbourne. The sewage at this plant is domestic in nature with little or no industrial component, and is collected via a short sewerage system leaving little time for degradation of the sewage in the pipe network. A schematic diagram of the CDS pilot plant is shown in Figure 2.

Raw sewage was screened through a CDS screen with a 1.2 mm aperture. Coagulant was added and flashed mixed in a centrifugal pump. There was no pH correction and the final pH varied between 6.4 and 6.7. Polyelectrolyte was added and floc maturation occurred in either mixed tanks or a pipe reactor for 0.5 to 4 min. Sewage entered the second CDS unit where the flocs were captured and clarified effluent was produced. The second CDS unit was 900 mm in diameter and operated at 4-10 L/s (23-57 $m^3m^{-2}h^{-1}$). Higher flowrates are possible through the unit, but the capacity of the feed pump limited the experiments to this rate. The aims of the pilot plant tests were to prove that screening of flocculated sewage was possible and to compare the performance of different polymers in the plant to their predicted performance from laboratory tests.

Laboratory Tests

Laboratory tests were performed with sewage from Melbourne Water's Eastern Treatment Plant at Carrum, Melbourne. Raw sewage was collected each morning for use throughout the day. The sewage was coarse screened through a 1 mm screen before use and the total solids (TS) measured. The variability in the TS concentration was 870 ± 50 mg/l.

A series of polyacrylamide polyelectrolytes were used. The polyelectrolytes were selected from the results of jar tests using the final residual turbidity and a subjective estimate of floc strength as the basis for comparison. The characteristics of the polyelectrolytes as stated by the manufacturers are shown in Table 1.

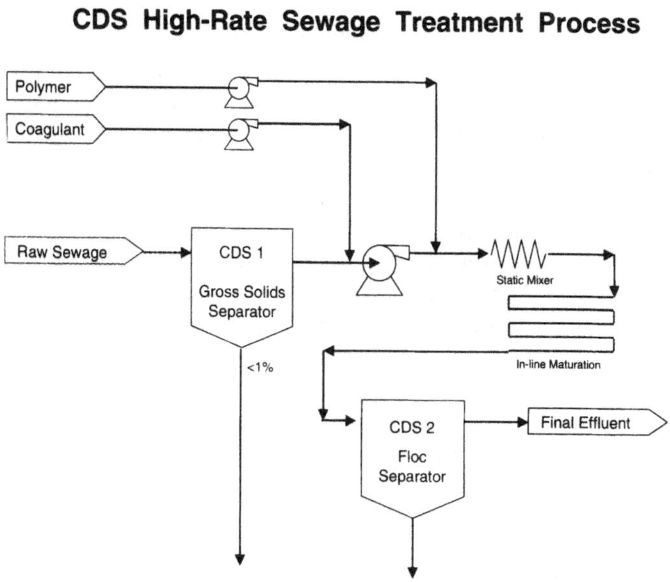

Fig. 2. Schematic diagram of the CDS pilot plant

Table 1. Polyelectrolyte characteristics

Polyelectrolyte	Charge	Charge density	Molecular Weight	Form
A	Cationic	Low	High	Powder
B	Anionic	Low	High	Powder
C	Cationic	Low	High	Emulsion*
D	Cationic	Medium	High	Emulsion*
E	Cationic	High	High	Emulsion*
F	Cationic	High	High	Emulsion*
G	Cationic	Medium	Very high	Powder

*The concentrations of emulsion polyelectrolytes given in this paper refers to the concentration of active ingredient.

Critical Mixing Speed

The strength of flocs in laboratory tests were estimated based on determining a critical mixing speed in jar tests where floc breakage occurred. The critical mixing speed of the flocs was compared by using a standard low speed mixing regime (jar test) to form flocs, following which the flocs were subjected to incrementally higher mixing speeds until an abrupt increase in turbidity was measured. In the jar tests sewage was rapidly mixed with 15 mg/l Al^{3+} (jar test minimum for lowest turbidity) and pH corrected to 5.9. Polyelectrolyte was added and flash mixed for 60 seconds, after which the mixing speed was lowered to 100 rpm for 15 minutes. The flocs were allowed to settle for 2 minutes, a sample of the clarified supernatant taken and the turbidity measured. The sample was returned to the jar test, the mixing speed incrementally increased and the sample mixed for 5 minutes. The sewage was again left to settle for 2 minutes, a sample taken and its turbidity measured. This process was repeated until the turbidity increased abruptly, and the speed at which this occurred was taken to be the speed at which the flocs began to break and is termed the critical mixing speed. This term can be misleading, as the floc size decreased for all increases in speed although this did not always result in an increase in turbidity. The abrupt increase in turbidity indicates that the flocs are too small to settle in the two minute settling period. Variation of the settling period between 1 and 5 minutes, however, did not significantly alter the measured critical mixing speed. This is probably because the critical mixing speed also corresponds to the speed where floc breakage is occurring by surface erosion rather than floc splitting. The critical mixing speed was reproducible to approximately ±100 rpm, and sewage age did not have a large effect on the critical mixing speed. The results are presented as graphs of the critical mixing speed versus polyelectrolyte dose for the different polyelectrolytes tested.

Residual Polyelectrolyte

Residual polymer in solution following flocculation was measured by coagulating clay with the settled sewage solution. Jar tests with the appropriate chemical conditions were conducted and the solids allowed to settle. Clay (2 g/l, Merck Bole white powder, kaolin) was added to the solution, mixed and allowed to settle for 5 minutes in a 100 ml settling cylinder. The top 30 ml of solution was sampled and the turbidity measured. This turbidity was compared with a standard curve of turbidity versus polyelectrolyte dose and the residual polyelectrolyte concentration estimated. The standard curve was prepared using sewage that had been coagulated (15 mg/l Al^{3+}, pH 5.9), settled and filtered (GA55 paper). Clay and known doses of polyelectrolyte were added to the clarified sewage and the turbidity measured after 5 minutes of settling. A graph of turbidity versus polyelectrolyte dose was then obtained. The turbidity in the standard curve was sensitive to polyelectrolyte doses between 0 and an upper limit. Therefore, concentrations above the upper limit were diluted to bring the polyelectrolyte concentrations into range. The upper limit varied between 0.5 and 2.0 mg/l depending upon polyelectrolyte type. The residual polyelectrolyte concentration was shown to decrease at higher shear rates than used in the jar test. Hence, the results reported in this paper represent the maximum residual polyelectrolyte concentrations likely to be encountered.

Results and discussion

Initial pilot plant trials were conducted to determine the operating requirements of the CDS unit. Polyelectrolyte A doses of approximately 7-10 mg/l or greater were required to obtain good effluent quality as shown in Figure 3.

Typical effluent TSS concentrations from the CDS unit operating with polymer A were 20 mg/l. By comparison, polymer B (an anionic polymer) was inferior to polymer A with final effluent TSS concentrations of 40 mg/l, while polymer C produced slightly better effluent quality than polymer A at similar doses.

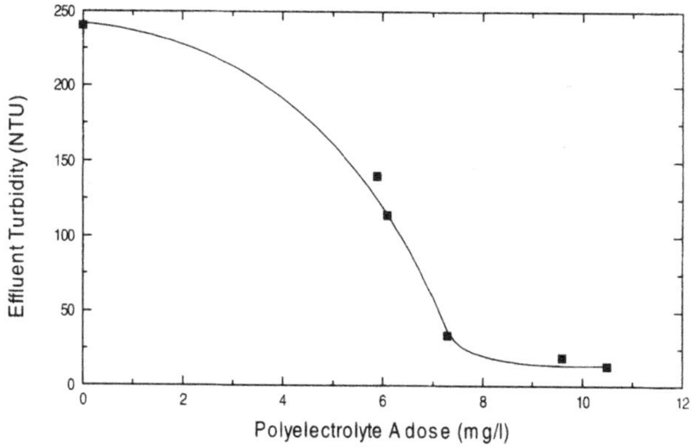

Fig. 3. Effluent turbidity from the CDS pilot plant as a function of polyelectrolyte A dose (20 mg/l Al^{3+}, flow rate = 4 l/s, 23 $m^3m^{-2}h^{-1}$, screen = 1200µm)

The best performance of the CDS unit was obtained with 20 mg/l Al^{3+} and 10 mg/l polymer C at a loading rate of 57 $m^3m^{-2}h^{-1}$, and produced an effluent quality of turbidity < 5 NTU and < 20 mg/l TSS. The BOD_5 was reduced by 95% and the total coliforms by 99.5%. Less than 1% of the flow went to sludge, which had a solids concentration of approximately 3% w/w. Optimisation of sludge production has not yet been studied.

The poor performance of polymer B in comparison to polymer A was thought to be the result of polymer B producing weaker flocs which were broken in the CDS unit. Laboratory tests were then conducted to verify this hypothesis and to determine if flocs of equivalent or greater strength could be made using lower concentrations of alternative polyelectrolytes. Lower doses of polyelectrolyte would potentially reduce operating costs and residual polyelectrolyte concentrations in the effluent.

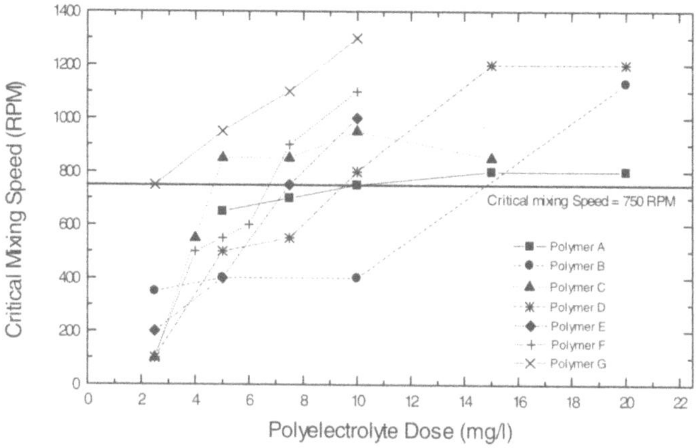

Fig. 4. Effect of polyelectrolyte dose for different polyelectrolyte types on critical floc shear rate

Figure 4 shows the comparison of critical mixing speeds determined in laboratory tests as a function of polyelectrolyte dose and type of polyelectrolyte.

Polymer A shows a critical mixing speed of 750 rpm at 10 mg/l; the dose required in the CDS unit. The critical mixing speed for polymer A only increased slightly at doses above 10 mg/l, suggesting that there would be little benefit in using higher doses than 10 mg/l. The critical mixing speed curve of polymer B lies below polymer A at low doses. Figure 4 suggests that doses of 15 mg/l polymer B are required for good performance in the CDS unit. This is consistent with the observations in the CDS unit, where doses of up to 20 mg/l polymer B were needed to produce treated effluent turbidities of < 20 NTU.

Polymer C, the emulsion equivalent of polymer A, reaches a critical mixing speed of 750 rpm at a dose of 5 mg/l compared with 10 mg/l for polymer A. At doses higher than 5 mg/l, the critical mixing speed only increases slightly for polymer C suggesting that little improvement in performance could be gained above this dose.

Polymers D, E and F were similar polyelectrolytes except that the charge density increased from polymer D to polymer F. The critical mixing speed for these polyelectrolytes increased linearly from 0-10 mg/l, and the ultimate critical mixing speeds of these polymers were above 1000 rpm. Thus flocs produced with these polymers may be able to withstand higher shear rates than those produced with polymer A. The dose required to attain a critical mixing speed of 750 rpm varied between 7.5 and 10 mg/l, with lower doses required as the polymer charge density increased.

Polymer G was a very high molecular weight polyelectrolyte and produced the highest critical mixing speeds at all polyelectrolyte doses. The dose required to reach 750 rpm was 2.5 mg/l and the ultimate critical mixing speed was above 1300 rpm. Thus this polyelectrolyte would be predicted to require the lowest dose in the CDS unit and to be capable of withstanding higher shear rates which may occur if the throughput of the CDS unit was increased.

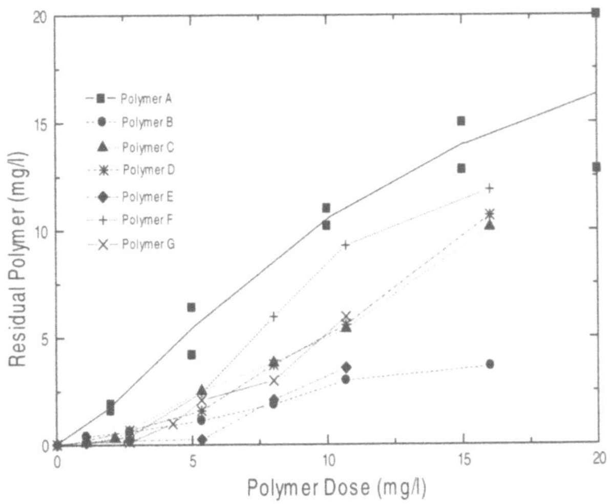

Fig. 5. Effect of polymer dose and type on residual polyelectrolyte concentration

Table 2. Polyelectrolyte dose and residual polyelectrolyte dose at a critical mixing speed of 750 RPM.

Polyelectrolyte	Polyelectrolyte dosed (mg/l)	Residual polyelectrolyte (mg/l)
A	10	10
B	15	4
C	5	2
D	10	6
E	8.5	2.2
F	7.5	6.4
G	2.5	0.15

However, the presence of high concentrations of residual polyelectrolyte may result in treated effluent being unsuitable for discharge to the environment. Figure 5 shows the residual polyelectrolyte concentration as a function of dose for each polyelectrolyte. All polyelectrolytes show a flat or low slope region at low doses, where most of the polyelectrolyte is adsorbed onto solids and little is left in solution. However, a critical concentration is reached, where further increases in concentration result in substantially greater residual polyelectrolyte doses. This behaviour is most pronounced for Polymers B and E and least pronounced for Polymers A and F

The polymer concentration and residual polyelectrolyte dose at a critical mixing speed of 750 rpm (that estimated to represent the shear rate in the CDS unit) are shown in Table 2. Polymer G has the lowest dose required to reach the critical mixing speed along with the lowest residual polyelectrolyte concentration. Hence, this polyelectrolyte appears to be the most suitable for use in applications where

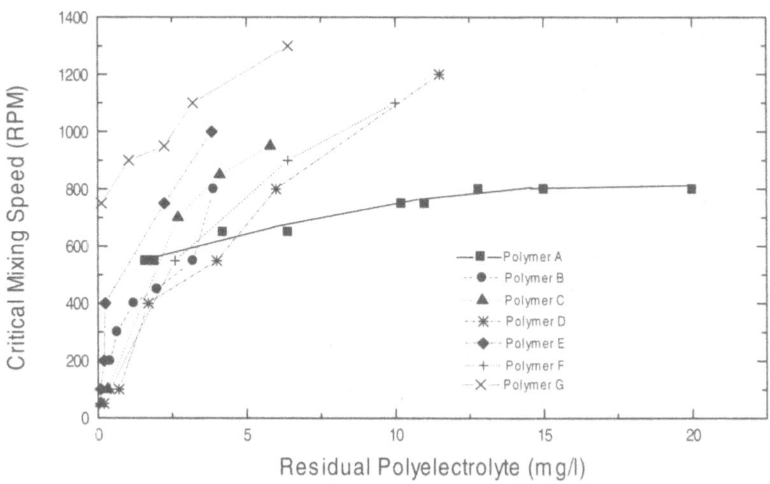

Fig. 6. Relationship between critical mixing speed and residual polyelectrolyte dose for various polyelectrolytes

high floc strength is required. This is further demonstrated in Figure 6, where polymer G has the lowest residual polyelectrolyte dose at any given critical mixing speed as well as the highest critical mixing speed.

Further pilot plant testing is planned to determine the performance of polymer G in the CDS separator, as well as to measure residual polyelectrolyte concentrations in the pilot plant effluent.

Conclusions

A high loading rate (57 $m^3m^{-2}h^{-1}$) CDS screen was capable of producing high quality effluent from raw sewage in a very short time. The CDS screening process produced an effluent with a turbidity of 5 NTU, TSS of 20 mg/l, BOD_5 of 20 mg/l, and coliform concentrations of 1 x 10^5 cfu/100ml. Chemical doses of 20 mg/l Al^{3+} and 10 mg/l polymer C were required and a 3% w/w sludge was produced.

Effluent of this quality is suitable for disinfection, and therefore the CDS process may find application treating peak sewer loads or sewer overflows. These applications require intermittent use, and a low capital cost – high operating cost process may prove to be economically viable.

Polyelectrolyte dose and type has a strong influence on the shear resistance of flocs. Polymer G, a medium charge density, very high molecular weight polyelectrolyte produced the strongest flocs at all polyelectrolyte doses. At a critical mixing speed of 750 rpm, which indicated good performance of the CDS unit, a dose of 2.5 mg/l polymer G was predicted to be required. All other polymers required 2-6 times this dose to reach a critical mixing speed of 750 rpm. Use of polymer G resulted in a residual polymer concentration of 0.15 mg/l when dosed for a critical mixing speed of 750 RPM, 10 times smaller than all other polyelectrolytes when dosed at concentrations required for a critical mixing speed of 750 RPM. Hence, polymer G was the most suitable polymer for applications were high strength flocs are required.

Current research aims to test this polymer in pilot plant trials and to determine whether there is an upper limit to floc strength that can be obtained using ultra high molecular weight polymers.

References

1. Booker, N.A., Cooney, E., Öcal, G., Priestley, A.J.: The Sirofloc sewage treatment process: A high rate process for sewage clarification. In Chemical Water and Wastewater Treatment III, R. Klute and H.H. Hahn (Eds), Springer-Verlag, Berlin Heidelberg, (1994) pp. 231-242
2. Booker, N.A., Cooney, E.L., Priestley A.J.: Innovative physico-chemical wastewater treatment research down-under. In Chemical Water and Wastewater Treatment IV, H.H. Hahn, E. Hoffmann and H. Ødegaard (Eds), Springer-Verlag, Berlin Heidelberg New York, (1996) pp. 277-286
3. Guibelin, E., Delsalle, F., Binot P.: The Actiflo® process: A highly compact and efficient process to prevent water pollution by stormwater flows. Wat. Sci. Tech. 30 (1994) 87

4. Zeghal, S., Boissonnade, G., Sztajnbok, P.: Innovative wastewater treatment in Europe, 18th Federal Convention, Australian Water and Wastewater Association, Adelaide, 11-14th April, (1999)
5. Gray, S.R., Booker N.A. Arld, R.: Effect of floc characteristics on high rate filtration of sewage. In Chemical Water and Wastewater Treatment V, H.H. Hahn, E. Hoffmann and H. Ødegaard (Eds), Springer-Verlag, Berlin Heidelberg New York, (1998) pp. 205-217
6. Tanaka, Y., Miyajima, K., Funakosi, T., Chida, S.: Filtration of municipal sewage by ring shaped floating plastic net media. Water Research 29 (1995) 1387
7. Tambo, N., Hozumi, H.: Physical characteristics of flocs-II. Strength of floc. Water Research 13 (1979) 421
8. Yeung, A.K.C., Pelton, R.: Micromechanics: A new approach to studying the strength and breakup of flocs. J. Coll. Interface. Sci. 184 (1996) 579
9. Yeung, A., Gibbs, A., Pelton, R.: Effect of shear on the strength of polymer induced flocs, J. Coll. Interface. Sci. 196 (1997) 113
10. Mühle, K., Domasch, K.: Stability of particle aggregates in flocculation with polymers. Chem. Eng. Process. 29 (1991) 1
11. Moudgil, B.M., Springgate, M.E., Vasudevan, T.V.: Characterisation of flocs for solid/liquid separation processes, Solid/liquid separation: waste management and productivity enhancement, Maralidhaus, H.S. (Ed) International Symposium, Battelle Press, Columbus, Richmond (1990) pp 245-253
12. Gehr, R., Henry, J.G.: The adsorption behaviour of cationic polyelectrolytes in dissolved air flotation. Wat. Sci Tech. 14 (1982) 689
13. Murgatroyd, C., Barry, M., Bailey, K., Whitehouse, P.: A review of polyelectrolytes to identify priorities for EQS development. WRc Technical Report P21, WRc plc, Buckinghamshire, UK (1996)

Direct Wastewater Membrane Filtration for Advanced Particle Removal from Raw Wastewater

A.F. van Nieuwenhuijzen, H. Evenblij and J.H.J.M. van der Graaf

Delft University of Technology, Department of Sanitary Engineering - PO Box 5048, NL2600 GA Delft , TheNetherlands
a.vannieuwenhuyzen@citg.tudelft.nl

Abstract

Direct filtration techniques were applied for fast and advanced particle removal in the pre-treatment of municipal wastewater, in order to optimise the total wastewater treatment system. Pilot-scale research into direct membrane ultrafiltration of raw wastewater indicated that this technique might be feasible. There was an average normalised production flux of 300 $l/m^2 \cdot h \cdot bar$ at a TMP of 0.2 to 0.4 bar and a cross flow rate of 2.4 m/s. Removal performances were excellent with all suspended and colloidal substances being removed by the membranes with pore size of 30 nm. Furthermore the influence of the cross flow rate, the applied TMP and the cleaning procedure on the filtration performance was investigated. It was concluded that direct membrane filtration shows potential for wastewater treatment.

Introduction

Physical-chemical pre-treatment is based on the separation of mainly suspended and colloidal particles from wastewater in a first process step. Since a major part of the influent COD consists of particulate material [1], advanced particle removal in the physical-chemical pre-treatment results in a lower load on the subsequent treatment steps. This makes it possible to design the total treatment plant to be more energy efficient and more compact [2].

With this knowledge, a research program is being conducted in The Netherlands with the objective to develop new and more sustainable wastewater treatment scenarios based on physical-chemical pre-treatment [3]. The research is initiated and funded by STOWA, the Dutch Foundation for Applied Research on Water Management, and is carried out by the Delft University of Technology and the Wageningen University. Based on this evaluation study [4], different kinds of physical-chemical pre-treatment techniques, including direct raw wastewater filtration with rough media filters and membranes, were recommended for further experimental research.

Literature on filtration techniques for pre-treatment of wastewater is rather scarce and application in practice is hardly done. Only a small number of research activities into direct influent filtration, mainly with rough media up-flow floating filters, have been reported, e.g. [5] and [6]. Besides a South Korean small-scale paper on hotel wastewater [7] and trial set-ups in Sydney (WWTP Malabar and Cronulla) [8], investigations into direct membrane filtration of raw municipal wastewater have hardly been described so far.

Nevertheless, direct filtration of wastewater may have specific advantages over the commonly used settling process [9]. Firstly, it is possible to achieve an extensive particle removal even without the use of chemicals. Secondly, filtration techniques can be applied in a very compact way, which may be an important aspect in highly populated areas like The Netherlands.

After investigations into direct rough media filtration [10], a further step was taken with the application of membranes. The aim of this experimental research was to explore the characteristics of direct membrane filtration as a pre-treatment technique for advanced particle removal. Removal performances and operational process conditions were investigated for different filter configurations.

Experimental Set-up

Material

The investigations were carried out with a membrane unit (Figure 1), consisting of two, 1 inch wide and 1 metre long, membrane modules, each containing 12 membrane tubes with a diameter of 5.2 mm.

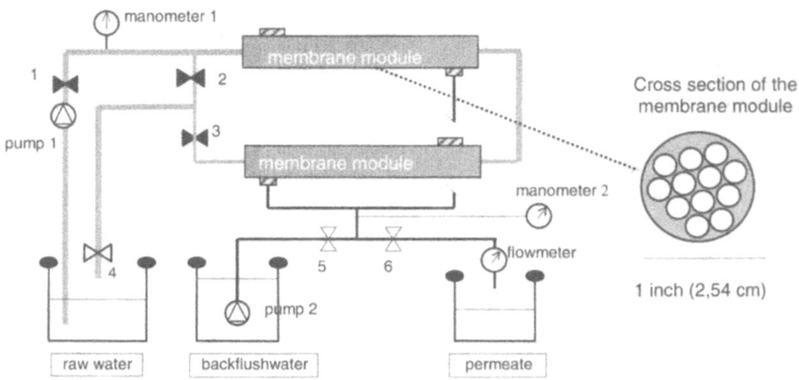

Fig. 1. Schematic diagram of the membrane test unit

The total effective membrane area was 0.17 m^2 per module; the membrane pore size was 30 nm. The applied membranes were of the type STORK FX4385, a hydrophilic polyvinylidenefluoride membrane on a supporting layer of polyester/polyolefine fabric. According to the technical specifications, the initial flux (distilled water at 25 °C) at 1 bar should be higher than 1,000 $l/m^2{\cdot}h$.

The two modules were coupled in series as shown in Figure 1. The cross flow velocity and the transmembrane pressure (TMP) were induced by pump 1. The TMP could be adjusted by operating valves 1 and 3. Pump 2 was used for cleaning the membranes by backflushing at a pressure of 0.8 bar.

The membrane unit was operated in the cross flow mode at different flow rates. Every ten minutes the production was interrupted for a backflush, using filtered tap water. The backflush time varied from 30 s to 120 s. The experiments were performed at a TMP ranging from 0.2 to 0.5 bar.

Feed water characteristics

As feed water for the membrane unit, wastewater was taken from the inlet of the municipal wastewater treatment plant Bennekom (18,000 p.e.), in the Netherlands. The temperature of the wastewater in the 80 litre feed water tank was kept constant. During the testing period the feed water was analysed; the characteristics are summarised in Table 1.

Table 1. Characteristics of the feed water

Parameter	Average concentration
Suspended solids	130 mg/l
Turbidity	130 NTU
COD	680 mg O_2/l
BOD_5	210 mg O_2/l
P_{total}	11 mg P/l
$N_{Kjeldahl}$	72 mg N/l
Temperature	21 °C

Testing procedure

During the experiments, flux, TMP and temperature were measured and samples were taken off the feed and permeate for analysis. All fluxes and TMP were recalculated to a standard temperature of 15 °C.

Each filtration run started with measuring the permeability (normalised flux in $l/m^2{\cdot}h{\cdot}bar$) of the membranes by using clean tap water. The permeability characterises the condition of the membrane.

After this first permeability test, the filtration of raw wastewater began. For several hours the wastewater was fed through the membrane unit, producing permeate. The membrane filtration unit was operated at a specific cross flow velocity and a constant TMP and the changes of the flux during the filtration run time were measured. At certain intervals the feed and permeate were sampled and analysed with respect to COD, N, P, turbidity and suspended solids.

Immediately after an experiment with raw wastewater, the permeability of the membrane was measured again using tap water, in order to determine the condition of the membranes.

During a chemical cleaning the membranes were exposed to a chlorine solution of 350 mg/l, this being 75% of the maximum concentration allowed for this specific membrane.

Results and discussion

General overview

The first result from this project was that membrane filtration of raw wastewater is possible. Over the three month testing period, the membranes did not get damaged and produced a clear, constant quality permeate. A steady-state configuration was approached for the applied membrane unit at a TMP of 0.2 to 0.4 bar with a production time of 10 minutes and a 1-minute backflush. With this configuration the average normalised flux over the three month testing period was calculated to be 300 $l/m^2 \cdot h \cdot bar$, decreasing from about 600 $l/m^2 \cdot h \cdot bar$ at the start of the tests to 180 $l/m^2 \cdot h \cdot bar$ three months later.

With direct membrane ultrafiltration all suspended and colloidal material was removed and only dissolved pollutants stayed in the permeate (see Table 2).

Table 2. Average removal performances during testing period

Analysed parameter	Feed	Permeate	Removal
Turbidity (NTU)	140	0.2	99.9%
Suspended solids (mg SS/l)	130	n.d.	99.9%
COD_{total} (mg O_2/l)	680	210	69%
$N_{Kjeldahl}$ (mg N/l)	72	60	17%
P_{total} (mg P/l)	11.0	7.1	36%

TMP = 0.2 bar, production = 10 min., backflush = 1 min., n.d. = not detectable

Cross flow rate versus flux

In the process of cross flow membrane filtration, the cross flow velocity is one of the most important parameters. The cross flow causes a turbulent flow in the membrane tube and a shear stress [11] on the membrane surface to hinder or remove the cake layer built up by particles and flocs. This (partially) prevents or limits fouling of the membrane.

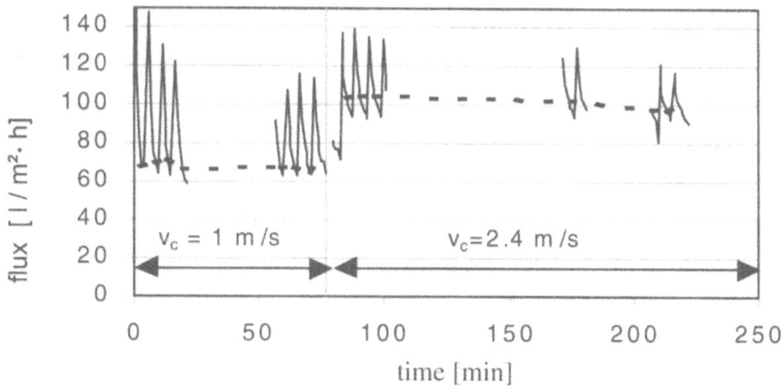

Fig. 2. Effect of increasing cross flow velocity on the flux at a TMP of 0.5 bar

Figure 2 shows the effect of an increase of the cross flow velocity on the permeate flux. Raising the cross flow velocity (v_c) from 1 m/s to 2.5 m/s increases the average flux from 70 $l/m^2{\cdot}h$ to 110 $l/m^2{\cdot}h$.

The increase in flux is probably caused by a more intensive removal of fouling or cake layer from the membrane. Because the flow conditions in the membrane tube of 5.2 mm diameter were already turbulent at a cross flow rate of 1 m/s ($Re \geq 4{,}000$ at 0.88 m/s), the effect was not caused by the turbulence criterion. The decisive parameter is supposedly the shear stress on the membrane surface, which becomes efficient above 20 Pa [11]. In the case of the applied membrane tubes, the minimum shear stress was reached at a cross flow rate of 2.4 m/s (22 Pa at 2.5 m/s). The relationship between critical flux and cross flow was not investigated, but is assumed to play a role in the optimisation of this membrane filtration application.

Production time versus backflush time

In the experimental set-up, the production of permeate was interrupted every ten minutes for a backflush, in order to clean the membranes. To investigate the influence of backflush cleaning on the membrane filtration process, the flux was recorded just before the start of each backflush and immediately after the backflush. In this way a curve of start-fluxes (Js) and a curve of end-fluxes (Je) was obtained. In Figure 3 these curves are plotted as a function of the number of production runs.

In the first three sections of the graph, the backflush time (bf) increased from 30 to 60 to 120 seconds at a standard TMP of 0.2 bar and a constant production time of 10 minutes. By doubling bf from 30 to 60 seconds, the start-fluxes increased as well. The end-fluxes hardly changed, resulting in higher permeate production per filtration run. A further increase of bf up to 120 seconds did not improve the value of the permeate fluxes.

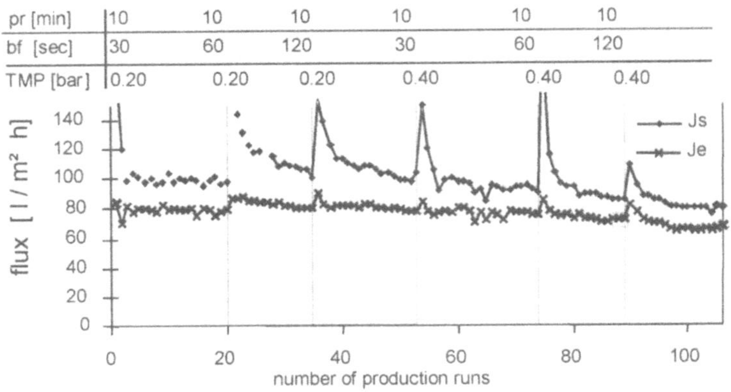

Fig. 3. Effect of backflush (bf), production time (pr) and TMP on the start-flux (Js) and end-flux (Je)

By raising the TMP from 0.2 to 0.4 bar (see difference between section 3 and 4), the start- and end-fluxes hardly changed. When these fluxes were converted into pressure normalised fluxes, it could be seen that by doubling the TMP, the permeability or so-called permeance [12] of the membrane was halved. The increase in backflush time did not influence the production fluxes at a TMP of 0.4 bar, which may indicate the formation of a more resistant (compact) fouling or cake layer on the membrane due to the higher pressure.

With the applied backflush-TMP of 0.8 bar, a 1-minute backflush was enough to reach the maximum effect, preferably at a TMP of 0.2 bar.

Figure 3 also shows that the end-fluxes are almost constant for all experiments. At the beginning of an experiment, the start-flux decreased rapidly within a couple of production runs to an almost constant level. After these first few production runs clogging of the membranes apparently decreased and the permeate flux reached a constant value. This indicates that independently of the process parameters, the flux after ten minutes reaches a specific value. Fouling of the membranes manifested itself as a decrease in start-fluxes, but less as a decrease in end-flux.

Flux deterioration without backflush

As described in the last section, the flux decrease occurred over the first minutes of the test run, and after that, the flux approached a constant value. Thus, at a certain flux value, in combination with the cross flow velocity and the TMP, permeate fluxes did not decrease further over time. This assumption was tested in a production run that lasted over 1.5 hours, without being interrupted by a backflush. The resulting flux pattern is presented in Figure 4.

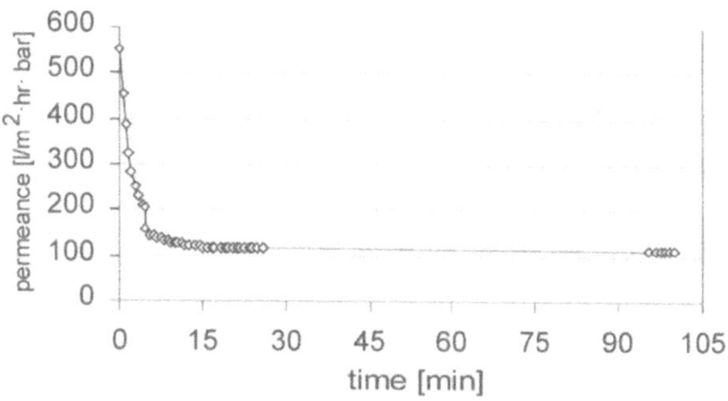

Fig. 4. Flux pattern during production run without backflushing

As in the normal test runs, the flux at the beginning of the permeate production decreased rapidly from 550 litre per m^2 per hour per bar to less than 200 $l/m^2 \cdot h \cdot bar$, but reached a constant permeability of 120 $l/m^2 \cdot h \cdot bar$ after 15 minutes. The flux stayed constant for more than one hour, without backflushing. Presumably, an equilibrium has been reached between cake layer formation and removal.

Fouling

During the filtering of wastewater, the membranes tend to foul. The easily removable fouling or cake layer was flushed away by the regular backflush, so formation and removal are in balance. In between the tests, the membranes were cleaned chemically. It was expected that backflushing and chemical cleaning would remove the total reversible fouling. However, after chemical cleaning, a certain amount of fouling still remained on the membrane: an irreversible type of fouling. This fouling could be quantified by determining the permeability of the membrane over the total experimental period. Figure 5 shows the permeability curve during the tests.

Before any experiments with wastewater were done, the membrane was only fed with clean tap water to test the membrane unit. During this period, the permeability of the membrane decreased considerably, from 1,100 $l/m^2 \cdot h \cdot bar$ to 650 $l/m^2 \cdot h \cdot bar$. This can be attributed to adsorption of ions to the membrane surface. It seems likely that this process went on during the whole series of experiments with wastewater, thus causing a more or less constant decrease in permeability from 650 $l/m^2 \cdot h \cdot bar$ to less than 200 $l/m^2 \cdot h \cdot bar$. This adsorption process caused a fouling that could not be removed effectively by the normal chemical cleaning procedure. Earlier research found that a high cross flow velocity resulted in a stagnation of the fouling [11]. This assumption could not be verified in this research, because the cross flow velocity could not be raised above 2.5 m/s. Others [13,14] found that the decrease in flux stopped when 20-30 % of the initial flux of the membrane was reached. These results agree with those of our own experiments, as does the result that with a TMP of 0.2 bar a constant flux could be reached.

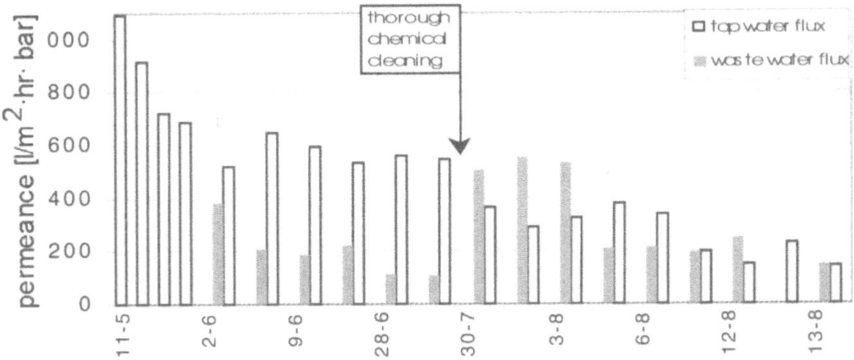

Fig. 5. Permeability of the membranes during the testing period

The result of filtering without backflush showed that under certain conditions an equilibrium could be reached between fouling and cleaning by the cross flow. Bouhabila *et al.* [13] states that the adsorption process could even be eliminated and fluxes could become constant over time. This phenomenon is assumed to be related to a (secondary) critical flux, as suggested by Field *et al.* [15].

Removal performances and wastewater characterisation

As presented in Table 2 turbidity and suspended solids were totally removed by the applied membranes due to a complete particle (> 30 nm) removal. The particle-related fraction of the individual contaminant can be derived from the removal performances of the membrane unit for COD, nitrogen and phosphorus. For the Bennekom wastewater 69% of the incoming COD is related to particulate material. Nitrogen is, as expected, present in a more soluble form, with only 17% of Kjeldahl-nitrogen related to particles. 36% of the total phosphorus was bound on or incorporated into particles.

By characterising wastewater according to particle size, it might be possible to predict the possibilities and performances of the physical and chemical treatment techniques applied in wastewater treatment. In the present investigation, the COD removal performance indicated that only about one-third of the total incoming COD still has to removed by biological treatment or other treatment techniques that can deal with dissolved organic components.

A major topic of discussion on the possible application of membrane filtration as advanced pre-treatment is the post-treatment of the produced effluent. As described above, the primary effluent consists of only soluble contaminants requiring specific post-treatment techniques. Biofilm systems and membrane bioreactors are more suitable than suspended activated sludge systems for this kind of water, especially with regard to suspended solids content and biological sludge production. One of the major problems in the post-treatment, due to a lack of carbon, seems to be nitrogen removal by nitrification-denitrification. The COD/N-ratio of the produced permeate was, for example about 3.5, indicating a rather difficult denitrification. These kinds of difficulties could also be predicted by characterising wastewater according to particle size.

Conclusions and recommendations

Taking into account the technical limitations of the tested membrane unit, the following conclusions can be drawn:

Pre-treatment of raw wastewater with membranes is possible and has remarkable potential. From the experimental research it can be said that the removal performances of direct membrane filtration are excellent; all colloidal and suspended material, 69% of COD, 36% of phosphorus and 17% of nitrogen were removed. The produced permeate is of a constant quality; only dissolved pollutants remained in the permeate. An average normalised flux of approximately 300 $l/m^2 \cdot h \cdot bar$ is possible. Practical values for TMP are between 0.2 and 0.4 bar, since higher values would cause a compact, more resistant cake layer on the membrane surface.

Without backflushing, it is possible to keep the flux at a constant level, which suggests only limited fouling of the membranes. The cross flow velocity plays an important role in this process, mainly because of the shear stress it causes along the membrane surface. There seems to be an optimal cross flow velocity, at a certain TMP, above which no flux-decreasing fouling occurs. This could be related to the (secondary) critical flux.

Future investigations should address scaling up of the facilities. A continuous pilot plant should be set up, in which parameters such as flux, TMP, cross flow rate, turbidity and COD are automatically monitored. The proposed existence of a (secondary) critical flux should also be verified.

With respect to the membranes, research should be done to investigate the efficiency of cleaning agents and procedures, and the necessity of cleaning at all.

Another field of research is the further treatment of the permeate and the retentate produced by the membrane process. It seems likely that for some applications, the permeate can be used in a regular treatment step. It should be kept in mind, that the permeate is free of particles and that it could easily get polluted again with particulate matter in a subsequent treatment step. Concerning the retentate, it can be said that it seems appropriate to concentrate it in further membrane processes, to achieve high dry solids concentrations. Thus, one may think of a cascade of membrane steps, in which the retentate is thickened with every step, and each step contributes a certain amount of particle-free permeate.

References

1. Levine, A.D., Tchobanoglous, G., Asano, T.: Size Distribution of Particulate Contaminants in Wastewater and their Impact on Treatability. Water Research 25 (8) (1991) 911-922
2. Van Nieuwenhuijzen, A.F. van, Mels, A.R., Graaf, J.H.J.M. van der, Klapwijk, B., Koning, J. de, Rulkens, W.H.: Identification and Evaluation of Wastewater Scenarios, Based on Physical-Chemical Pretreatment. In: Chemical Water and Wastewater Treatment V (H.H. Hahn, E. Hoffmann, H. Ødegaard, eds.), Springer Verlag, Berlin 1998, 351-362
3. Mels, A.R., Van Nieuwenhuijzen, A.F., Van der Graaf, J.H.J.M., De Koning, J., Klapwijk, A., Rulkens, W.H.: Sustainability criteria as a tool in the development of new sewage treatment methods. Water Science & Technology 39 (5) (1999) 243-250
4. Van Nieuwenhuijzen, A.F.: Physical-Chemical Pretreatment of Wastewater in The Netherlands. Technology & Environment UTA 8 (2) (1999) 73-79
5. Booker, N.A., Cooney E.L. ,Priestley, A.J.: Innovative Physico-Chemical Wastewater Treatment Research Down-Under. In: Chemical Water and Wastewater Treatment IV (H.H. Hahn and E. Hoffmann, eds.), Springer Verlag, Berlin 1996, 277-286
6. Ødegaard, H., Ulgenese, Y., Brevik, D. ,Liao, Z.: Enhanced Primary Treatment in Floating Filters. In: Chemical Water and Wastewater Treatment V (H.H. Hahn, E. Hoffmann, H. Ødegaard, eds.), Springer Verlag, Berlin 1998, 189-204
7. Inside information by Dr. N.A. Booker, CSIRO Molecular Science, Australia
8. Kyu Hong Ahn, Ji-Hyeon Song, Ho Young Cha, Kyung Guen Song, Hyungseok Yoo: Application of Tubular Ceramic Membrane for Building Wastewater Reuse. In: Proceedings IAWQ Biennial Conference, Vancouver, 1998, 136-143
9. Van Nieuwenhuijzen, A.F., Van der Graaf, J.H.J.M.: Application of direct filtration techniques for raw wastewater treatment. In: Proceedings World Filtration Congress 8, Brighton, UK, April 2000, 815-819
10. Van Nieuwenhuijzen, A.F., Van der Graaf, J.H.J.M., Mels, A.R.: Direct influent filtration as pretreatment step for more sustainable wastewater treatment systems. To be presented at the 1st IWA World Congress, Paris, France, 3-7 July 2000
11. Elmaleh, S., Abdelmoumni, L.: Experimental test to evaluate performances of an anaerobic reactor provided with an external membrane unit. Water Science & Technology 38 (4-5) (1998) 385-392
12. Koros, W.J., Ma, Y. Shimidzu, T.: Terminology for membranes and membrane processes - IUPAC Recommendations 1996. Journal of Membrane Science 120 (1996
13. Bouhabila, E.H., Ben Aïm, R., Buisson, H.: Microfiltration of activated sludge using submerged membrane with air bubbling (application to wastewater treatment). Desalination 118 (1998) 315-322
14. Megat Johari, M. M. N., Jusoh, A., Ghazali, A. H. Ahmadun, F. R.: Microfiltration of oxidation pond effluent using single flexible tubular fabric membrane and polyelectrolyte dosage. Water Science & Technology 34 (9) (1996) 181-187
15. Field, R.W., Wu, D., Howell, J.A., Gupta, B.B.: Critical flux concept for microfiltration fouling. Journal of Membrane Science 100 (1995) 259-272

High Rate Biological / Chemical Treatment Based on the Moving Bed Biofilm Process Combined with Coagulation

H. Ødegaard*, B. Gisvold**, H. Helness**, F. Sjøvold** and L. Zuliang*

*Norwegian University of Science and Technology, N-7491 Trondheim, NTNU, Norway
hallvard.odegaard@bygg.ntnu.no
**SINTEF Civil and Environmental Engineering, N-7465 Trondheim, Norway

Abstract

A high-rate wastewater treatment process for secondary treatment has been investigated. This process consists of a highly loaded moving bed biofilm reactor directly followed by a coagulation/floc separation step. It is demonstrated that the biofilm in such a process mainly deals with the soluble organic matter while coagulation deals with the particulate/colloidal matter. The bioreactor may, therefore, be designed based on the soluble COD loading only, resulting in a very compact plant, especially when a compact separation reactor (i.e. filtration) is used as well.

Introduction

Compact wastewater treatment processes are being looked for by cities all over the world as land available for treatment plants is becoming scarce. In many of the cities secondary treatment standards have to be met. Direct coagulation/-flocculation/floc separation, results not only in a very substantial removal of suspended and colloidal matter, but good removal of organic matter, bacteria and viruses, and micropollutants as well [1,2,3]. These compounds are either associated with particles and colloids or soluble, high molecular weight organic substances

In order to meet secondary treatment standards, however, it may be necessary to remove low molecular weight, soluble organic matter as well. For this purpose biological processes are most suitable from an economic point of view. Biofilm reactors are especially suitable if the goal is to make compact treatment plants. Many of the most compact biofilm reactors (like the granular media biofilters), however, cannot accept a high load of particulate matter, since this would easily clog the filter and result in too frequent filter washing. It is, therefore, normal to use a two-step process consisting of a pre-coagulation step followed by a biofilter step. By utilising lamella-separators for floc separation in the primary step, such treatment plants can be made very compact [4]. Nevertheless one is reluctant to use a very high organic load because of the fear of clogging of the biofilters. An alternative will be discussed in this paper, namely to combine the new moving bed biofilm process with direct coagulation. This process can accept both a high particulate load as well as a high soluble organic load.

The moving bed biofilm reactor (MBBR)

In the moving bed biofilm reactor (Figure 1) the biomass grows on carriers that move freely in the water volume of the reactor, kept within the reactor volume by a sieve arrangement at the reactor outlet. In aerobic processes, the biofilm carrier movement is caused by the agitation set up by the air, while in anoxic and anaerobic processes a mixer keeps the carriers moving (Figure 1a and b). The biofilm carrier (K1) is made of high density polyethylene (density 0,95 g/cm^3) and has the shape of a small cylinder with a cross on the inside of the cylinder and "fins" on the outside (Figure 1c). The cylinder has a length of 7 mm, and a diameter of 10 mm (not including fins). Lately a larger carrier (K2) of similar shape (length and diameter about 15 mm) has been introduced as well, intended for use in plants with coarse inlet sieves and especially for upgrading of activated sludge plants.

One of the important advantages of the moving bed biofilm reactor is that the carrier filling-fraction (% of reactor volume occupied with carriers in empty tank) may be subject to preferences. The standard filling fraction is 67 %, resulting in a total, specific carrier area of 465 m^2/m^3 with the K1 carrier. Since the biomass grows primarily on the inside of the carrier, the effective specific surface area is 335 m^2/m^3 for the K1 carrier and 210 m^2/m^3 for the larger K2 carrier at 67 % filling fraction. It is recommended that filling fractions should be below 70 %. One may, however, use as much as needed below this, which is convenient, especially when upgrading plants – for instance from activated sludge to moving bed reactors.

The moving bed biofilm process has been used for many different applications [5,6]. In this paper we shall only discuss the high-rate moving bed biofilm process. It has been shown that the carriers K1 and K2 performs equally well when comparisons are made based on the effective biofilm surface area [8].

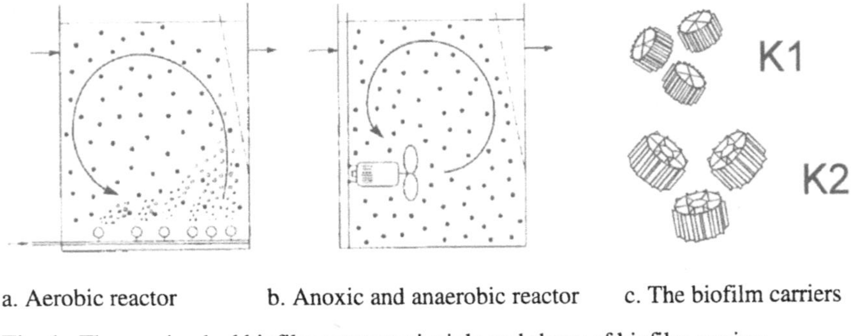

a. Aerobic reactor b. Anoxic and anaerobic reactor c. The biofilm carriers

Fig. 1. The moving bed biofilm reactor principle and shape of biofilm carriers

The high-rate moving bed process was also investigated earlier [7], but at that time, the focus was on an alternative process scheme based on the use of aluminium-nitrate as a coagulant and anoxic biodegradation.

The high-rate Kaldnes moving bed biofilm process

The fate of particles in biofilm reactors is not totally clear, but it is obvious that particulate matter is being degraded to a far lesser extent in a high-rate biofilm process than in a standard activated sludge process. When operating at such a high organic loading that the maximum COD degradation rate prevails, the COD-removal will primarily be due to consumption of soluble organic matter. The idea behind the high-rate Kaldnes moving bed biofilm process (Figure 2), is that the biofilm is supposed to deal with the soluble organic matter, while the coagulant is supposed to deal with the separation of the particulate matter, including colloids. If the new larger carrier (K2) is used, pre-treatment can consist of pre-sieving (3-4 mm sieve) only, which then results in an extremely compact plant, as shown in Figure 2, especially when a compact separation method (i.e. filtration) is employed.

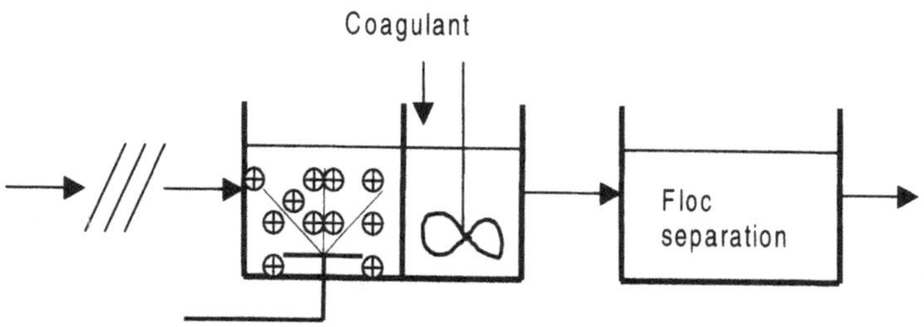

Fig. 2. Schematic flow diagram for the high-rate process discussed

When used for BOD-removal only, the process has traditionally been designed for a volumetric loading rate of 4-5 kg BOD_7/m^3d at 67 % carrier filling fraction (effective specific surface area: 335 m^2/m^3) and 15°C. This corresponds to ca 15 g BOD_7/m^2d, which is somewhat higher than for other biofilm processes (like RBC) for the same purpose. In the experiments reported here, the performance of the process was investigated at much higher organic loadings.

Biodegradation of soluble organic matter

The biodegradation experiments were primarily carried out in order to investigate the influence of carrier size and shape in the moving bed process. These results were published elsewhere [8]. Here we shall concentrate on the influence of organic matter loading on biodegradation and only report the results with the carriers K1 and K2. The experiments were conducted in pilot plants, each having

Table 1. Average, maximum and minimum influent values for the pilot plant

	Period 1			Period 2			Period 3		
	Ave. ± st.dev.	Max	Min	Ave. ± st.dev.	Max	Min	Ave. ± st.dev.	Max	Min
SS	136 ± 98	505	53	152 ± 55	232	58	88 ± 18	136	53
COD	323 ± 166	893	139	498 ± 235	915	125	219 ± 66	435	119
SCOD	123 ± 39	236	69	219 ± 128	431	36	100 ± 38	211	42
pH	7.3 ± 0.2	8.0	7.0	6.7 ± 0.3	7.1	6.5	7.5 ± 0.1	7.8	7.4

two parallel lines, consisting of one moving bed reactor and one settling tank. The volumes of the bioreactors were 20 l. The moving bed reactors were operated at organic loads in the range of 10-120 g COD/m^2d and 5-45 g SCOD/m^2d. The experiments were carried out in three different periods. In the first one both reactors were given the same volumetric load at a filling fraction of 60 %, while in the second period the filling fraction was varied to give the same effective area load at constant flow. The third period was devoted to a comparison between the two Kaldnes carriers (K1 and K2) at 70 % filling fraction. The two lines were operated in four sub-periods at close to constant flow in each period (e.g. the same residence time) and hence the same volumetric loading rate. The flow of the four periods corresponded to average residence times of 380, 52, 27 and 18 min. In Table 2 the wastewater characteristics for the various experimental periods are given.

The raw water temperature was in the range of 10 - 15 °C and the oxygen concentration in the range of 4 - 6 mg O_2/l. Within this range, variations in O_2-concentration are not expected to have any influence on the rate of COD-removal.

Results from bidegradation experiments

In order to evaluate degradation of organic matter independent of the biomass separation step, one may look at the removal rate of soluble/filtered COD (SCOD) versus the loading rate. Figure 3 shows that the maximum removal rate in this wastewater was found to be around 30 g SCOD/m^2d. This maximum rate was reached at a loading of around 60 g SCOD/m^2d. A line through data points up to this loading is close to linear, indicating that the degradation rate was limited by the availability of biodegradable organic matter at lower organic loads.

The difference between this line and the 100 % removal line represents the soluble COD that could not be biodegraded in this water within the actual residence time. The reason for the somewhat lower removal in the second part of the experiment (Figure 3b and 4b) stems mainly from the fact that the water was more dilute, with a greater portion of the total COD that was not biodegradable. Most of the results are also obtained at very low residence times.

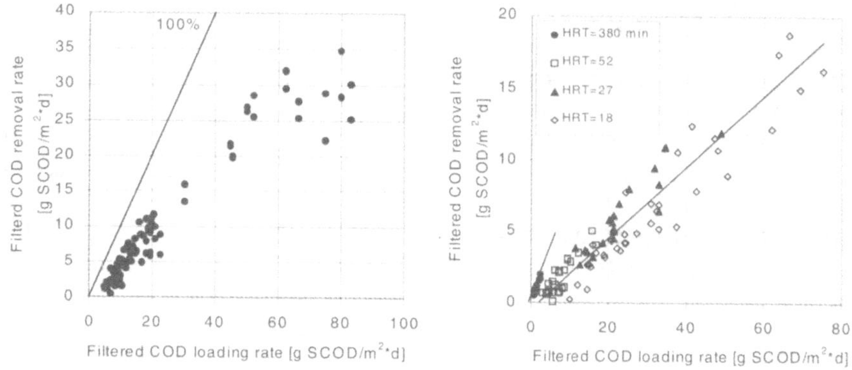

a. Period 1 and 2 b. Period 3

Fig. 3. Soluble COD removal rate versus soluble COD loading rate

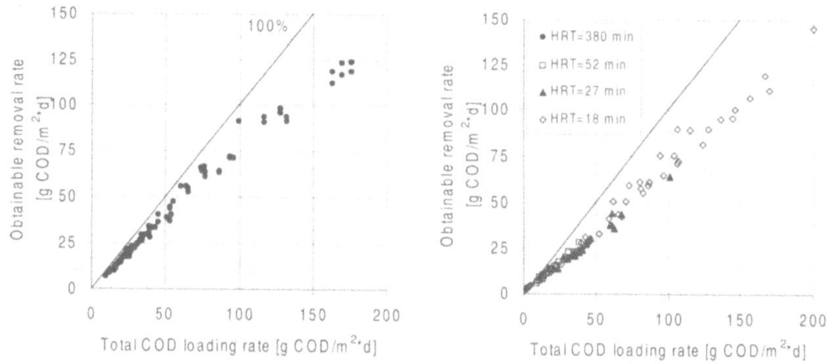

a. Period 1 and 2 b. Period 3

Fig. 4. "Obtainable" removal rate versus total COD loading rate

The results of period 3 show, however, that there is not much difference between the results at 18 or 27 min residence time and those at 52 min. This indicates that the removal of the biodegradable COD is rapid and that hydrolysis does not play an important role during these short residence times. It is interesting to note from Figure 3b, however, that at a very low load and long residence time (380 min - indicated by the line), the slope of the removal/loading rate relationship is significantly higher than that at the lower residence times (52, 27 and 18 min). This indicates that hydrolysis takes place at this long residence time.

Up to now we have concentrated on the removal of soluble (filtered) COD. It is not easy to analyse the total COD removal rates in the bioreactor, since the characteristics of both the soluble and the particulate organic matter change in the reactor due to hydrolysis, assimilation etc. In order to be able to take the particulate matter into account, we have analysed what one may call the "obtainable" COD removal rate defined as:

$(COD_{influent}-SCOD_{effluent}) \cdot Q/A$

where Q is the flow and A is the effective surface area of the carrier. This term illustrates the removal rate of organic matter if all particles larger than 1 µm were removed in a downstream separation step. Figure 4 shows that 85 - 90 % removal of COD could have been obtained all the way up to a loading rate of 100 g COD/m^2d, if the biomass downstream the bioreactor had been completely removed.

The results demonstrate that a much higher design load than normally used for secondary treatment may be accepted if efficient biomass separation is assured. In highly loaded plants clarification of the biomass may, however, represent a problem. Another pilot study was carried out to investigate this.

Separation of the biofilm by settling

Settleability experiments have been carried out in both jar tests and on pilot scale on effluents from high-rate moving bed reactors. In the jar-test experiments an apparatus based on continuous in-line mixing of the coagulants and pipe flocculation was used [9]. The pipe flocculators were operated at decreasing G-values (385 sec^{-1} for 21 sec and 135 sec^{-1} for 47 sec) and the flocculated suspension was introduced to the 2 l jars from the bottom at a flow of 1 l/min. All experiments were performed with 15 min of slow mixing (25 rpm) and 15 min settling except for the ferric chloride tests where 60 min of settling was used.

The pilot plant experiments were carried out in the same pilot plants as the biodegradation experiments (see above). When a coagulant was used, however, two flocculation chambers (with decreasing paddle speed and 40 min residence time) were introduced between the bioreactor and the settling tank. The settling tanks had a diameter of 0.38 m and a settling depth of about 1 m. Hydraulic bioreactor retention times in the range of 18 to 380 min were used, resulting in overflow rates in the range of 0.05 to 1 m/h in the settling tanks.

Results of settleability experiments

Figure 5 shows the results from the jar-test experiments on the effect of coagulation on separation of biofilm from a high rate moving bed reactor.

It is demonstrated in Figure 5a that even a relatively small amount of metal (about 0,2 mmol), dramatically improved settleability over no coagulant addition. Figure 5b demonstrates that reasonably good SS-removal(85 %) could also be obtained using of a relatively low dosage (2 mg/l) of a cross-linked, medium charged, cationic polymer.

Figure 6 and 7 show the settleability results from the pilot plant tests. Figure 6 shows the influence of organic loading on the settleability of the biomass when no coagulant was added. The SS-removal efficiency is given as a function of the total COD loading rate on the bioreactor at different overflow rates on the settling tank. One should be careful in interpreting the actual removal percentages, since these settling tanks were very small, but the overall picture with respect to influence of loading rate can be considered to be correct.

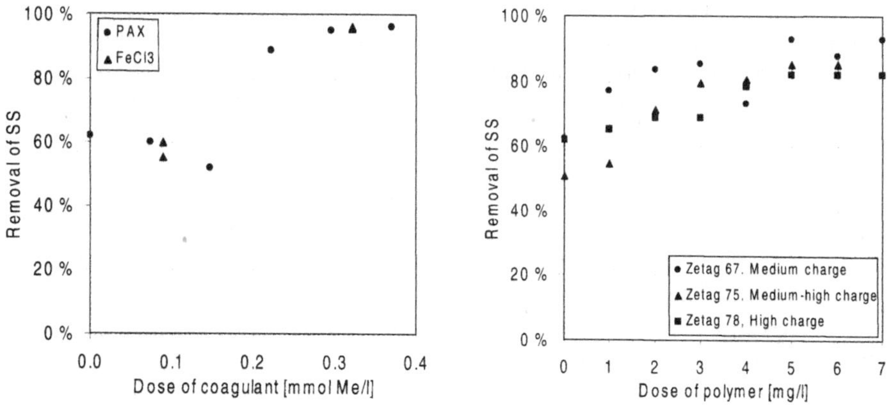

a: Addition of inorganic metal salts (PAX - prepolymerised $AlCl_3$)

b: Addition of cross-linked (25 %), high MW cationic polymer

Fig. 5. Removal of suspended solids in jar tests of a high-rate MBBR effluent

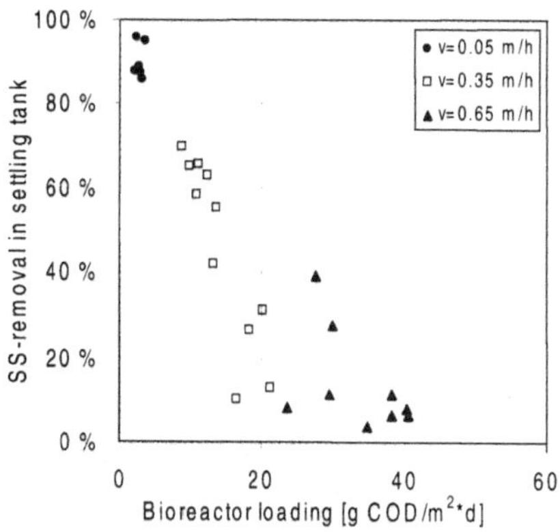

Fig. 6. Influence of organic loading rate in bioreactor on settleability

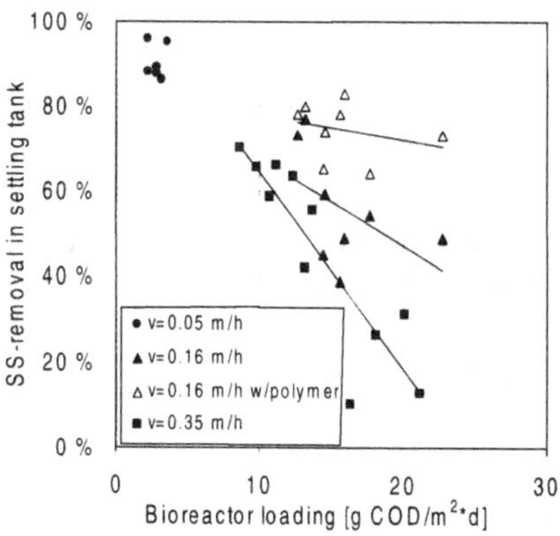

Fig. 7. Influence of polymer addition on settleability

The overflow rates correspond to three levels of flow and consequently different levels of organic loading on the bioreactor. This makes the analysis a little complicated since both loading rate and surface overflow rate were varied at the same time.

It is obvious, however, that not only the overflow rate but also the organic loading on the bioreactor has a pronounced effect on settleability. At a given overflow rate (for instance 0,35 m/h in Figure 6), there is a decrease in settleability with increasing organic load. This was the case both when total COD and soluble COD was considered, indicating that the organic loading regime experienced by the microorganisms, does influence the settleability of the biofilm that is sloughed off the carriers. The consequence of this, from a practical point of view, is that settling ought to be enhanced by coagulation when this flow scheme is operated at high organic loading rates. In Figure 7, results from experiments where two pilot plants have been run in parallel are compared in order to study the effect of adding a cationic polymer coagulant at a given overflow rate. 1,5-2 mg/l of the medium charged, high MW polymer was added. The regression lines in Figure 7 are drawn for illustration purposes only.

Even though there is considerable scatter in the data, the results for a given overflow rate indicate that a) settleability is better with polymer addition and b) settleability is less influenced by the organic loading on the bioreactor when a polymer is added. The probable reason for this is that the polymer can flocculate the smallest particles that are more abundant as organic loading increases. The poorer settleability at higher organic loading is therefore compensated by polymer coagulation.

Separation of the biofilm by direct filtration

Since the amount of sludge to be separated is quite low in the actual process scheme, direct filtration may be an alternative to flocculation/settling. If phosphate removal is required, a metal coagulant (Al or Fe) will be necessary, but if the goal is to remove only SS and BOD, a cationic polymer alone, or in combination with a low dose of metal salt, may be used. This would minimise sludge production and might make the use of direct filtration possible.

A filter for such an application should, of course, be built with a high sludge retaining capacity in order to achieve acceptable filter run times. In this project we used an up-flow filter with expanded clay aggregates (Filtralite) as filter medium with a wide range of grain sizes (1.5-4 mm). The filter grains arrange themselves from coarse to fine in the direction of flow after backwashing. This is ideal in terms of storage capacity and even distribution of particle deposits throughout the entire filter bed depth. Because of the low density of the lightweight expanded clay aggregates, this filter is operated with low filter backwash rates.

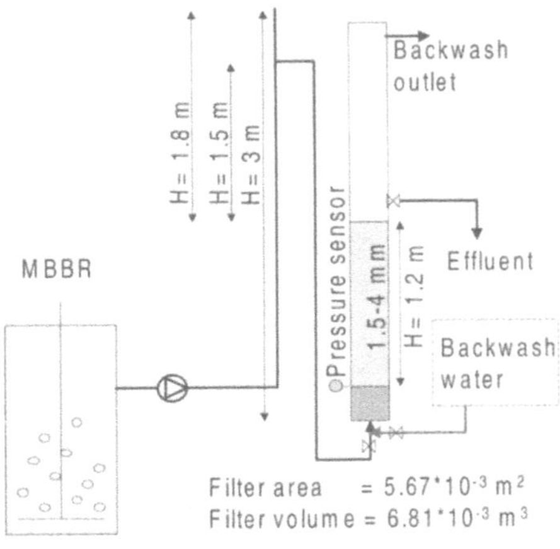

Fig. 8. Filter pilot plant

The filter of the pilot-plant used in our experiments (Figure 8) had a bed depth of 1.2 m, a 10-cm gravel layer (5-10 mm) as support layer, and 1.5-4 mm crushed Filtralite as filter medium. The inside diameter of the filter column was 85 mm and the total available head 1.8m. The head-loss could be determined by a pressure sensor at the base of the filter column.

Experiments were performed without coagulant addition as well as with the addition of various coagulants - ferric chloride at a dosage of 10 mg Fe/l, the low MW, highly charged polyDADMAC (Magnafloc 368) at a dosage of 1 mg/l and the high MW, highly charged poly-acrylamide (Floerger FO4440SH) at 1 mg/l. The organic loading on the bioreactors varied somewhat throughout the experiments but was on average 8.0 kg COD/m^3*d (36.5 g COD/m^2*d) and 4.5 kg SCOD/m^3*d (20.5 g SCOD/m^2*d).

Results of filterability experiments

Figure 9 shows the relationship between effluent SS-concentration and filtration rate. The lines are drawn for illustration purposes only. There is a close to linear relationship between the effluent SS-concentration and the filtration rate, no matter what kind of pre-treatment was used. The worst results were obtained when iron alone was used, while the two different cationic polymers gave about the same results, which were almost 10 mg SS/l lower that with iron alone

Even at filtration rates as high as 20 m/h, the effluent SS-concentration could be kept under 20 mg/l. In contrast, when iron alone was used, the filtration rate had to be decreased to 5 m/h in order to obtain the same effluent SS concentration. These results are probably due to the fact that iron alone produces small particles through coagulation in addition to those already present in the water. These small particles escape the filter more easily than those bound together through the action of the polymer. Iron alone caused weak flocs leading to filter breakthrough as the determining factor for filter run time.

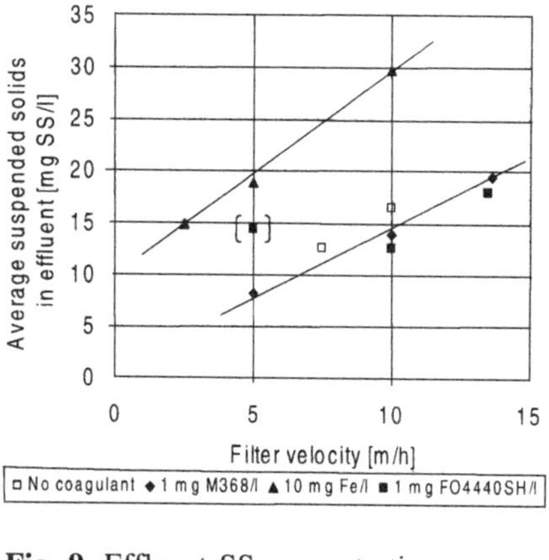

Fig. 9. Effluent SS-concentration versus filtration rate

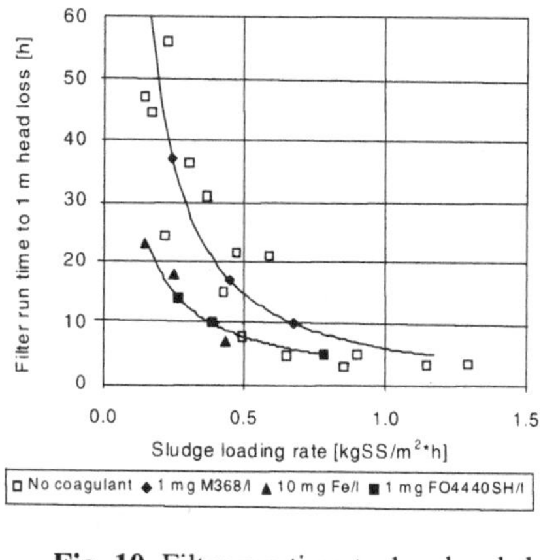

Fig. 10. Filter run time to 1 m head- loss versus sludge loading rate

The filter run was terminated either when the maximum allowable head-loss (set at 1m) was reached or when breakthrough occurred (when maximum allowable effluent SS concentration, set at 30 mg SS/l, was reached). In all runs with iron alone as coagulant, breakthrough determined the length of the filter run (max allowable effluent concentration of 30 mg/l was reached before max allowable head-loss). On the other hand, when a polymer was used as coagulant, maximim head-loss determined the length of the filter run.

In Figure 10 the filter run time is shown as a function of the sludge loading rate. The lines are drawn for illustration purposes only. It can be seen that the cationic polymer with the lower molecular weight (Magnafloc 368) gave longer filter runs at a given sludge loading rate than the one with high molecular weight (FO440SH) at the same dosage. This is caused by the fact the low MW polymer acts as a pure coagulant, resulting in relatively small, compact flocs, while the high MW polymer acts according to the bridging mechanism as well, resulting in larger flocs that can not penetrate equally far into the filter. The fastest head-loss build-up was experienced with iron alone as a result of the higher amount of sludge to be separated as a consequence of metal hydroxide precipitation.

When only 1 mg/l of the low MW, cationic polymer was added, about 16 hours filter run time was obtained at a sludge loading rate of about 0,5 kg SS/m^2h. A sludge loading rate as high as 0,75 kg SS/m^2h could be used if a filter run time of 10 hours was acceptable. The latter loading equals a filtration rate of about 5 m/h without presettling ($SS_{in\ filter}$~150 mg SS/l). At filtration rates below 10 m/h, the effluent SS-concentration could be expected to be ≤ 15 mg SS/l.

Conclusions

The following conclusions can be drawn from this study:

1. The combination of a high-rate moving bed reactor, possibly without primary settling, followed directly by a coagulation step will result in an extremely compact wastewater treatment plant, especially when a high-rate separation method (i.e. filtration) is also employed.
2. If the goal is to remove only SS and BOD (secondary treatment), a cationic polymer alone can be used, minimising sludge production. In this case, a low molecular weight, highly charged cationic polymer should be the chosen. The necessary dosage can be expected to be in the range of 1-2 mg/l.
3. If phosphate removal is required as well, a metal coagulant has to be used, but a low dosage is needed, normally < 0,2 mmol/l.
4. If direct filtration is used for floc separation, a sludge loading rate of about 0,5 kg SS/m^2h would result in a filter run time of about 16 hrs with the actual filter.
5. At filtration rates between 7 and 10 m/h, the effluent SS-concentration could be expected to be ≤ 15 mg SS/l. At filtration rates < 7 m/h, the effluent SS concentration could be ≤ 10 mg SS/l.

References

1. Ødegaard, H.: Particle separation in Wastewater treatment. Documentation 7'th European Water Pollution Control Association Symposium, Munich, May, 1987
2. Ødegaard, H.: Coagulation as the first step in waste water treatment. In : Hahn, H.H., Klute, R. (eds): Pretreatment in Chemical Water and Wastewater Treatment. Springer Verlag, Berlin/Heidelberg, 1988, pp. 249-250
3. Ødegaard,H.: Norwegian experiences with chemical treatment of raw wastewater. Wat. Sci. Tech. 25 (12) (1992) pp. 255-264
4. Pujol, R., Sagberg, P., Lemmel, H., Hamon, M.: The use of reagents in up-flow submerged biofilters In : Hahn, H. H. and Klute (eds): Chemical water and wastewater treatment III. Springer Verlag, Berlin/Heidelberg (1994) pp 221 - 230
5. Ødegaard, H., Rusten, B., Westrum, T.: A new moving bed biofilm reactor - Applications and results. Wat.Sci.Tech. 29, (10-11) (1994), pp 157-165
6. Ødegaard, H., Rusten, B., Siljudalen, J.: The development of the moving bed biofilm process - From idea to commercial product. European Water Management 2 (3) (1999) pp 36-43
7. Æsøy. A., Ødegaard, H., Sandberg, R.: Anoxic degradation of dissolved COD for enhanced organic matter removal in compact chemical treatment plants. In : Hahn, H. H., Hoffmann, E. and Ødegaard, H. (eds): Chemical water and wastewater treatment IV. Springer Verlag, Berlin/Heidelberg (1996) pp 387 - 398
8. Ødegaard, H.. Gisvold, B., Strickland, J.:The influence of carrier size and shape in the moving bed biofilm process Wat. Sci. Tech. 41 (4-5) (2000)
9. Ødegaard, H., Fettig, J., Ratnaweera, H.: Coagulation with prepolymerized metal salts". In Hahn, H.H. and Klute, R. (eds): Chemical Water and Wastewater Treatment. Springer Verlag, Berlin/Heidelberg (1990) pp.189-219

Particle Size Distribution (PSD) Obtained in Effluents from an Advanced Primary Treatment Process Using Different Coagulants

A. Chávez Mejía and B. Jiménez Cisneros

Institute of Engineering, UNAM, Circuito Escolar S/N, Ciudad Universitaria, México
acm@pumas.iingen.unam.mx

Abstract

The particle size distribution (PSD) obtained in effluents treated with two types of coagulants was compared. The objective was to maximize the removal of particles with a size between 0.75 and 1.5 µm, 1.5 and 5.0 µm, and 20 and 80 µm, corresponding to the sizes of fecal coliforms, salmonella and helminth ova, respectively. The coagulant doses tested were 20, 30, 40 and 50 mg/L for both anhydrous aluminum sulfate and PAX·XL-60. The raw wastewater had a PSD between 0.04 and 310 µm, of which 4.4% of the particles were found between 0.75 and 1.5 µm (by volume), 12.7% between 1.6 and 5.0 µm, and 39.4% of the particles were between 20 and 80 µm. It was determined that with 50 mg/L of aluminum sulfate 91% of the particles in the influent were eliminated, whereas with 30 mg/L of PAX XL-60 a 93% removal was achieved. The percentage of remaining particles with diameters between 0.7 and 5.0 µm was 48.9 and 53.8%, respectively, for the two coagulants. Particles between 20 and 80 µm were completely removed by the treatment system in both cases. Under these conditions the effluents had total suspended solids (TSS) concentrations between 39 and 37 mg/L, turbidity between 30 and 39 NTU, and COD between 218 and 233 mg/L, respectively. In terms of the removal of microorganisms, both coagulants removed helminth ova to values below 1 ova/L. Fecal coliforms were reduced from $2.82 \cdot 10^8$ to $5.45 \cdot 10^6$ with aluminum sulfate and to $4.85 \cdot 10^6$ NMP/100 mL with PAX XL-60; salmonella were reduced from $5.43 \cdot 10^6$ to $9.85 \cdot 10^5$ and to $4.85 \cdot 10^6$ NMP/100 mL with the respective coagulants. These levels would require an additional disinfection step to eliminate these organisms to acceptable levels for irrigation. Under the conditions in this study, the change in the PSD made it possible to determine the coagulant dose required to produce suitable water for agricultural reuse by an advanced primary treatment.

Introduction

In Mexico, due to the scarcity of water, the wastewater regulations are formulated to encourage reuse for agricultural purposes, and therefore the water quality standards are oriented towards pathogen control. For this reason, Mexican legislation (NOM-001 ECOL/96 [1]) adopted the recommendations of the World Health Organization [2], which stipulate less than 1000 MPN/100 mL of fecal coliforms (FC) and less than one helminth egg/L in treated water used for unrestricted irrigation. Additionally, for crops that are not consumed raw, water containing up to 5 ova/L, but still less than 1000 MPN/100 mL of fecal coliforms may be used for irrigation.

Helminth ova and fecal coliforms are part of the total suspended solids (TSS) of the water and, consequently, the treatment processes that eliminate particles between 0.7 to 80 μm are those of primary interest. Physical-chemical processes are effective for removing TSS and the colloidal fraction. Biological treatment processes, on the other hand, may not be appropriate, given that in agricultural irrigation the presence of organic material, nitrogen and phosphorus is actually a benefit. The objective of this study, therefore, was to maximize the removal of helminth ova and FC as general indicators of water quality, and salmonella as a specific concern in Mexico. In particular, the removal of particles between 0.7 and.5.0 μm (FC and salmonella), and between 20 and 80 μm (helminth ova) was evaluated.

Background

In Mexico, the incidence of helminth infections is high, with the national average for Ascariasis being 33.3 % [3]. This value varies drastically depending on whether the population is urban or rural, and is markedly higher for children between 0 to 4 years old. Helminths are human intestinal parasites, and helminth infections are one of the most common illnesses worldwide, with approximately one billion persons infected [4]. Helminth ova are ingested, and the worms reside in the intestine. To detect the ova in wastewater a complex technique is required. In this process, helminth ova are separated from the rest of the particles present so that they can be identified under a microscope. The use of raw wastewater for irrigation is the principal cause of helminth infections [5]. In fact, the risk of Ascariasis in children from 0 to 4 years old is 16 times higher than the national average in areas using raw wastewater for irrigation. It should be mentioned that this effect is not observed with the transmission of other illnesses such as giardiasis (*Giardia lamblia*) or amoebiasis (*Entamoeba histolytica*). According to Ayres [6], helminth ova are between 20 and 80 μm, and FC vary between 0.7 and 1.5 μm [7]. Salmonella vary between 0.7 to 1.5 μm in width and 2.5 to 5 μm in length.

The information on the type of contaminants present in wastewater in relation to the size of suspended solids is very limited [8]. From the available data it is not possible to determine whether it is the small or large particles that are the most important carriers of contaminants. It is known that the contaminants that must be separated from the wastewater are complex mixtures of particles, organic constituents, and soluble inorganics. The size of impurities in the water varies within approximately four orders of magnitude (< 0.001 μm for dissolved material, 0.001 to 1.0 μm for colloids, 1 to 100 μm for supracolloids, and > 100 μm for settleable solids [8 9]. The removal efficiency of nutrients and contaminants by different processes could be improved with a better understanding of the particle size distribution (PSD) in wastewater [10]. In addition, this distribution provides useful information about the nature of the particles and their characteristics and how they might be removed in treatment processes.

For example, coagulation-flocculation processes are designed to change the size of the particles in suspension, however PSD is rarely measured [11]. In physical-chemical treatment, the efficiency of coagulation and flocculation is measured by means of turbidity and the concentration of total suspended solids (TSS). Nevertheless, the measurement of the PSD provides knowledge of the contaminants in a direct and exact way [12,13]. In fact, Ødegaard and Lawler et al. [13,14] reported that an analysis of the PSD is one of the main parameters that should be considered in the design of treatment plants. In addition, Cairns et al. [16] showed that this parameter has a strong impact on the efficiency of disinfection with UV if the particles are greater than 7 μm.

In this paper the PSD of an effluent from an advanced primary treatment (APT) process using two types of coagulants was analyzed. The wastewater treated by this system had to comply with the Mexican regulation for agricultural reuse (NOM-001 ECOL/96 [1]).

Methods

The wastewater used in this research came from one of the sewers in Mexico City with a flow rate of 35 m^3/s. According to Jiménez et al. [17, 18], this influent was characterized by a high and variable concentration of suspended solids (60 to 1500 mg/L), helminth ova, (6 to 98 HO/L), fecal coliforms ($6.7 \cdot 10^8$ to $5.2 \cdot 10^9$ NMP/100 mL including species such as *Y. enterocolitica, S. typhi, C. freundii, K. pneumoniae* and *K. oxytoca, E. cloacae* and *E. agglomerans,* and *A. hydrophila),* salmonella ($5.4 \cdot 10^6$ to $2.2 \cdot 10^9$ NMP/100 mL), *Pseudomonas aeruginosas* ($1.76 \cdot 10^4$ to $1.1 \cdot 10^7$ NMP/100 mL), and protozoan cysts (978 to 1814 organisms/L including mainly *E. histolytica, E. coli, Giardia lamblia* and *Balantidum coli*). It is clear that the presence of all these contaminants would have a great impact on public health if this wastewater were to be used for irrigation.

As a first step, it was necessary to determine the optimum dose of coagulant. A series of jar tests was performed, in which anhydrous aluminum sulfate and PAX XL–60 (dry basis) were applied in parallel in doses of 10, 20, 30, 40 and 50 mg/L; 1.0 mg/L of a high molecular weight anionic polyelectrolite (Prosifloc A-252) was also added. For each test there was also a control sample that received no coagulant and flocculant. The apparatus used for the jar tests is shown in Figure 1.

The conditions for each step (coagulation, flocculation, and sedimentation) are listed in Table 1. These conditions were based on another study [19], which determined the effect of the mixing velocity and contact time for each stage of APT.

The parameters analyzed in the influent and effluent were determined according to Standard Methods [20]. The size and distribution of particles was measured using a particle counter (Coulter LS 230) with a measurement interval from 0.04 to 2000 μm; the count was performed in triplicate for each treatment.

Table 1. Mixing conditions used for the jar tests

Stage	Revolutions per minute, rpm	$G\ s^{-1}$	Time
Coagulation	300	336	30 s
Flocculation	50	23	5 min
Sedimentation	0	0	5 min

Table 2. Physico-chemical parameters in the influent and effluent

TSS, TS, TDS, pH, turbidity, temperature, redox potential, real and apparent color, PSD, COD total, alkalinity, helminth ova, fecal coliforms and *Salmonella* spp.

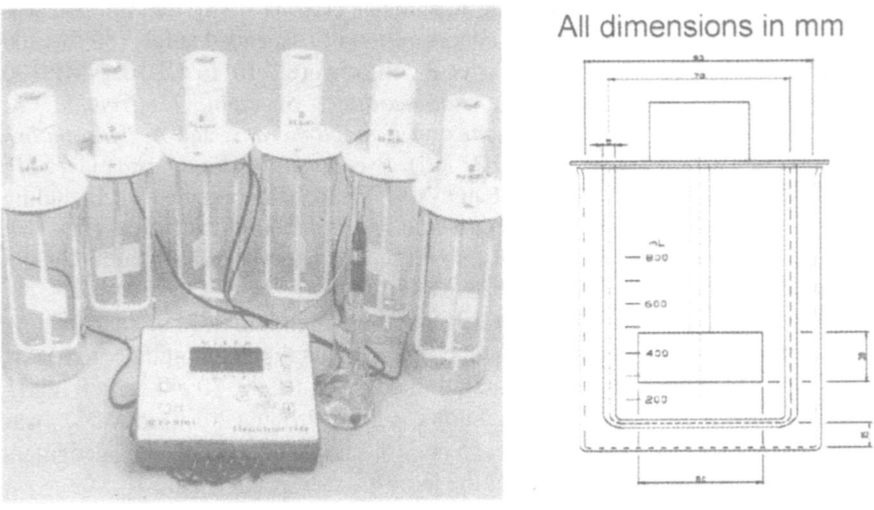

Fig. 1. Apparatus used for the jar tests and sketch with dimensions

Results and Discussion

Particle size distribution in wastewater

The particles in the wastewater used in this study varied between 0.04 and 310 μm and the typical distribution was as follows (see Figure 2): 4.7% (as volume) of the particles were between 0.04 and 0.74 μm, 4.4% were between 0.75 and 1.5 μm, 12.7% between 1.5 to 5.0, 37.0% between 5.1 and 19.0 μm, 39.4% between 20 and 80 μm, and finally only 1.8% were greater than 80 μm. The TSS varied from 156 to 964 mg/L, the turbidity from 139 to 461 NTU, COD from 309 to 941 mg/L, and in terms of microbiological characteristics helminth ova ranged from 5.0 to 22.5 HE/L (14.7 HE/L average), fecal coliforms from $5.5{\cdot}10^7$ to $1.1{\cdot}10^{10}$ NMP/100 mL ($2.82{\cdot}10^8$ NMP/100 mL average) and salmonella from $6.0{\cdot}10^5$ to $1.7{\cdot}10^7$ NMP/100 mL ($5.43{\cdot}10^6$ NMP/100 ml average).

The goal of the treatment was to eliminate helminth ova (20 to 80 μm), fecal coliforms (0.7 to 1.5 μm) and salmonella (0.7 to 1.5 μm width, 2.5 to 5 μm length) from the treated water. As shown in Figure 3, 95.3% of the particles were larger than 0.75 μm, which translates into a high probability that these microorganisms were present in the raw wastewater. Primary sedimentation without addition of coagulant and flocculant removed only the particles greater than 80 μm and 4% of the supracolloidal particles. Therefore coagulation/flocculation is needed to improve these removal rates.

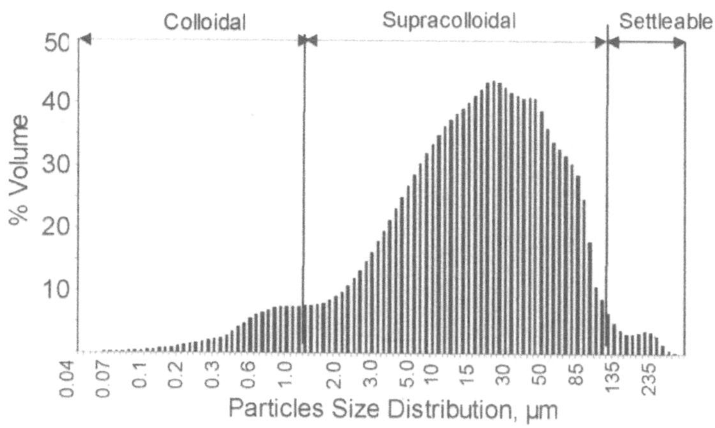

Fig. 2. Particle size distribution in the raw wastewater

The effect of adding coagulant to the raw wastewater is shown in figure 4. It was found that the efficiency of the coagulants was determined largely by the TSS. The average concentration of TSS in the wastewater was 350 mg/L, and in order to achieve good removal, 50 mg/L of anhydrous aluminum sulfate was required. Under these conditions, the residual TSS fluctuated between 28 mg/L and 50 mg/L, with an average of 39 mg/L, corresponding to a removal of 88.8%. When PAX XL-60 was used as coagulant, only 30 mg/L was required to achieve a similar TSS removal. The resulting effluent TSS was between 26 mg/L and 54 mg/L, with an average of 37 mg/L and a removal efficiency of 89.4%. Increased doses of PAX XL-60 did not increase TSS removal significantly (90.5 and 90.9% for 40 and 50 mg/L of PAX XL-60, respectively).

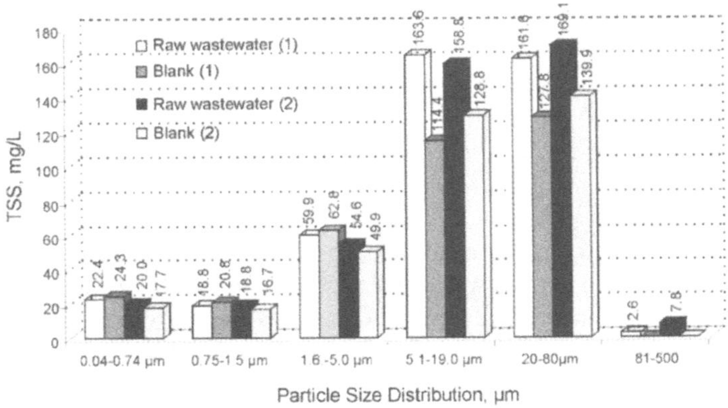

Fig.3. Particle size distribution intervals for the raw wastewater and blank

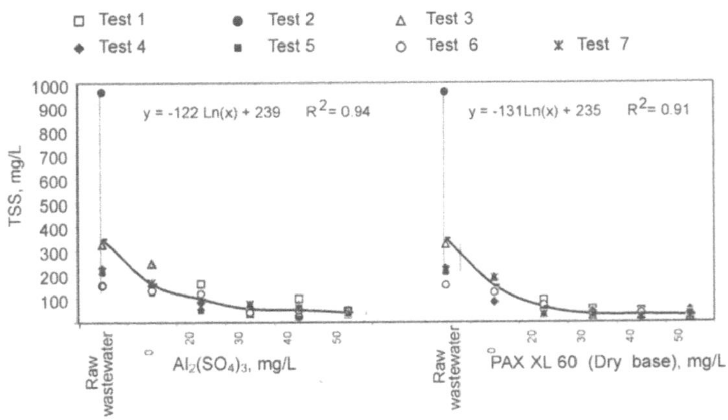

Fig. 4. Concentration of TSS achieved with different doses of anhydrous aluminum sulfate and PAX XL 60

PAX-XL 60

Table 3 shows other physical-chemical characteristics of the effluent for the selected doses of the two coagulants (50 mg/L of aluminum sulfate and 30 mg/L of PAX XL-60). With aluminum sulfate the turbidity was reduced to 89.7%, COD_t 62.5%, real color 67%, apparent color 53.6%, and TS 44.8%. In fact, the water contained a large quantity of dissolved compounds, including calcium, magnesium, and sodium, which is reflected in the high conductivity measured (1298 mS/cm). When 30 mg/L of PAX-XL 60 was applied the efficiency of removal was 86.6% for turbidity, 60% for COD, 65% for real color, 46.3% for apparent color, and 43.5% for TS. In terms of sludge, the production rates were 0.37 and 0.35 g/L, respectively, for the two coagulants.

As shown in Figure 5, as the coagulant dose of aluminum sulfate was increased the percentage of particles present in each size interval was reduced. The remaining 9% showed the following distribution. In the size range 0.7 to 1.5 μm (size of fecal coliforms), only 2.6 to 7.6% of the particles were present. For particles between 1.6 to 5.0 μm removal was from 36% to 51.2% depending on the dose; and in the size range 20 to 80 μm (size of helminth ova) all of the particles were removed except in the case where 20 mg/L was applied (0.3 mg/L of the particles remained).

Table 3. Constituents removal using 50 mg/L of $Al_2(SO_4)_3$ and 30 mg/L of PAX XL 60

Parameter	Number of Samples Inf	Number of Samples Eff	Mean Inf	Mean Eff (1)	Mean Eff (2)	Standard Deviation Inf	Standard Deviation Eff (1)	Standard Deviation Eff (2)	η (%) (1)	η (%) (2)
COD_t mg/L	7	7	582	218	233	194	101	102	62.5	60.0
Turbidity, NTU	7	7	292	30	39	150	6.6	11	89.7	86.6
TSS, mg/L	6	7	350	39	37	306	8	10	88.8	89.4
Color (R), Pt-Co	7	7	> 550	182	193	0	40	74	67.0	65.0
Color (A), Pt-Co	7	7	179	83	96	41	30.8	40	53.6	46.3
TS, mg/L	6	6	1318	723	720	996	227	215	44.8	45.4
pH	7	7	7.7	7.4	7.4	0.2	0.2	0.2	----	----
Temperature, °C	7	7	21.7	22.2	22	3.0	1.50	1.50	----	----
Redox potential	7	7	-64.1	-41.4	-43.7	15	18	15		----
Conductivity, mS/cm	7	7	1298	1211	1252	0.3	0.2	31	6.7	3.5
TDS, mg/L	4	6	743	686	714	269	228	212	7.6	4.0
Alkalinity, $CaCO_3$/L	1	1	318	268	292				15.7	8.2
Sludge production , g/L		3		0.37	0.35	---	----	---		

(R) = real; (A) apparent, (1) $Al_2(SO_4)_3$, (2) PAX XL 60

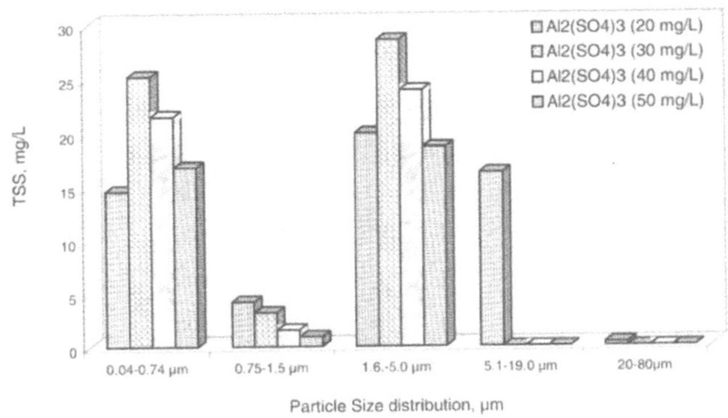

Fig. 5. Particle size distribution of effluent treated with different doses of $Al_2(SO_4)_3$

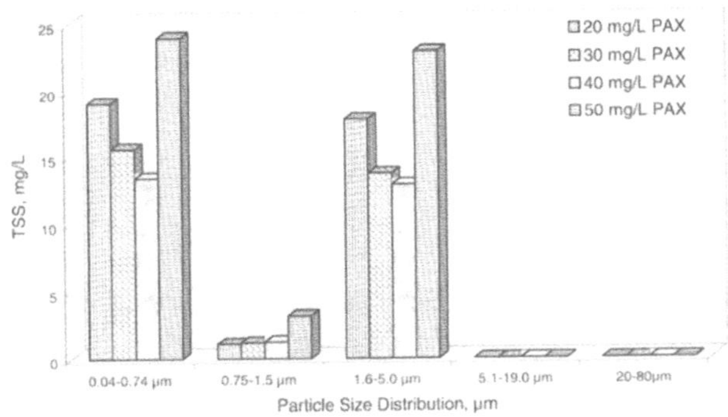

Fig. 6. Particle size distribution of effluent treated with different doses of PAX- XL 60

On the other hand, PAX XL-60 removed more particles than aluminum sulfate, but as shown in Figure 6, increasing the dose up to 50 mg/L caused a restabilization of the particles. The remaining particles within the treated WW (7%) showed the following distribution. For all the applied doses, no particles were detected between 20 and 80 µm, whereas in the size range 0.75 to 1.5 µm, the percentage varied between 2.9 and 6.3%. For the particles between 1.6 to 5.0 µm, 45 to 46.9% remained. Furthermore, since the highest dose of 50 mg/L PAX XL-60 gave the highest remaining particle value, it appears that too high a dose actually has a negative effect on physical-chemical treatment performance.

The concentration of TSS in each size interval is shown for each coagulant in Table 4. Particle removal improved with an increase in coagulant dose. The application of 50 mg/L of aluminum sulfate and 30 mg/L of PAX produced a similar PSD. 91% and 93%, respectively, of the particles in the raw wastewater were removed by the two coagulants. In both cases, the remaining particles in the effluent were within the size range from 0.04 to 3.8 µm.

Table 4. Behavior of PSD using different coagulant doses (TSS in mg/L and removal in percent)

	Raw	Coagulant dose							
	waste	**20 mg/L**		**30 mg/L**		**40 mg/L**		**50 mg/L**	
Particle size, µm	water mg/L	TSS mg/L	Rem. %	TSS mg/L	Rem. %	TSS mg/L	Rem. %	TSS mg/L	Rem. %
0.04 – 0.74	20.0								
$Al_2(SO_4)_3$		14.6	27 %	25.2	-26%	21.5	-7.5 %	16.9	16 %
PAX XL-60		19.1	4 %	15.6	22%	13.4	33 %	23.9	-20 %
0.75 – 1.5	18.8								
$Al_2(SO_4)_3$		4.2	78 %	3.2	83%	1.6	91%	0.9	95%
PAX XL-60		1.1	94 %	1.2	94%	1.2	93%	3.2	83%
1.5 – 5.0	62.4								
$Al_2(SO_4)_3$		20.0	68 %	28.5	54 %	23.9	62 %	18.7	70 %
PAX XL-60		17.8	71 %	13.7	78 %	12.9	79 %	22.9	63 %
5.1 – 19.0	130.6								
$Al_2(SO_4)_3$		0.3	99.8%	0.0	100 %	0.0	100 %	0.0	100 %
PAX XL-60		0.0	100 %	0.0	100 %	0.0	100 %	0.0	100 %
20 - 80	163.6								
$Al_2(SO_4)_3$		0.3	99.8%	0.0	100 %	0.0	100 %	0.0	100 %
PAX XL-60		0.0	100 %	0.0	100 %	0.0	100 %	0.0	100 %
> 80	2.6								
$Al_2(SO_4)_3$		0.0	100 %	0.0	100 %	0.0	100 %	0.0	100 %
PAX XL-60		0.0	100 %	0.0	100 %	0.0	100 %	0.0	100 %
Total removal of particles									
$Al_2(SO_4)_3$			90 %		86 %		88 %		91 %
PAX XL-60			90 %		92 %		93 %		87 %
Size range of residual particles									
$Al_2(SO_4)_3$		0.04 – 25.0		0.04 – 4.3		0.04 – 3.8		0.04 – 3.8	
PAX XL-60		0.04 – 4.2		0.04 – 3.8		0.04 – 3.5		0.04 – 3.8	

Removal of pathogenic microorganisms

The physical-chemical process removed a significant number of helminth ova (HO). With 50 mg/L of aluminum sulfate the initial concentration of 14.7 HO/L was reduced to 0.6 HO/L (96% removal); 30 mg/L of PAX XL-60 resulted in a reduction to 0.30 HO/L (94.5% removal). The treated effluent met the standards for wastewater reuse in terms of HO. On the other hand, only minimal removal of fecal coliforms and salmonella was observed for both types of coagulants and at all doses (less than 2 logarithmic units, see Table 5). This means that an additional disinfection step is necessary. Based on the PSD of the effluent (< 3.8 µm), UV disinfection may be appropriate.

The high removal of HO is directly related to the particle size. In fact, all TSS in the HO size range (20-80 µm) were removed. In the case of FC and salmonella, the removal efficiency of this size range is considerably lower, and does not even take into account the reproductive capabilities of these microorganisms.

Table 5. Microbiological quality of influent and effluent

Dose mg/L	Helminth ova HO/L	Removal %	Fecal coliforms MPN/100 mL	Removal %	*Salmonella* spp. MPN/100 mL	Removal %
Raw wastewater						
0.0	14.7	-----	2.82E+08	------	5.43E+06	-----
$Al_2(SO_4)_3$						
20	1.60	89.1	1.50E+07	94.7		
30	1.40	90.5	6.85E+07	75.7	4.30E+06	20.8
40	1.40	90.5	6.85E+07	75.7	4.30E+06	20.8
50	0.60	96.0	5.45E+06	98.0	9.85E+05	82.0
75	0.30	98.0	4.85E+07	82.0	1.00E+06	81.6
100	0.32	98.0	2.00E+07	93.0	4.50E+05	91.7
PAX-XL 60						
20	0.60	96.0	1.60E+08	43.2	3.00E+06	44.8
30	0.80	94.5	4.85E+07	83.0	4.80E+06	11.6
40	0.7	95.2	5.75E+07	79.6	1.30E+06	76.1
50	0.4	97.3	1.71E+07	94.0	1.80E+06	67.0
75	0.6	96.0	9.50E+06	96.6	1.60E+06	70.5
100	0.5	96.6	2.50E+07	91.1	1.60E+06	71.0

Conclusions and Recomendations

The wastewater used in this study had a typical PSD ranging in size from 0.04 to 310 μm, including colloids, supercolloids, and settleable solids. Given that the size of helminth ova, fecal coliforms and salmonella falls within this particle size range, it makes sense to use a wastewater treatment process that eliminates particles based on size.

The optimal dose of 30 mg/L of PAX-XL 60 removed all but 7% of the particles; these were within the size range from 0.04 to 3.8 μm. The resulting effluent contained 37 mg/L TSS, 39 NTU turbidity, and 233 mg/L CODt. With 50 mg/L of anhydrous aluminum sulfate, 91.5% of the particles were eliminated from the system, and as in the previous case the effluent particles were within the size range 0.04 to 3.8 μm; the effluent contained 39 mg/L TSS, 30 NTU turbidity, and a final COD_t of 218 mg/L.

The use of PSD to monitor water quality made it possible to determine the required doses for an advanced primary treatment system, and at the same time, defined the removal capacity of the system. With both coagulants, particle removal increased with increased dose. A dose of 50 mg/L of anhydrous aluminum sulfate left only 2.6% of the particles having a diameter from 0.75 to 1.5 μm (FC); 51.2% with a diameter between 1.51 and 5.0 μm remained (salmonella). All of the particles between 20 and 80 μm were removed (helminth ova). A dose of 30 mg/L of PAX XL-60 achieved similar removals.

All the doses of both coagulants achieved excellent removal of helminth ova, with effluent concentrations less than 1 HO/L. Removal of fecal coliforms and salmonella, however, was less satisfactory, so that it is necessary to treat the effluent with an additional disinfection step to reduce the concentration of these organisms. The high removal of HO is directly related to the particle size. In fact, all TSS in the HO size range (20-80 μm) were removed. The removal efficiency of particles in the FC and salmonella size range, however, is considerably lower, and this does not even take into account the reproductive capabilities of these microorganisms.

Rerferences

1. Official Mexican Standard. NOM-001-ECOL/1996. (Establishes the maximum permitted limits of pollutants in the wastewater discharges into water and onto national property) The Official Federal Gazette. January 6, 1997, 67-81
2. World Health Organization: Health guidelines for the use of wastewater in agriculture and aquaculture, Technical Report Series No. 778. (1989) WHO, Geneva
3. Biagi, F.: Enfermedades parasitarios. La Prensa Mexicana. México (1990) pp 376
4. Tay, Z., Aguilera, L., Quiróz, G., Castrejón, V.: Parasitología Médica. Francisco Méndez Cervantes. México, (1991) pp 6-327
5. Cifuentes, E., Blumenthal, J., Ruiz-Palacios, G., Beneth, S.: Health Impact Evaluation of Wastewater in Mexico. Public Health Revue. 19 (1992) 243-250
6. Ayres, R.: The Enumeration of Human Intestinal Nematode Ova in Raw and Treated Wastewater. ODA Research Scheme R. 4336. Final report. University of Leeds, Department of Civil Engineering. Leed, U.K. (1989) 639-646
7. Jawetz, Melmick, Adelberg´s : Microbiología Médica. El Manual Moderno, Mexico. (1995)
8. Levine, D., Tchobanoglous, G., Takashi, A.: Size Distributions of Particulate Contaminants in Wastewater and Their Impact on Treatability. Wat Res. 25 (8) (1991) pp 911-922
9. Richert, D. A., Hunter, J.V.: General Nature of Soluble and Particulate Organic in Sewage and Secondary Effluents. Water Research. 5 (7) (1971) pp 421
10. Thiem, A., Herwing, V., Neis, U.: Particle size analysis for improved sedimentation and filtration in wastewater treatment, Wat. Sci. and Tech. 39(8) (1999) pp 99-106
11. American Water Woks Association: Committee Coagulation as an Integrated Water Treatment Process. J. Amer. Water Works Assoc. 81 (10) (1990) 72-78
12. Lawler, F.: Particle Size Distribution in Treatment Processes; Theory and Practice. Water Science and Technology. 36(4) (1997) pp 15-23
13. Ødegaard, H.: Optimized Particle Separation in the Primary Step of Wastewater Treatment. Wat. Sci. Tech. 37(10) (1998) pp 43-53
14. Ødegaard, H.: Chemical Floc Formation in Wastewater Treatment – An Introduction. Prog. Wat. Tech. (1979) pp 103-110
15. Lawler, F., Willkes, D.R.: Flocculation Model Testing Particles Sizes in a Softening Plant. Research & Technology Journal AWWA. 76 (7) (1984) pp 90-97

16. Cairns, W. L.: Comparing Disinfection by Ultraviolet Light and Chlorination the Implications of Mechanism for Practice. Water Environmental Federation Specialty Conference. May 1993. Whippany, N. Y. (1993) pp 555-565
17. Jiménez, B., Chávez, A.: Treatment of Mexico City Wastewater for Irrigation Purposes. Environ. Tech., 18 (1997) 721-730
18. Jiménez, B., Chávez, A., Hernández, C.: Alternative Treatment for Wastewater Destined for Agricultural Use. Wat. Sci. Tech., 40 (4-5) (1999) 355-362
19. Chávez, A., Jiménez, B.: Distribución del Tamaño de Partícula (DTB) obtenida con la aplicación de PAX XL 60 y Sufato de Aluminio. Compromiso Nacional con el Medio Ambiente, Tomo I, FEMISCA. March 21 to 24, 2000. Morelia, Mich. Mexico (2000) pp 421- 431
20. Standard Methods for the Examination of Water and Wastewater 19th edn. American Public Health Association/American Water Works Association/Water Environment Federation. (1995) Washington D.C. USA

Enhanced Pre-Precipitation while Retrofitting the Biological Tanks at Bromma WWTP, Stockholm, Sweden

J. Öman and B.G. Hellström

Stockholm Water Co, 106 36 Stockholm, Sweden
Johanna.oman@stockholmvatten.se; bghellstrom@stockholmvatten.se

Abstract

In the spring of 2000 the biological treatment at Bromma WWTP (160 000 m^3/d) was retrofitted to accommodate all-year-round nitrification and denitrification and thereby harmonize with the new council directive (98/15/EEC) regarding nitrogen removal. In order to retain nitrification and a high organics removal during the construction period a full-scale experiment using ferric chloride (Kemwater PIX-111) for pre-precipitation was performed in May 1999. This paper presents the results of full-scale test and for the first three months of the construction period.

By dosing 10-13 g Fe/m^3 as PIX-111 the load of organic matter and suspended solids (SS) in pre-settled water was reduced by 15-30% and 40-45%, respectively. As a consequence of the enhanced organics removal the F/M ratio did not rise from its normal value of 0.25 kg BOD_7/kg VSS·d and the BOD_7/TKN ratio decreased from 3.1 to 2.4. Moreover, the organic fraction in the activated sludge rose from 67% to 78%. Hereby nitrification was preserved throughout the construction period, in spite of a reduced aerobic volume of 33%.

Introduction

Bromma wastewater treatment plant (WWTP) was built in 1935 and has been extended and retrofitted a couple of times throughout the years. Since the mid-nineties the plant has been operated with nitrogen removal and a total nitrogen (Tot-N) reduction of about 45%. In order to harmonize with the new council directive (98/15/EEC) and the limit of 10 g Tot-N/ m^3, construction work on the biological tanks was started in January 2000. Over a construction period of six months the old ceramic aeration system is going to be replaced by new membrane fine bubble diffused aerators to allow better aeration and dissolved oxygen control. Baffles will be erected, dividing the tanks into different zones allowing a flexible aerobic/anoxic volume control. Moreover, the plant will be fitted with step-feed possibilities and nitrate recirculation.

During the construction period, two out of six biological tanks are taken out of operation. The aerated SRT is equal to or less than four days and there is a limitation in raising the MLSS above 3000 mg/l due to problems with filamentous bacteria, i.e. bulking sludge. In order to avoid overloading the tanks in operation, either the flow to the tanks or the organic load has to be decreased. It is possible to by-pass the pre-settled water directly to the sand filters, but then there is a risk of only moving the problem a step ahead, with overloaded filters and poor effluent quality as the result.

With pre-precipitation using ferrous sulphate, 10 g Fe/m^3, the organics removal (COD_{Cr}) in the primary sedimentation equals 40%. In order to enhance the pre-precipitation, laboratory tests with different products were conducted in the spring of 1999. Partly based on the results from the tests, but mainly based on the desire to stick to an iron product, ferric chloride was chosen. During May 1999 a full-scale test with ferric chloride (Kemwater PIX-111) was performed.

This paper presents the results of the full-scale test with PIX-111 in May 1999 and from the first 14 weeks of the construction period in the year 2000.

Hypothesis and objective

If the pre-precipitation is enhanced so that the organic load on the biological tanks does not increase during the construction period, the requirements from the authorities will be met, and the nitrification will be retained during the construction period.

The objective was to decrease the load of COD_{Cr} and SS in the pre-settled water by one third so that the load on the individual biological tanks would not increase during the construction period when the volume was reduced by one third.

Bromma WWTP technical data

Bromma WWTP is designed for a flow of 160 000 m^3/d, and 270 000 persons from the western district of Stockholm are connected. The Bromma plant is one of three WWTPs operated by Stockholm Water Co. Since the treated wastewater from the three plants is discharged to the Baltic Sea by a common pipe, the effluent demands are common: BOD 10 g/m^3; Tot-P 0.3 g/m^3 and Tot-N 10 g/m^3.

Bromma WWTP consists of mechanical, chemical and biological treatment as well as a final separation step consisting of two-media sand/clay filters (Figure 1). For pre-precipitation 10 g Fe/m^3 of ferrous sulphate is added to the incoming wastewater before the step screens. Before the filters an additional 2 g Fe/m^3 is dosed to the wastewater. The biological treatment consists of six aerated tanks where the first eighth of the volume is anoxic, i.e. a pre-denitrification process.

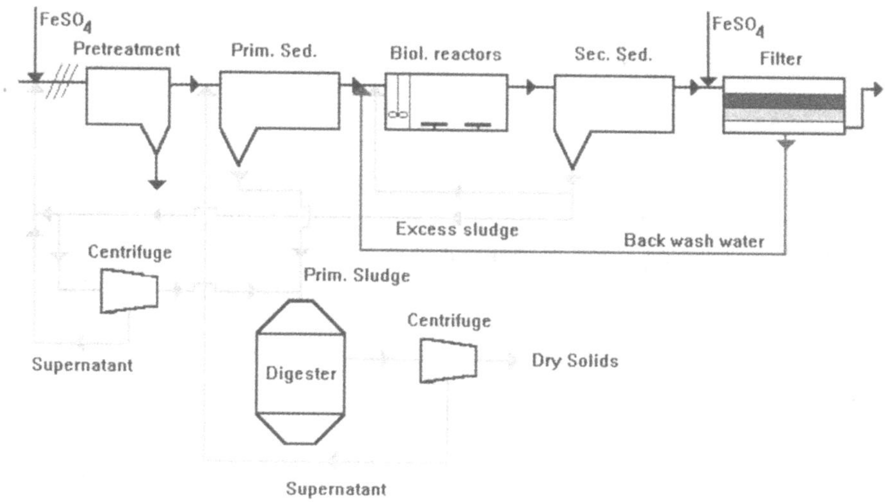

Fig. 1. A schematic of Bromma WWTP

When flow is high and sludge characteristics poor a part of the pre-settled water flow can be by-passed directly to the sand/clay filters. By experience and tradition the maximum flow to the biological treatment step has been set to 2.0 m^3/s (170 000 m^3/d).

The sludge from the primary clarifiers is sent to five parallel digesters (HRT 24 d). The waste activated sludge from the biological tanks is first dewatered by a centrifuge and is thereafter sent to a digester (HRT 14 d). The digested sludge, 2 % DS, is dewatered by centrifuges, 30-35% DS.

Averages of some characteristic parameters calculated from weekly composite samples from the first 14 weeks in 1999, are presented in Table 1.

Table 1. Average values of incoming (in), pre-settled (ps) and effluent (out) wastewater at Bromma WWTP, weekly composite samples from weeks 1 – 14, 1999 (reference period)

Parameter	in	ps	out	Unit
Flow	156 500	161 000		m^3/d
COD_{Cr}	261	156	32	g/m^3
BOD_7	118	74	4	g/m^3
SS (1 μm)	163	91	3	g/m^3
Tot-P	3.1	2.6	0.1	g/m^3
Tot-N	22	25	14	g/m^3

Methods and Materials

Dosage of PIX-111

The ferric chloride was injected through a channel just before the grit chambers where the velocity of the water was high. The dose was high in the beginning of the full-scale test, 16 g Fe/m^3, in order to get a quick answer if the chemical was sufficiently well mixed. After only a couple of days it was obvious that the dosage had the intended effect and that spot was used for feeding the ferric chloride throughout the test and later during the construction period.

The dosage was controlled by the incoming wastewater flow, which corresponds fairly well to the variations of organic load. In order to avoid excessively high doses, when flow was high and the wastewater diluted, a maximum dose limit was introduced at 1.5 m^3/s. The max-dose limit was raised to 2.0 m^3/s during the construction period, which started off in the high flow season.

Table 2. An overview of studied periods and coagulants

	Ref	**May**	**2000**
Period	04.01.99 - 11.04.99	29.04.99 - 29.05.99	03.01.00 - 09.04.00
Coagulant	$FeSO_4 \cdot 7\ H_2O$	$FeCl_3$	$FeCl_3$
Dose (g Fe/m^3)	10	10	12

Table 3. An overview of sampling and analysis frequency

Period Sample	**Ref**	**May** Number of analyses a week	**2000**	Sample	Parameters
Wastewater	1	7	1	Composite	A
Wastewater	1	1	1	Composite	B
Wastewater	1	5	5	Composite	C
Primary sludge	1	4	1	Composite	E
Activated sludge (AS)	1	1	1-2	Grab	D
Waste AS	1	1	1	Grab	E
Digested sludge	1	1	1	Grab	E
Dewatered sludge	1	1	1	Composite	E

A) Tot-P, COD_{Cr}, Fe, TKN, Tot-N, NH_4-N, NO_3-N (preserved samples) **B**) BOD_7, alkalinity **C**) Suspended solids (SS) **D**) SS, VSS, sludge characteristics, i.e. sludge quality index (SQI) and studies under microscope (filamentous bacteria and protozoa) **E**) Dry solids (DS), volatile solids (VS)

From earlier experiments two tanks with a volume of 35 m^3 each and controlling apparatus was available at the plant. After the installation of three feed pumps and some pipes the dosage of ferric chloride could start.

The dose was 10 g Fe/m^3 during the full-scale test, with the exception of the first week when the dose was 13-16 g Fe/m^3. During the construction period the dose was varied between 8 and 16 g Fe/m^3 with an average of 12 g Fe/m^3 (Table 2).

Sampling and Analyses

The analysis program was based on 24-hour composite samples, which were collected seven days a week from three spots at the plant; incoming (in), pre-settled (ps) and effluent wastewater (out). An overview is shown in Table 3.

Results and discussion

The results from the two periods with ferric chloride, PIX-111, the full-scale test in May 1999 and the first 14 weeks of the construction period in 2000 are presented. The results from the periods with PIX-111 were compared to the results from the first 14 weeks in 1999 (reference period) when 10 g Fe/m^3 of ferrous sulphate was used as pre-precipitation chemical.

Enhanced pre-precipitation

Based on the analysis of SS and COD_{Cr}, there was a significant difference between the load in the pre-settled water during the full-scale test (May) and the reference period (Ref) (Fig 2). The average dose was in both cases 10 g Fe/m^3. The load of SS in the pre-settled water was 45% lower and the organic load (COD_{Cr}) 30% lower than in the reference period. The dashed line (Figure 2) shows the objective to reduce the load in the pre-settled water by one third. The conclusion was that PIX-111 could by far meet the goal for SS removal and almost meet the goal regarding organics removal. When the dose was raised to 13 g Fe/m^3 for a couple of days at the start of the test, an enhanced removal of almost 40% was actually reached, indicating that a higher dose would result in an even higher organics removal. It can be noted that the load of incoming water was lower in May than in the reference period due to lower flows.

Moreover, the reduction of Tot-P and BOD_7 was remarkably improved with ferric chloride (Table 4).

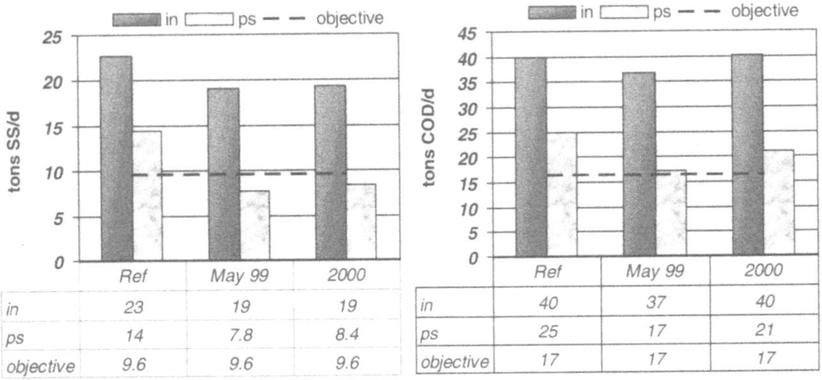

Fig. 2. Incoming and pre-settled SS and organic load in the reference period (Ref), during the test in May (May) and during the first 14 weeks of the construction period (2000)

The results from the construction period (2000) are in the same range regarding SS, BOD_7, and Tot-P removal, but not as impressive when looking at COD removal. The organic load in pre-settled water was only 15% lower than the load in the reference period. In fact, the reduction of SS and organic matter (COD_{Cr} and BOD_7) was overall higher in May than in the winter of 2000. However, the reduction in winter 2000 was still higher than that during the reference period (Table 4).

To try to improve the removal of organics, the dose was raised to 13 g Fe/m^3 and after one week to 15 g Fe/m^3. The higher dose did not improve the reduction, however, so the dose was lowered to 11 g Fe/m^3.

The reason for the better results in May was not yet determined when writing this paper. It was, however, established that the COD removal reached the same levels as in May 1999 at the end of April. Therefore, a reasonable explanation could be differences in wastewater characteristics, for example, charge, soluble organic matter, and temperature between the winter and the spring period. There was, however, no significant difference between the alkalinity in the pre-settled water during these periods; the average value was 225-230 g HCO_3/m^3.

Table 4. The reduction (%) of COD_{Cr}, SS, BOD_7 and Tot-P calculated from the loads into the plant and the loads in the pre-settled water.

Parameter	Ref	May	2000
COD_{Cr}	37	53	47
SS	43	58	57
BOD_7	37	51	49
Tot-P	13	64	65

Effects on the biological processes

When the construction work started in January 2000 and 33% of the tank volume was taken out of operation it was considered important to get a high fraction of bacteria (VSS/SS) in the tanks and to avoid an increase of the food-to-microorganism ratio (F/M). If these objectives could be met, the same volume of pre-settled water could be biologically treated and hence the quality of the outlet water would be maintained. With an unchanged BOD_7/TKN ratio in the pre-settled water the nitrification process could most likely be preserved.

The organics ratio of the suspended solids (VSS/SS) was 67% in the activated sludge when ferrous sulphate was used for pre-precipitation. After the change to PIX-111 for the test in May the VSS/SS ratio went up to 76% in less than a week. The average VSS/SS ratio was 78% during the first 14 weeks in 2000. This is in agreement with the fact that the concentration of ferrous was 8 g Fe/m^3 in pre-settled water when 10 g Fe/m^3 as ferrous sulphate was used. With the same dose of ferric, the average concentration was 3.5 g Fe/m^3. This is due to the fact that a dose of 10 g Fe/m^3 as ferrous sulphate left a residual 8 g Fe/m^3 in the pre-settled water, whereas the same dose of ferric chloride (PIX-111) left only 3.5 g Fe/m^3 on average. As a consequence of the decreased content of inorganic sludge, the wasting rate (Q_w) could be reduced by 20-30%.

The F/M ratio calculated as kg BOD_7/kg SS·d (*F/M* in Figure 3) was slightly higher during the construction period. When the F/M ratio was calculated with respect to kg VSS and day, however, the parameter ended up at the same value as the year before, 0.25 kg BOD_7/kg VSS·d (*F/M** in Figure 3). This indicates that the organic load on the biological tanks did not rise, due to the enhanced pre-precipitation, when one third of the volume was taken out of operation.

At Bromma WWTP the aerobic sludge retention time (SRT) is very low, 3-4 days, regarding the fact that it operates with all-year-round nitrification (Figure 4). Due to a lower production of activated sludge and hence a lower wasting rate, the aerobic SRT did not decrease during the construction period.

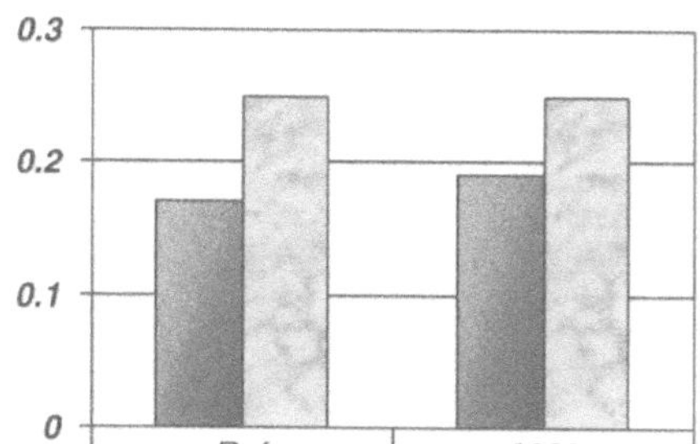

	Ref	2000
F/M	0.17	0.19
F/M*	0.25	0.25

Fig. 3. The F/M ratio (kg BOD_7/kg SS·d) and F/M* ratio (kg BOD_7/kg VSS·d) during the reference period and during the construction period

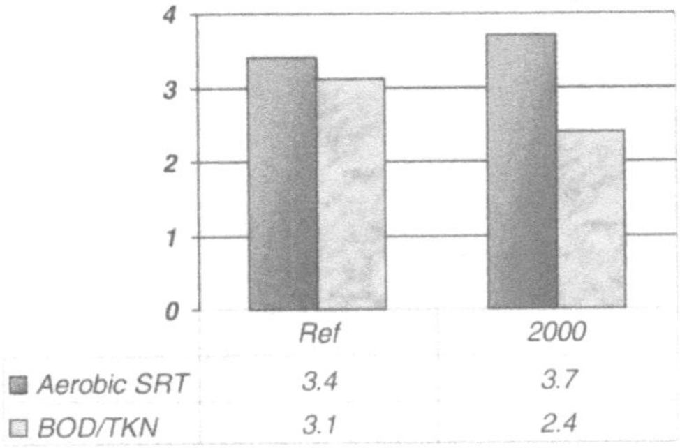

Fig. 4. The aerobic SRT and the BOD_7/TKN ratio during the reference period and during the construction works

The BOD_7/TKN ratio decreased from 3.1 in the reference period to 2.4 during the construction period. The lower ratio favored the nitrifiers, and the concentration of ammonia in the effluent water (NH_4-N_{out}) did not differ from the year before (Figure 5). It can be noted that the average temperature, 12 °C, was one degree warmer in the winter of 2000 than in the winter the year before. A higher temperature has, of course, a positive effect on the nitrification process. The measured value of NH_4-N_{out} includes the wastewater that bypassed the biological treatment step. At Bromma WWTP the sludge characteristics are normally poor during the winter period and good in the summer/autumn period. The SQI-value usually rises in December, when the first snow melts, and remains high until June. In wintertime 1999 the SQI equaled 105 ml/g and during the construction period it was 150 ml/g on average due to filamentous bacteria. The inferior sludge characteristics during the construction period could probably be attributed to the high removal of organics by PIX-111, as a lack of substrate is known to favor the growth of filamentous bacteria.

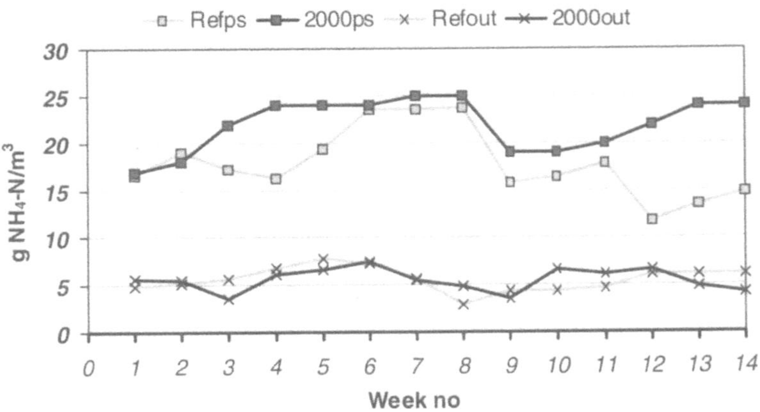

Fig. 5. The NH_4-N-concentration in the pre-settled water and in the effluent

Effects on the sludge treatment

Due to an enhanced pre-precipitation the mass flow of primary sludge (PS) to the digesters was about 15-20% higher when PIX-111 was dosed. As mentioned earlier, the wasting flow was lower: the flow from the centrifuge was 22% lower than that in the reference period. The PS was fed to five digesters and the dewatered waste-sludge (DWS) to one digester during 1999 and 2000.

The organic load on the tanks digesting PS raised from 1.2 to 1.4 kg VS/m^3 of digester volume and day. This resulted in a higher production of gas, which increased by 19% (Figure 6).

Neither the specific production of gas nor the degree of digestion changed markedly when PIX-111 was dosed; the specific production of gas stayed at 0.7 m^3 gas/ kg VS fed, while the degree of digestion was about 50%.

A lower wasting flow resulted in a lower production of gas from the digester for waste activated sludge; 73 m^3/h compared to 92 m^3/h in the reference period. The organic load was slightly higher, 2.4 compared to 2.0 kg VS/m^3·d, but an HRT of 14 days was probably too short to digest the extra organic matter.

The overall production of gas from all the digesters increased by 11%.

In December 1998 two new centrifuges for dewatering the digested sludge were installed. It is not possible to evaluate the effect of the new pre-precipitation chemical since the operation of the centrifuges has been subject to changes after the installation as the personnel has optimized the process continuously. It can, however, be pointed out that the ferric chloride has probably not had an adverse effect on the dewatering process since the average DS was 35% from January to April this year and 30% in 1999.

The total sludge production was 17 tons of DS per day from January to March this year and 16 tons of DS per day for the same months in 1999. The small increase could be due to a higher dose of iron during the construction period.

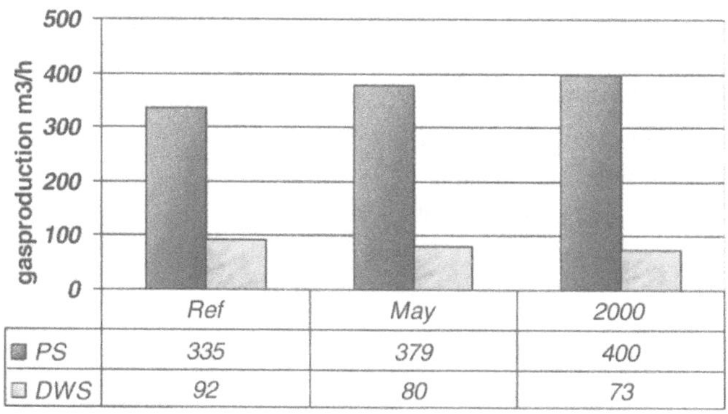

Fig. 6. The production of gas in digesters fed with primary sludge (PS) and from the digester fed with dewatered waste-sludge (DWS)

Conclusions

The objective was to reduce the load of organic matter and SS in the pre-settled water by one third in order to retain nitrification during the construction period. The reduction due to pre-precipitation with ferric was higher during the full-scale test than during the construction period and therefore the results are presented in an interval.
With 10-13 g Fe/m^3 ferric chloride (PIX-111)

☐ the load (tons/d) in pre-settled wastewater was reduced by:

- 40-45% regarding SS
- 15-30% regarding COD_{Cr}
- 25-35% regarding BOD_7 and
- 45-55% regarding Tot-P

The enhanced removal had the following effects:

☐ the production of digester gas increased by

- 19% for the digestion of primary sludge
- 11% regarding digestion of both primary and waste activated sludge (total production of gas)

☐ the VSS/SS ratio in the activated sludge rose from 67% to 78%,
☐ the BOD_7/TKN ratio decreased from 3.1 to 2.4 and
☐ the SQI was 45 units higher (105 ml/g in wintertime 1999 and 150 ml/g in 2000).

During the construction period, when one third of the aerobic volume was taken out of operation, the consequences of a more organic activated sludge were that:

☐ the aerobic SRT did not decrease
☐ the F/M ratio (kg BOD_7/kg VSS·d) did not increase
☐ the nitrification was retained and
☐ the requirements from the authorities were met.

Removal of Phosphonates and Polyphosphates in the Coagulation Process

J. Fettig, P. Seydel and C. Steinert

University of Paderborn, Dept. of Environmental Engineering, An der Wilhelmshöhe 44, D-37671 Höxter, Germany
j-fettig@moellinger.hx.uni-paderborn.de

Abstract

The interaction of both phosphonates and tripolyphosphate with three coagulants, PAX 20, alum, and ferric chloride, was studied in laboratory experiments. Besides coagulation of P-components in synthetic test water, adsorption on preformed hydroxide flocs was evaluated. Phosphonates were removed more efficiently by the Al coagulants, however, adsorption on flocs did not seem to be the only removal mechanism. Alum was the best-suited coagulant for polyphosphate removal. Also in this case several mechanism must be assumed. Water hardness improved the removal of both P-species.

Introduction

Phosphonates are a group of organic compounds that contain one or more phosphonic acid groups $-C\text{-}PO_3H_2$. They are applied for different purposes, e.g. as sequestrants of bivalent metal ions in the textile industry, as inhibitors of corrosion and scale formation in cooling systems, as stabilisers for bleaching chemicals and enzymes in cleaning agents, and as deflocculants for pigments and other solids in suspensions. According to an estimate by a Dutch group of experts, more than 11.000 t/a of phosphonates were used in Western Europe in 1996, while the corresponding amount in the US was about 9.500 t/a in 1992 [1]. Their predominant field of application is in the industries although they are also used in households as components of detergents and dish washing products.

Phosphonates are characterised by high water solubilities and a good thermal and chemical stability. However, in many applications they cannot be kept in a closed-loop process but will become a compound of an industrial or domestic wastewater stream after use. In activated sludge plants they must be considered as hardly biodegradable [2]. Therefore there is some concern about potential negative effects in the environment [3], and studies on their environmental concentrations have been conducted in several European countries.

In Sweden, total phosphonate concentration in raw sewage in cities was estimated to about 350 μg/l, resulting in effluent concentrations of 190 μg/l in two-stage treatment plants due to sorption on activated sludge flocs, and 90 μg/l in three-stage plants due to additional removal by coagulants [4]. In Belgium, phosphonate concentrations in raw sewage were calculated to 170-290 µg/l and in treated sewage to 17-145 µg/l [5]. In Switzerland, the fate of three phosphonates in seven treatment plants was investigated [6]. The total influent concentrations varied between about 60 and 1220 µg/l, the effluent concentrations were below detection limit (15-30 µg/l) except in one case where still 70 µg/l were found. However, in all plants tertiary treatment was applied, and phosphonate elimination was attributed to sorption processes on metal hydroxide containing flocs.

In Germany, the behaviour of six phosphonates in different drinking water treatment processes was studied at concentrations of 50 µg/l [7]. Complete removal could only be reached by coagulation, where alum performed better than ferric chloride. Also in this study sorption processes were considered being the major removal mechanisms.

Polyphosphates produced by condensation of orthophosphates are compounds with –P-O-P- bondings. At least 750.000 t-P/a of condensed phosphates were applied world-wide in 1996 [8]. They are mainly used in detergents and cleaners, as food additives, for water conditioning, and in ceramics, leather and textile industries. The specific properties which are utilised here, include complexation of polyvalent cations for water softening, adsorption on solid surfaces for dispersion of particles, and alkalisation of cleaning liquids.

The most often applied condensed phosphate, sodium tripolyphosphate $Na_5P_3O_{10}$ (STPP), was the main constituent in powder laundry detergents for many years. In the seventies ca 250.000 t-P/a were used for that purpose in Western Europe. After 1986, phosphorous in detergents was banned in several countries including Norway and Switzerland, while the level was limited by law or voluntary agreements in other states. As a result, the phosphorous consumption for detergents in Western Europe had decreased to about 140.000 t-P/a already in 1989 [9].

Condensed phosphates undergo hydrolysis to orthophosphate within a certain time frame. The reactions are enhanced by high temperatures, alkaline pH values, and enzymatic activities in the water [10]. It is assumed that STPP from households is almost completely hydrolysed in the sewer system and in the treatment plant, respectively [11]. However, under non-favourable conditions, e.g. in wastewater from certain industries, higher concentrations of condensed phosphates might be encountered in the treatment stage. Furthermore biologically treated sewage was found to contain several 0.1 mg/l of polymeric phosphates that may be released from dead bacteria [12]. It has been shown that condensed phosphates are removable by precipitation, however, the dosages required for efficient elimination are reported to be higher than for orthophosphate [11,13].

In this study, the behaviour of phosphonates and STPP in the coagulation process was investigated. The conventional jar test procedure was used, and three different coagulants were compared with respect to phosphorous elimination and turbidity removal. In addition, the same phosphorous compounds were added to systems containing freshly precipitated metal hydroxides in order to assess the contribution of sorption processes to overall removal efficiencies. Finally, jar tests were conducted with textile factory wastewater containing both phosphonates and polyphosphates.

Materials and Experimental Methods

A synthetic test water was used in order to conduct the study at constant boundary conditions. The test water's background composition is given in Table 1. Kaolinite representing the solid matter was a commercial product with a particle size range of 0-100 μm and a mean diameter of 19.5 μm. Initial pH was adjusted by adding appropriate amounts of HCl.

Table 1. Synthetic background test water prepared with deionized water

Compound		Soft water	Hard water
$NaHCO_3$	[mmole/l]	5	5
$CaCl_2$	[mmole/l]	0.2	2
Kaolinite	[mg/l]	100	100
pH_0		7.5	7.5

The phosphonates added to the test water originate from the complexing agent Securon 540 (Henkel, Germany) that is frequently used in the textile industries. This liquid product is free of other organic surfactants and reacts slightly acidic, i.e. the pH of a 1% solution is about 6. The phosphorous content of the product was determined to 83 g P/l. A powder laundry detergent used for cleaning of working clothes in industries was taken as the source for STPP. In this case the phosphorous content of the product was determined to 8.8 %. A stock solution of 10 g/l was prepared freshly prior to each experimental series from which the amounts required were taken.

Biologically treated wastewater from a textile factory in Northrhine-Westphalia (Germany) was used in an additional series of tests. Both phosphonates and polyphosphates are applied in the factory, and at the time of the study, these compounds made up for about 65 % of the total P-concentration of 7.7 ± 2.5 mg P/l in the raw wastewater. Since no coagulants were added in the plant at that time, total-P removal was only on the order of 5 %. However, during the wastewater's long detention time of about seven days in the two-stage treatment plant, non-orthophosphates were converted to orthophosphate, and the effluent contained 6.3 mg P/l of the latter and only 1.0 mg P/l of other P-components on an average.

A jar test apparatus (IKA-Werke, Germany) equipped with 1 l jars was used that were filled with 0.8 l of test water each. After addition of the coagulant, 1 min of rapid mixing (200 min^{-1}), 15 min of slow mixing (20 min^{-1}) and 30 min of settling were used in the coagulation experiments. The adsorption studies were conducted as follows: Constant amounts of coagulant were added to 4 jars filled with the background test water. A fifth jar was taken as a reference without coagulant addition. After 1 min of rapid mixing (200 min^{-1}) and 15 min of semi-rapid mixing (80 min^{-1}) when the coagulants had completely been precipitated as hydroxides because of solution pH, different amounts of phosphonates or STPP, respectively, were added while pH was kept constant by adding HCl or NaOH. The systems were mixed (20 min^{-1}) for another 15 min and then allowed to settle for 30 min.

After sedimentation the supernatant was analysed with respect to residual turbidity (Ratio/XR turbidimeter, Hach Co, US) and pH (pH-meter 539, WTW, Germany). Samples filtered through 0.45 µm membranes were taken for total-P analysis, except in the series with textile wastewater where non-filtered samples were used. The phosphorous species were converted to orthophosphate by microwave digestion using potassium persulphate. Orthophosphate was measured according to the German Standard Method DIN 38 405 part 11.

Aluminium sulphate (ALG) and PAX 20 provided by Kemwater, Sweden, and ferric chloride obtained as a reagent-grade product $FeCl_3$*6 H_2O from Merck, Germany, were used as coagulants. Stock solutions of 1 g Al/l (ALG) and 5 g Fe/l were prepared for dosing. Pax 20 which contains 5.9 % Al and 0.6 % Fe, was considered having an Al equivalent of 6.2 % as a simplified formula.

Results and Discussion

Removal of phosphonates by coagulation

The results on P-removal from soft test water by PAX 20 are shown in Figure 1 for three different initial concentrations of the phosphonates. Similar results were obtained with ALG and $FeCl_3$, respectively. Residual concentrations well below 1.0 mg P/l could be reached in all of the series, however, the amount of coagulant required increased clearly with increasing P-concentration.

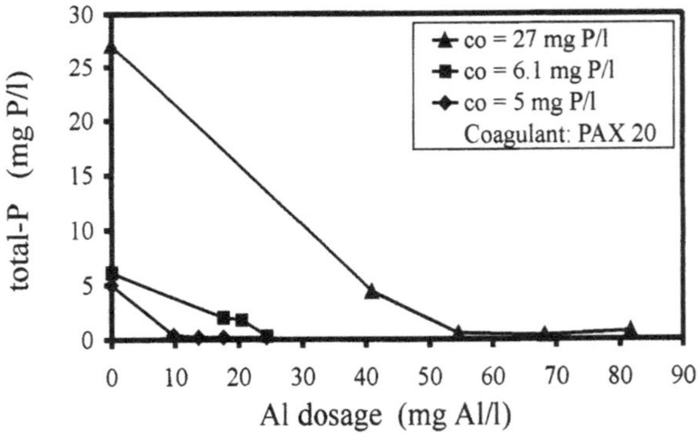

Fig. 1. Phosphonate removal by PAX 20 in the coagulation process

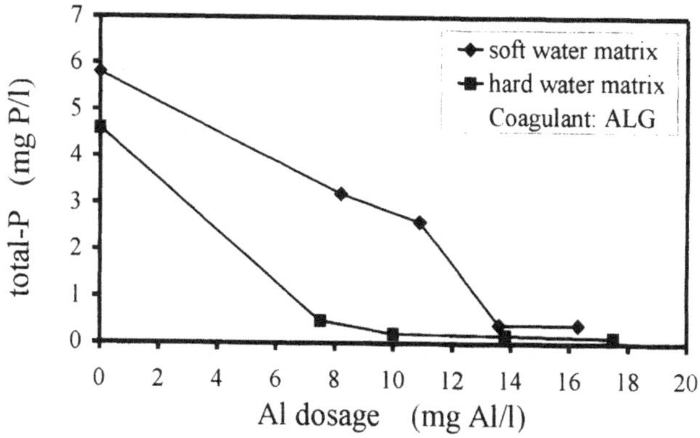

Fig. 2. Phosphonate removal by ALG from soft and hard test water

The effect of increased water hardness on phosphonate removal is illustrated in Figure 2. The results obtained with the other coagulants were similar. It is obvious that the P-species are more easily removed when 2 mmole/l of Calcium ions are present instead of 0.2 mmole/l. Since alkalinity in both types of test water was the same, this effect cannot be attributed to differences in pH, but rather to increasing formation of Calcium phosphonate complexes in hard water prior to coagulation

The data were evaluated by relating the Al or Fe dosage required in order to obtain residual concentrations of 1 mg P/l, to the initial P-concentration on a molar base. This ratio corresponds to the parameter β used in Germany for the estimation of coagulant demand in wastewater treatment [14]. The results given in Table 2 show that PAX 20 and ALG have similar efficiencies in soft water while $FeCl_3$ is much less efficient. In hard water, all of the coagulants perform better, although the relative improvement is higher for ALG and $FeCl_3$ than for PAX 20. Compared with coagulant dosages typical for precipitation of orthophosphate, i.e. β = 1.5–2 with monomeric metal salts [14], the values obtained here are considerably higher.

These data are in agreement with observations made by Metzner and Nägerl [15] who needed an alum dosage corresponding to β = 4.7 in order to remove 74 % of initially 1.9 mg P/l of phosphonates from an aqueous solution. Horstmann and Grohmann [16] reported that precipitation of orthophosphate was inhibited by 0.2–1 mg/l of phosphonates, but this influence could be overcome by higher coagulant dosages. Held [17] found that alum removed a particular phosphonic acid much better than ferrous sulphate. Also in these studies, phosphonate removal was attributed to sorption processes on metal hydroxide flocs.

Table 2. $\beta = c(Me)/P_0$ (mole/mole), derived from dosages required for $P_{res} = 1$ mg/l

Coagulant	Soft test water	Hard test water
PAX 20	2.81 ± 1.24	2.36
ALG	2.84 ± 0.39	1.62
$FeCl_3$	5.34 ± 2.11	2.86

Removal of phosphonates by adsorption

The adsorption tests were only conducted with soft test water. Based on the initial and residual concentrations measured, the amounts of phosphonate adsorbed were calculated from a mass balance. The results are presented in Figure 3 as adsorption isotherms, together with "P-capacities" obtained in the coagulation tests that were calculated by dividing the amounts of P removed by the corresponding metal ion dosages. In that way, removal by adsorption onto preformed flocs can be directly compared with removal efficiencies in the coagulation process.

The adsorption data show almost identical behaviour of the $Al(OH)_3$-precipitates formed by alum and PAX 20 application, respectively, while smaller amounts of phosphonate are adsorbed on the $Fe(OH)_3$-precipitates. However, on a molar base phosphonate adsorption is on the order of 0.2 mole P per mole Me and thus very similar in all cases. In studies of orthophosphate adsorption on preformed metal hydroxide flocs, the influence of parameters like pH, contact time, water hardness and floc age has been evaluated [18, 19, 20]. When data from these studies and from this work obtained under similar conditions are compared, a fairly good agreement can be stated between the adsorption of orthophosphate and phosphonates on preformed flocs.

Meanwhile the coagulation data, although scattering to some extent, correspond to capacities that are almost completely higher than the isotherms. The effect is more pronounced with the Al coagulants than with the iron salt. Therefore it is concluded that additional mechanisms for phosphonate removal must be accounted for in the coagulation process. If one assumes that reactions of the phosphonates with metal ions or positively charged metal hydroxo complexes occur simultaneously with the precipitation of $Me(OH)_3$, one possible explanation would be that the reaction products are either incorporated in part into the flocs or that they contribute to an improved phosphonate binding to the floc surfaces. However, a verification of this hypothesis would require a thorough investigation of complex formation reactions of the phosphonate species which was beyond the scope of this study. The observation that aluminium species are obviously more efficient than iron species in this respect, agrees with the findings of Klinger et al. [7].

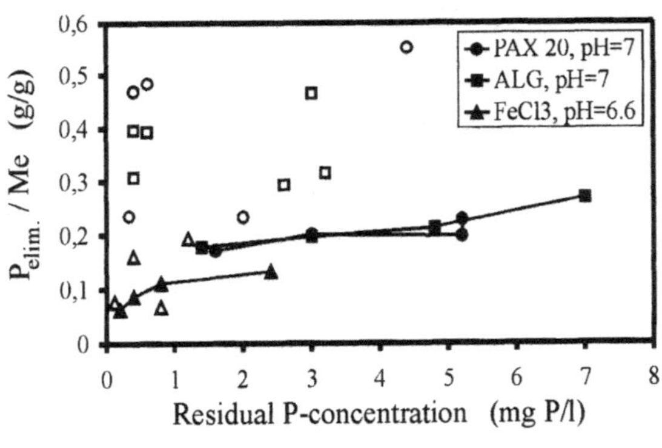

Fig. 3. Adsorption data of phosphonates on freshly precipitated metal hydroxides in soft test water compared with "capacities" derived from coagulation results (open symbols)

Removal of polyphosphates by coagulation

The results on P-removal from soft test water by ALG are shown in Figure 4 for two different initial concentrations of STPP. Similar results were obtained with PAX 20 and $FeCl_3$, respectively. Residual concentrations below 1.0 mg P/l were reached in all of the series, however, the amount of coagulant required was strongly dependent on the initial P-content.
The effect of increased water hardness on polyphosphate removal is illustrated in Figure 5 for PAX 20 as a coagulant. It can be seen that STPP is more easily eliminated when 2 mmole/l of Calcium ions are present instead of 0.2 mmole/l. Similar results were obtained with the other coagulants.

The data were again evaluated by relating the Al or Fe dosage required in order to obtain residual concentrations of 1 mg P/l, to the initial P-concentration on a molar base. The results given in Table 3 show that ALG is the most efficient coagulant in soft water. In hard water, all of the coagulants perform better, although the relative improvement is higher for PAX 20 than for $FeCl_3$ and ALG.

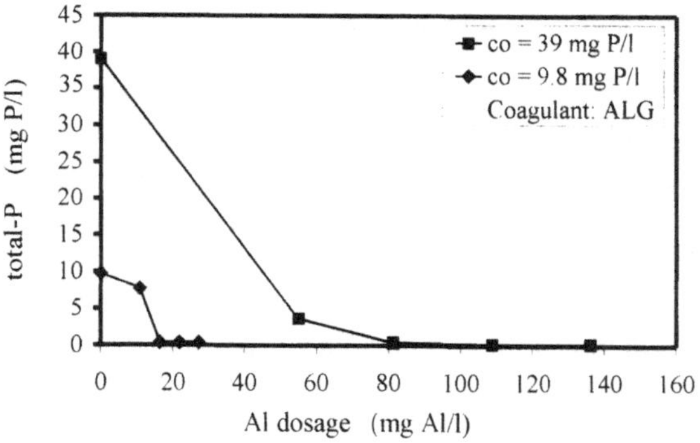

Fig. 4. Polyphosphate removal by ALG in the coagulation process

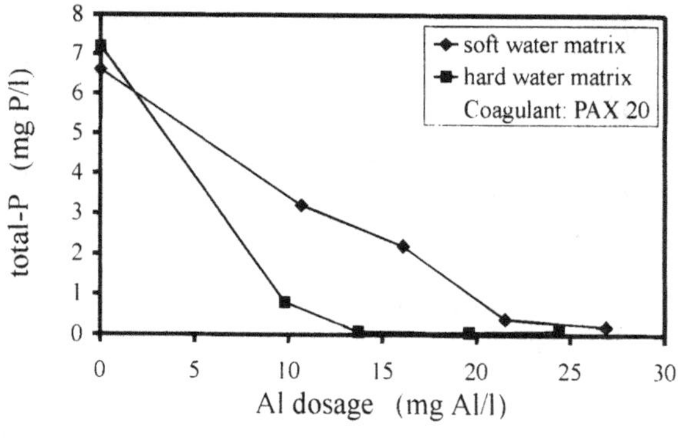

Fig. 5. Polyphosphate removal by PAX 20 from soft and hard test water

Table 3. $\beta = c(Me)/P_0$ (mole/mole), derived from dosages required for $P_{res} = 1$ mg/l

Coagulant	Soft test water	Hard test water
PAX 20	3.79 ± 0.67	1.51
ALG	2.18 ± 0.28	1.55
$FeCl_3$	3.31 ± 0.23	2.20

According to these data polyphosphates can efficiently be removed in the coagulation process, but the coagulant demand is significantly dependent on water hardness. This effect must be accounted for when data from the literature are compared. The strong influence of Calcium was not observed for the precipitation of orthophosphate by aluminium salts [21]. Since a higher amount of Calcium polyphosphate complexes are likely to be formed in hard water, it can be assumed that these complexes are more easily removable than the polyphosphate anions.

Removal of polyphosphates by adsorption

The adsorption tests were only conducted with soft test water. Based on the initial and residual concentrations measured, the amounts of tripolyphosphate adsorbed were calculated from a mass balance. The results are presented in Figure 6 as adsorption isotherms, together with "P-capacities" obtained in the coagulation tests that were calculated by dividing the amounts of P removed by the corresponding metal ion dosages.

The adsorption data show a similar behaviour of the $Me(OH)_3$-precipitates formed from alum and $FeCl_3$, respectively, while larger amounts of STPP are adsorbed on

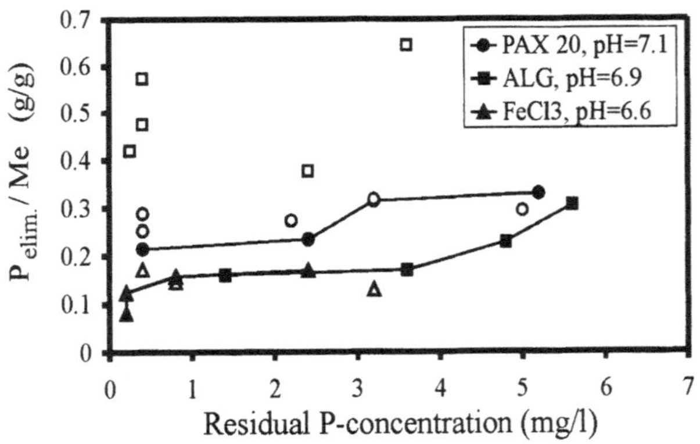

Fig. 6. Adsorption data of polyphosphates on freshly precipitated metal hydroxides in soft test water compared with "capacities" derived from coagulation results (open symbols)

the precipitates formed from PAX 20. On a molar base, adsorption capacities for STPP are in the range of 0.15–0,27 mole P per mole Me and thus in the same order as the phosphonate adsorption results. They also agree with adsorption capacities on $Al(OH)_3$ flocs of about 0,28 mole P per mole Me that were found by Galarneau and Gehr [22] for another condensed phosphate at pH = 7.

A comparison of adsorption and coagulation data shows no clear difference for the coagulants PAX 20 and $FeCl_3$, respectively. Therefore, removal of STPP in the coagulation process could be attributed here exclusively to adsorption on the flocs. However, for ALG the picture is quite different: The capacities related to coagulation are much higher than the isotherm data. Hence there must be an additional mechanism for polyphosphate removal by alum coagulation. If one compares the properties of the coagulants, this effect seems to be Al-specific (since it is not observed with Fe(III) ions), and it is related to the presence of monomeric Al species. Further studies will be needed in order to better understand the interactions between polyphosphates and trivalent cations.

Coagulation of textile wastewater

Residual total-P concentrations and residual turbidity in textile wastewater are shown in Figure 7 as a function of coagulant dosage. PH in treated water was between 6.8 and 7.3. Accordingly, residual P-concentrations of 0.5 mg P/l were reached with both Al coagulants, while $FeCl_3$ was less efficient. This can be attributed in part to poor turbidity removal by $FeCl_3$. However, about 1 mg P/l in the wastewater was due to organic phosphorous components. There is evidence that it may have played a certain role as well, since the specific Al dosages related to 1 mg/l of residual P corresponded to β-values as high as 3.2-3.4.

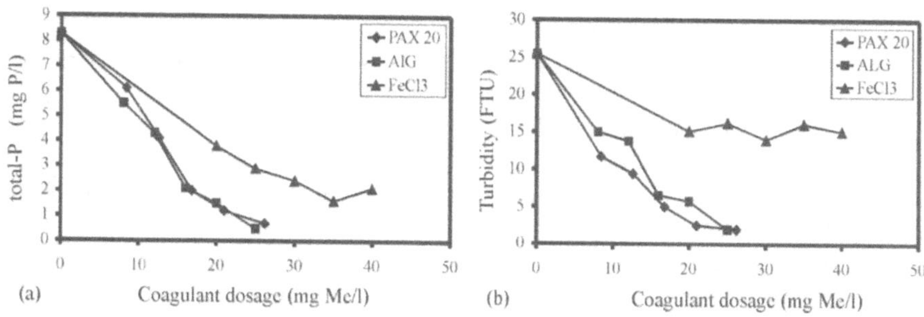

Fig. 7. Removal of (a) total-P and (b) turbidity in biologically treated textile wastewater

Conclusions

1. Alum and PAX 20 are more efficient for phosphonate removal by coagulation than $FeCl_3$. Adsorption on flocs does not seem to be the only relevant mechanism here. High water hardness results in lower coagulant demand.
2. Alum is also quite efficient for the removal of tripolyphosphate, while PAX 20 and $FeCl_3$ are less effective coagulants in soft water. With respect to the latter systems, adsorption on flocs might be the predominant mechanism.
3. These findings can be used to estimate P-removal from industrial wastewaters.

References

1. Overleggroep: Phosphonates in Domestic Laundry- and Cleaning Agents. Nota 4, Consultative Expert Group Detergents - Environment, Zeist, The Netherlands (1997)
2. Horstmann, B., Grohmann, A.: Untersuchungen zur biologischen Abbaubarkeit von Phosphonaten. Vom Wasser 70 (1988) 163
3. Koch, M.: Komplexchemisches und bakterientoxisches Verhalten ausgewählter Schwermetallphosphonatkomplexe. Oldenbourg Verlag, München, Germany (1995)
4. Landner, L., Walterson, E.: Phosphonates Used for Detergent Applications. Swedish Environmental Research Group, Stockholm, Sweden (1993)
5. Vandepitte, V.: Eco-physiological aspects of phosphonates and their biodegradation in the environment. Ph.D. Thesis, University of Gent, Gent, Belgium (1994)
6. Nowack, B.: The Behavior of Phosphonates in Wastewater Treatment Plants of Switzerland. Wat. Res. 32 (1998) 1271
7. Klinger, J., Sacher, F., Brauch, H.-J., Maier, D., Worch, E.: Verhalten organischer Phosphonsäuren bei der Trinkwasseraufbereitung. Vom Wasser 91 (1998) 15
8. Staffel, Th.: Production and Application of Sodium phosphates and potassium phosphates. Product Information, BK Giulini GmbH, Ladenburg, Germany (2000)
9. Morse, G.K., Lester, J.N., Perry, R.: The Environmental and Economic Impact of Key Detergent Builder Systems in the European Union. Selper Publ., London, UK (1994)
10. Jolley, D., Maher, W., Cullen, P.: Rapid Method for Separating and Quantifying Orthophosphates and Polyphosphates. Wat. Res. 32 (1998) 711
11. Gleisberg, D.: Kontrolle des Phosphors in der Umwelt – zu Vorkommen, Elimination und Algenverfügbarkeit. Z. Wasser-Abwasser-Forsch. 24 (1991) 157
12. Bendt, T., Pehl, R., Pullmann, R., Rolfs, C.H.: Möglichkeiten und Grenzen des on-line-Monitoring von Phosphor in Kläranlagen. Vom Wasser 84 (1995) 79
13. Smith, R.S., Cohen, J.M., Walton, G.: Effects of Synthetic Detergents on Water Coagulation. J. AWWA 48 (1956) 55
14. ATV Abwassertechnische Vereinigung e.V.: Verfahren zur Elimination von Phosphor aus Abwasser. Arbeitsblatt A 202, Hennef, Germany (1992)
15. Metzner, G., Nägerl, H.D.: Umweltverhalten zweier Wasserkonditionierungsmittel auf Phosphonat- und Polyacrylatbasis. Tenside Detergents 19 (1982) 23
16. Horstmann, B., Grohmann, A.: Der Einfluß von Phosphonaten auf die Phosphatelimination. Z. Wasser-Abwasser-Forsch. 19 (1986) 236
17. Held, S.: Zum Umweltverhalten von Komplexbildnern auf Phosphonsäurebasis. Textilveredlung 24 (1989) 394

18. Lijklema, L.: Interaction of Orthophosphate with Iron(III) and Aluminum Hydroxides. Environm. Sci. Technol. 14 (1980) 537
19. Thole, S., Borho, M., Voswinckel, G., Jekel, M.: Einfluß der Wassermatrix auf die Adsorption von Phosphat an Eisenoxidhydratschlämmen. Vom Wasser 79 (1992) 313
20. Boisvert, J.P., To, T., Berrak, A.: Phosphate Adsorption in Flocculation Processes of Aluminum Sulphate and Polyaluminum-Silicate-Sulphate. Wat. Res. 31 (1997) 1939
21. Fettig, J., Miethe, M, Kassebaum, F.: Coagulation and Precipitation by an Alkaline Aluminium Coagulant. In: Chemical Water and Wastewater Treatment IV, H. H. Hahn, E. Hoffmann, H. Ødegaard (Eds.). Springer, Berlin Heidelberg New York (1996) pp. 107-117
22. Galarneau, E., Gehr, R.: Phosphorus Removal from Wastewaters: Experimental and Theoretical Support for Alternative Mechanisms. Wat. Res. 31 (1997) 328

Industrial Wastewater Treatment

Photochemical and Photocatalytic Detoxification of Reactive Dyebath Wastewater by the Fenton's Reagent and Novel TiO_2 Powders

I. Arslan*, I. Balcioglu*, T. Tuhkanen** and D.W. Bahnemann***

*Bogaziçi University, Institute of Environmental Sciences, 80815 Bebek - Istanbul, Turkey
idilarslan@superonline.com
**Tampere University of Technology, Institute of Water and Environmental Engineering, P.O. Box 541, FIN-33101 Tampere, Finland
***ISFH GmbH, Sokelantstrasse 5, D-30165 Hannover, Germany

Abstract

The study described herein aimed to investigate the homogeneous (Fenton/UV-C) and heterogeneous (TiO_2/UV-A) advanced oxidation of commonly used reactive dyestuffs in simulated dyebath effluents. For the Photo Fenton (Fe^{2+}/H_2O_2/UV-C) reaction the effect of the initial H_2O_2 dose on the oxidative reaction rates was examined. Further, the photocatalytic treatment efficiency of two novel anatase TiO_2 powders (Millennium PC 500 and Sachtleben Mikroanatas) and platinized Degussa P25 TiO_2 type were compared with that of the more conventional P25 at different reaction pH's. In the case of the Photo Fenton reactions, the treatment performance was strongly affected by the initial H_2O_2 concentration, whereas for the TiO_2/UV-A process only the PC 500 TiO_2 powder exhibited a significantly pH dependent reaction efficiency. The decolourization rates followed the decreasing order of Fenton/UV-C > Pt-P25/UV-A > Mikroanatas/UV-A > P25/UV-A > PC 500/UV-A, whereas the decreasing order for TOC removal efficiency was Fenton/UV-C > Mikroanatas/UV-A > Pt-P25/UV-A > P25/UV-A > PC 500/UV-A for the selected reaction conditions. Removal of optical density at 254 nm wavelength ranged between 75 % (Mikroanatas/UV-A) and 96 % (P25/UV-A).

Introduction

Wastewater from the textile industry is highly colored and of a complex and variable nature depending upon the fibers processed as well as the associated dyestuffs and auxiliary chemicals employed [1]. The problem of colored effluent has become particularly identified with the dyeing of cellulose fibers and the use of reactive dyestuffs of which as much as 50 % may be lost to the effluent and cannot be reused [2]. Many of these synthetic dyestuffs are resistant to conventional treatment processes (biotreatment, coagulation, activated carbon adsorption etc.) due to their complex polyaromatic structure and hence cause aesthetic and environmental problems [3]. There is also concern that the aromatic amines, which are formed as metabolites by reductive cleavage of the -N=N- bonds in azo dyes under anaerobic conditions, could pose a more serious toxic hazard than the intact dye molecules [4]. Hence removal of effluent colour has become the integral part of effective textile wastewater treatment.

The destruction of hazardous and refractory organic pollutants by so-called advanced oxidation processes (AOPs), where strong oxidizing agents (e.g. O_3, H_2O_2) are combined with UV light ($\lambda < 300$ nm) to generate hydroxyl radicals ($OH^{\bullet}$), have been thoroughly and comparatively studied in the last three decades [5]. Recent investigations have proven that organic compounds like dyestuffs can be oxidized more quickly and thoroughly by the Fenton's reagent ($Fe^{2+} + H_2O_2$) as compared to the H_2O_2/UV-C process [6-8]. However, the Fenton's oxidation process requires high amounts of both reagents for effective oxidation and Fe(III)-iron sludge has to be removed after treatment [9]. Another drawback is that the Fenton's reagent can not completely mineralize organic pollutants. The rate of degradation of organic compounds via Fenton's reaction can be strongly accelerated by the introduction of UV light [10,11]. The increased efficiency of the Photo Fenton reaction is attributable to the photoreduction of the ferric ions formed during the Fenton's reaction [12];

$$Fe^{3+} + h\upsilon + H_2O \xrightarrow{\lambda<400\,nm} Fe^{2+} + H^{+} + OH^{\bullet} \qquad (1)$$

During the photoreduction of ferric ion to ferrous iron an additional $OH^{\bullet}$ is generated and the reproduced ferrous ion (Fe^{2+}) reacts with H_2O_2 to make Fe^{3+} and $OH^{\bullet}$ so that the cycle can continue. Consequently, significantly smaller amounts of Fe^{2+} can be used for treatment and less iron sludge produced.

On the other hand, heterogeneous photocatalysis has emerged as an effective alternative method for purifying water and air [13]. Up to now, titania (TiO_2) under the shape of mainly anatase has always been found as the most efficient photocatalyst. The decomposition of organic model pollutants in TiO_2 suspensions is due to photogenerated electron/hole pairs formed in this material under near-UV irradiation (300 nm > λ > 400 nm) [14];

$$TiO_2 + h\upsilon \xrightarrow{300\,nm>\lambda>400\,nm} h_{vb}^{\ +} + e_{cb}^{\ -} \qquad (2)$$

The holes that migrate to the semiconductor surface then react with electron donors (H_2O, OH^-) in the electrolyte to produce $OH^{\bullet}$ according to following reaction;

$$h_{vb}^{\ +} + OH^{-}_{ads} \rightarrow OH^{\bullet} \qquad (3)$$

Several studies have attempted to increase photoactivity of TiO_2 by noble metal deposition or ion doping. On the other hand, novel photocatalysts with better photocatalytic and adsorptive properties as a consequence of their smaller particle size, higher BET surface area or pure anatase crystal phase, were expected to posses superior photoactive properties and hence to increase heterogeneous photocatalytic treatment efficiency [15]. Surface platinization is known to prevent recombination of holes and electrons and to increase the $OH^{\bullet}$ production rates [16].

More recent studies dealing with the photocatalytic degradation of commercial dyes [17] and real dyehouse effluents [18] have revealed that it is possible to degrade dyes completely or to some extend via photocatalytic oxidation. In addition, the solution matrix can influence in particular the photocatalytic reaction rate in several ways. Specially inorganic ions whose presence in textile dyeing wash water at high concentrations cannot be ignored, may inhibit AOPs drastically. For instance the inhibitory effect of Cl^- and CO_3^{2-} ions typically found in textile wastewater at considerably high concentrations is of a rather pH-dependent nature. They can be active as $OH^{\bullet}$ scavengers, compete for active surface sites and absorb the UV-A irradiation necessary for photocatalytic excitation [19]. As a consequence the influence of background matrix requires special consideration.

Considering the above mentioned points, this paper is concerned with the applicability of AOPs for the treatment of various reactive dyestuffs in simulated effluents from the cotton dyeing and rinsing process stages. For this purpose, treatment efficiency of UV-A driven TiO_2-mediated heterogeneous photocatalysis was compared with the UV-C light assisted Fenton (Fe^{2+}/H_2O_2/UV-C) treatment processes. The effect of the initial H_2O_2 dose was investigated for the Photo-Fenton AOPs, while in the case of TiO_2/UV-A oxidation the individual effects of reaction pH and TiO_2 type on the processes efficiency were studied. Treatment performances of the above mentioned AOPs were assessed in terms of global environmental parameters such a first order decolourization rate constants (k_d values), TOC (total organic carbon) and $UV_{254\ nm}$ (aromatic carbon content) reduction.

Experimental Procedure

Simulated reactive dyebath effluent

The simulated dyehouse effluent was prepared as a mixture of four dyeing formulations most frequently used by a local integrated textile manufacturing plant in Turkey for dyeing cotton fabrics with reactive dyes. The synthetic spent dyebath consisting of six reactive dyestuffs (manufacturers Dystar and Clariant) widely applied in this industry and proper amounts of their associated auxiliary chemicals are presented elsewhere [20]. The environmental characterization of the 40 fold diluted reactive dyebath effluent that was studied in the experiments is given in Table 1. This dilution ratio was chosen since it represents the typical strength of effluents from the washing and rinsing stages in the above mentioned textile manufacturing plant. Deionized water used for the preparation of all reagent solutions was obtained from a Milli-Q purification system. H_2O_2 (30 % w/w) and $FeSO_4.7H_2O$ were purchased from Merck. Unreacted H_2O_2 was destroyed by catalase enzyme from bovine liver (780,000 U/mL) to quench the Fenton's reaction and to prevent its interference with the analytical measurements. Necessary pH adjustments were made with 0.1 N NaOH or HCl solutions.

Table 1. Characterization of the 40 fold diluted reactive dyebath effluent

Parameter	Value
COD	44.00 mg/L
TOC	30.55 mg/L
Inorganic Carbon (IC)	12.00 mg/L
CO_3^{2-}	220.83 mg/L
Cl^-	1062.00 mg/L
pH (Units)	10.5
Absorbance at 254 nm	47.10 1/m
Absorbance at 600 nm	38.14 1/m

TiO_2 photocatalysts

Two commercial anatase TiO_2 powders (Mikroanatas and PC 500) and 0.5 % w/w platinized titania (Pt-P25) were tested and compared with ordinary Degussa P25 (P25) for their photocatalytic treatment performance. Physicochemical and technical properties of the selected photocatalysts received from the supplier companies are summarized together with that of Pt-P25 in Table 2.

Table 2. Description of the different TiO_2 photocatalysts used in this study

Photocatalyst Powder	Supplier	BET Surface Area m^2/g	Particle Size nm
Degussa P25	Degussa AG	50	30
Pt-P25	ISFH	45	30
Mikroanatas	Sachtleben Chemie GmbH	271	10
PC 500	Millennium Inorganic Chemicals	287	5 - 10

The photochemical UV-C reactor

UV-C light assisted experiments were carried out for 1 h in an 4000 mL capacity truncated cone shaped Pyrex glass photoreactor (Kreil Corp.) equipped with an immersion type 25 W low pressure mercury lamp. The reactor was specially designed to improve the quantity of absorbed UV-C light [21]. The actual incident radiation flux was determined by hydrogen peroxide actinometry [22] as 2.26 W/L. The photoreactor was thoroughly washed and rinsed with concentrated HNO_3 solution before each use to prevent iron deposition on the reactor walls. Reactions were performed at ambient temperature (20 - 22 °C) and no cooling was provided. In the Fenton/UV-C experiments oxidant (H_2O_2) and catalyst (Fe^{2+} - salt) were only added at the beginning of the reaction directly after pH adjustment of the reaction solution to pH = 2.8 - 3.0.

The photocatalytic UV-A reactor

The experimental setup for the photocatalytic runs consisted of a UV-A light source, a UV-A transmitting Plexiglas reactor, a light homogenizer, a reflector and an air sparging unit. During all experiments the photoreactor was continuously sparged with air at a rate of 35 L/h to guarantee sufficient O_2 in the reacting medium and to keep the reaction solution in suspension. With all 16 UV-A lamps (40 W each) on, the light intensity was measured as 22 W/m^2 by a Dr. Hönle UV-meter directly at the irradiated side of the photoreactor wall. In the TiO_2/UV-A experiments, 150 mL dyehouse effluent was filled into the photoreactor and treated for 1 h. The 16 UV-A lamps were turned on 30 min before the photocatalytic reaction to obtain a constant light output. Prior to photocatalytic treatment, the pH of the reaction samples was adjusted to the desired value and then the TiO_2 powder was added to give a final concentration of 1 g/L in the effluent. Thereafter, the reaction solution was sonicated for 15 min to obtain a homogeneous suspension and magnetically stirred for another 20 min in the dark to allow equilibration of the dyehouse effluent + TiO_2 mixture. As in the case of the UV-C photoreactor, reactions were executed at room temperature and no cooling was provided.

Analytical methods

15 mL samples taken at various time intervals from the UV-C and UV-A photoreactors were analyzed for TOC, absorbance at 254 nm representing the aromatic carbon content of the wastewater as well as the absorbance at 600 nm, which was the wavelength of maximum absorbance of the samples and recorded by a Shimadzu UV1601 model spectrophotometer to follow up the oxidative decolourization kinetics. During the photocatalytic experiments, sample aliquots were first centrifuged in a Heraeus Sepatech Labofuge at 4000 rpm for 40 min to obtain a clear supernatant. TOC of the treated samples was measured with a Shimadzu TC-5000 analyzer equipped with an autosampler.

Results and discussion

The Fenton/UV-C Process

For the Photo-Fenton process, the reaction pH was chosen as 2.8, since ferrous iron precipitated as ferric hydroxide even at the lowest tried Fe(II)-ion dose when the pH was above 3.5. At Fe(II)-ion concentrations exceeding 0.5 mM, Fe(III)-iron sludge formation was observed even at pH = 2.5 - 3.0. Therefore, a pH of 2.8 and an Fe(II)-ion dose of 0.5 mM was chosen for the Fe^{2+} only (control), Fenton and Photo Fenton reactions.

The effect of the initial applied H_2O_2 dose on apparent first order decolourization rate constants, i.e. k_d values (measured as 1/min), overall per cent UV_{254nm} and TOC removal efficiencies obtained for the treatment of reactive dye wastewater via Photo Fenton's reaction with 0, 0.5, 5 and 10 mM H_2O_2 and a dark Fenton control experiment are displayed all together in Figure 1.

From the figure it is evident that the introduction of Fe^{2+}-ion catalyst increased reaction rates more drastically as compared to the H_2O_2/UV-C oxidation at the same pH. For instance, even when the applied H_2O_2 dose was only 0.5 mM during Photo Fenton reaction in the presence of 0.5 mM Fe(II)-ion, the k_d value (= 0.151 1/min) and ultimate TOC removal (= 19.9 %) were significantly higher than for applying 10 mM H_2O_2 in the H_2O_2/UV-C (k_d = 0.087 1/min and 13.3 % overall TOC

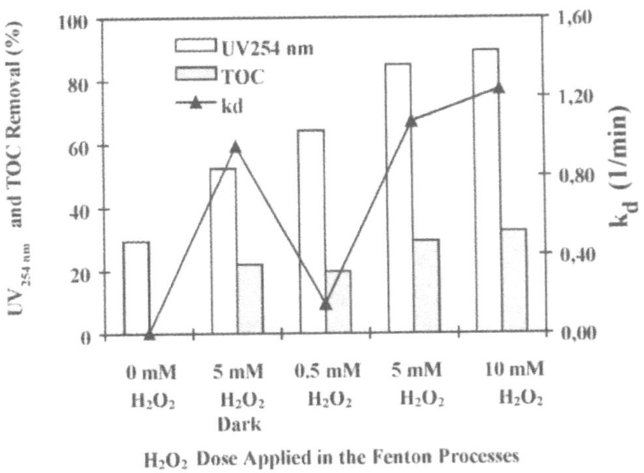

Fig. 1. k_d values, per cent overall UV_{254nm} and TOC removal efficiencies for the Fenton/UV-C at varying initial H_2O_2 doses, Fe^{2+}/UV-C and the dark Fenton (Fe^{2+}/H_2O_2) control process

removal) oxidation process. The dark Fenton's reaction in the presence of 5 mM H_2O_2 resulted in treatment efficiencies close to the corresponding Photo-Fenton reaction, owing to fact that this reagent produces $OH^{\bullet}$ even in the absence of UV-light [23]. The advanced oxidation rates were markedly enhanced by introducing UV-C light and elevating the oxidant dose since H_2O_2 was then consumed by three simultaneous reactions namely the direct Fenton's reaction, photoreduction of Fe(III)-ion to Fe(II)-ion (photo-regeneration of Fenton's reagent) and the direct H_2O_2-photolysis by UV-C irradiation [10].

The TiO_2/UV-A process

The influence of initial pH was studied with the relatively well studied P25 TiO_2 powder. Obtained k_d values, overall per cent $UV_{254\ nm}$ and TOC removal efficiencies are shown in Figure 2 for the TiO_2/UV-A processes obtained for the photocatalytic treatment of the simulated dyehouse effluent at pH = 4, 7, 9 and 10.5 (the actual pH of the reactive dye wastewater) with the 1 g/L P25 type of TiO_2. The figure also contains a mere UV-A control experiment conducted at pH = 4 in the absence of TiO_2. Decolourization was practically complete after a treatment time of 30 min for the TiO_2/UV-A process at all tested pH values, whereas no reaction was observed for the mere UV-A irradiation of the synthetic dyehouse effluent.

As it is obvious from Figure 2, no significant difference in the treatment efficiencies in terms of the studied process parameters was evident throughout the entire selected pH range. This was not in contrast to former studies implying that although the reaction mechanism might change with the reaction pH, degradation rates are quite insensitive to this parameter [14]. Additionally, carbonate and bicarbonate become effective $OH^{\bullet}$ scavengers at alkaline pH, whereas chloride reacts readily with $OH^{\bullet}$ at lower pH values. Thus the photocatalytic dye degradation is expected to be hindered at the whole studied pH range. Furthermore, slightly higher overall $UV_{254\ nm}$ removals were achieved by the P25/UV-A reaction as compared with the Photo Fenton oxidation system. From the above results it can be inferred that the overall treatment efficiency was not so seriously affected by changing the reaction pH that makes the P25-mediated heterogeneous photocatalytic treatment processes practically more attractive than the Photo Fenton oxidation system which has to be conducted at acidic conditions ($2 < pH < 5$).

Besides, at pH = 4, 52.5 % of the absorbance at 600 nm was removed prior to the photocatalytic degradation by initial dark adsorption of the anionic dyestuffs on the positively charged photocatalyst surface, together with an initial 54.9 % $UV_{254\ nm}$ and 5.4 % TOC at the same acidic pH (not shown in the figure). As expected, the preliminary removal due to pre-equilibration adsorption was not observed for the other examined pH values that were above the ZPC (zero point of charge) of the selected semiconductor ($TiO_{2,\ ZPC}$ = 6.3) due to electrostatic repulsion between the anionic dyestuffs and the negatively charged TiO_2 surface above $pH > 6.3$.

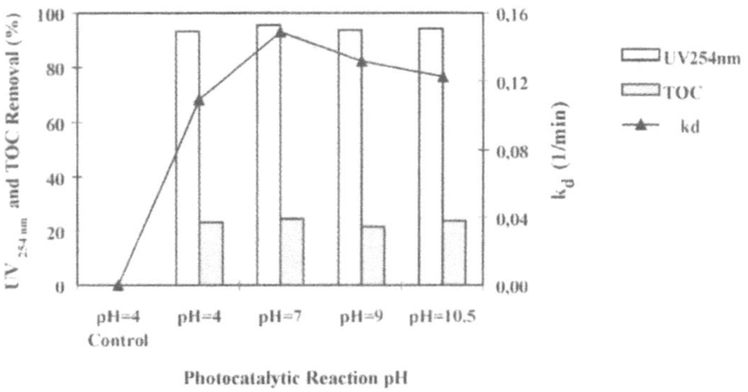

Fig. 2. Calculated k_d values, per cent overall $UV_{254\ nm}$ and TOC removal efficiencies for the TiO_2/UV-A and UV-A control process at varying pH values

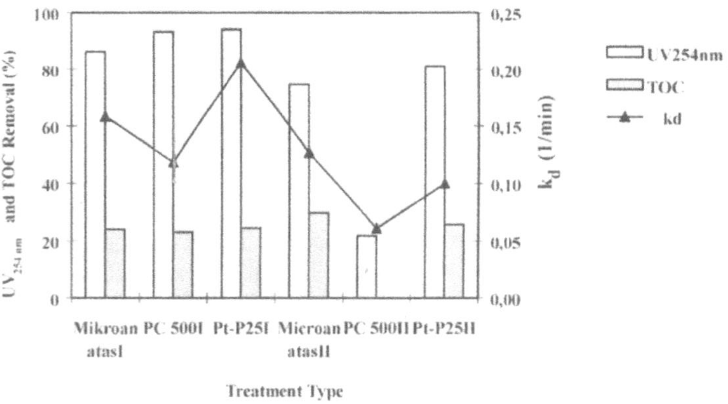

Fig. 3 . Calculated k_d values, per cent overall $UV_{254\ nm}$ and TOC removal efficiencies for the TiO_2/UV-A processes at varying TiO_2 types and pH values

For comparative purposes, the reaction kinetics of Pt-P25 TiO_2 and two other TiO_2 photocatalysts with anatase crystal structures were examined at acidic (pH = 4.0) and natural (pH = 10.5) pH values of the synthetic reactive dye wastewater. The acidic pH was investigated since at this pH distinct dark initial adsorption of the negatively charged dyehouse effluent on the selected TiO_2 photocatalysts took place. It is generally believed that a high BET surface area coupled with small particle size or surface platinization makes semiconductors additionally photoactive as they are thought to be important factors influencing the photocatalytic treatment efficiency [24,25]. However, in Figure 3 depicting the obtained photocatalytic treatment rates in terms of the investigated process parameters for different TiO_2 powder types (I refers to pH = 4.0 and II to pH = 10.5), it can be clearly seen that the obtained removal efficiencies were not higher than for the P25/UV-A treatment system and relatively higher reaction rates were only achievable in the case of Pt-25 at pH = 4.0 (with k_d = 0.206 1/min) and for Mikroanatas at pH=10.5 (with an overall TOC removal = 29.7 %). Moreover, the removal of $UV_{254\ nm}$ was markedly low for PC 500 at alkaline pH and no mineralization was obtained at that pH for the simulated dyehouse effluent. More than half of the reduction in colour and $UV_{254\ nm}$ as well as an average of 12.3 % TOC removal were attributable to the preliminary dark adsorption at pH = 4.0 on the positively charged photocatalyst surface.

Table 3 summarizes k_d values, overall $UV_{254\ nm}$ and TOC removal rates obtained for the investigated heterogeneous and homogeneous AOPs at optimized reaction conditions. Here the superiority of the Photo-Fenton's oxidation reaction in terms of decolourization rate and overall TOC reduction is evident, whereas the TiO_2/UV-A processes were slightly better in overall per cent $UV_{254\ nm}$ removal efficiency possibly as a consequence of the photoadsorption and photodegradation of the aromatic compounds that was taking place simultaneously.

Table 3. Pseudo first order decolourization rate constant k_d, overall % $UV_{254\ nm}$ and % TOC removals achieved for 1 h treatment with the studied AOPs

AOP Type	pH	k_d 1/min	Removal (%) $UV_{254\ nm}$	TOC
0.5 mM Fe^{2+}/10 mM H_2O_2/UV-C	2.8	1.238	89.5	32.5
P25/UV-A;	7.0	0.149	95.7	24.6
Pt-P25/UV-A;	4.0	0.206	94.3	24.5
Mikroanatas/UV-A;	10.5	0.127	74.6	29.7
PC500/UV-A;	4.0	0.119	93.4	23.1

Conclusions

The treatment efficiency and reaction kinetics of homogeneous (Fenton/UV-C) and heterogeneous (TiO_2/UV-A) AOPs for the oxidation of simulated reactive dyebath effluent containing various dyestuffs and their corresponding dye assisting chemicals were comparatively studied. Generally spoken, all investigated AOPs were capable of completely decolourizing and partially mineralizing the simulated reactive dye wastewater within a feasible oxidative treatment period of 1 hour once optimized for specific critical process parameters (e.g. H_2O_2 dose and reaction pH). For the Fenton/UV-C reactions, fastest colour and highest overall TOC removal rate were obtained. In case of TiO_2/UV-A treatment, the overall destruction of the aromatic carbon content expressed as $UV_{254\ nm}$ abatement was slightly better than for the homogeneous AOP. Results clearly delineated that when the decolourization of the dyehouse effluent is sufficient, the reaction duration can be reduced dramatically by applying Fenton/UV-C oxidation, the best treatment option among the investigated AOPs, that would lower the treatment costs tremendously. To decide which AOP is the most suitable for treatment of the wastewater in question, one should carefully outweigh costs for semiconductor powder separation after photocatalytic treatment of the effluent against fouling of the quartz sleeve housings and the pH adjustment to acidic values.

Acknowledgement

Idil Arslan wishes to express her thanks to CIMO (Center of International Mobility), Finland, and TUBITAK BAYG (NATO A2 scholarship), Turkey for their financial support.

References

1. Correia, V. M., Stephenson, T., Judd, S. J.: Characterization of textile wastewaters-a review. Environ. Technol., 15 (1994) 917-929
2. Easton, J. R.: The dye maker's view. In: Colour in Dyehouse Effluent, Peter Cooper (Ed.), the Society of Dyers and Colorists, Alden Press, Oxford (1995) pp. 9-21
3. Pagga, U., Brown, D.: The degradation of dyestuffs: Part II; Behaviour of dyestuffs in aerobic biodegradation tests. Chemosphere 15 (1986) 479-491
4. Reife, A., Dyes, environmental chemistry. Kirk-Othmer Encyclopedia of Chemical Technology, Fourth Ed., Vol. 8, John Wiley & Sons, Inc. (1993) 753-784
5. Glaze, W. H., Kang, J. W., Chapin, D. H.: The chemistry of water treatment using involving ozone, hydrogen peroxide and ultraviolet radiation." Oz. Sci. & Engrg., 9 (1987) 335-342
6. Kuo, W. G.: Decolorizing dye wastewater with Fenton's reagent. Wat. Res., 26 (1992) 881-886
7. Lin, S. H., Peng, C. F.: A continuous Fenton's process for treatment of textile wastewater. Environ. Technol., 16 (1995) 693-699
8. Arslan, I., Balcioglu, I. A. Tuhkanen, T.: Oxidative treatment of simulated dyehouse effluent by UV and near-UV light assisted Fenton's reagent. Chemosphere, 39(15) (1999) 2767-2783
9. Safarzadeh-Amiri, A., Bolton, J. R., Cater, S. R.: The use of iron in advanced oxidation processes. J. Adv. Oxid. Technol. 1 (1996) 18-26
10. Pignatello, J. J.: Dark and photoassisted Fe^{3+} - catalyzed degradation of chlorophenoxy herbicides by hydrogen peroxide. Environ. Sci. Technol., 26 (1992) 944-951
11. Ruppert, G., Bauer, R.: The photo-Fenton reaction-an effective photochemical wastewater treatment process. J. Photochem. Photobiol. A: Chem. 73 (1993) 75-78
12. Zepp, R. G., Hoigné, J.: Hydroxyl radical formation in aqueous reactions (pH 3-8) of iron(II) with hydrogen peroxide. Environ Sci. Technol., 26 (1992) 313-319
13. Ollis, D. F.: Process economics for water purificaiton: a comparative assessment, In: Photocatalysis and Environment. Trends and Applications, Schiavello, M. (Ed.), NATO ASI Series C: Mathematical and Physical Series 237, Kluwer, London (1988) pp 663-667
14. Bahnemann, D. W., Cunningham, J., Fox, M. A., Pelizetti, E., Pichat, P., Serpone, N.: Aquatic and Surface Photochemistry., In: Photocatalytic Treatment of Waters, Helz, G. R., Zepp, R. G., and D. G. Crosby (Eds.), CRC Press Inc., Florida (1994) pp. 261-316
15. Lindner, M., Bahnemann, D. W., Hirthe, B., Griebler, W. D.: Solar water detoxification: novel TiO_2 powders as highly active photocatalyts. J. Sol. Energy Eng., 119 (1997) 120-125
16. Kraeutler, B., Bard, A. J.: Heterogeneous photocatalytic preparation of supported catalysts; photodecomposition of platinum on TiO_2 powder and other substrates. J. Am. Chem. Soc. 100 (1978) 4317-4318

17. Arslan, I., Balcioglu, I. A.: Degradation of commercial reactive dyestuffs by heterogeneous and homogeneous advanced oxidation processes: a comparative study. Dyes Pigments, 43(2) (1999) 189-195
18. Davis, R. J., Gainer, J. L., O'Neal, G., Wu, I. W.: Photocatalytic decolourization of wastewater dyes. Water Environ. Res., 66 (1994) 50-53
19. Schmelling, D. C., Gray, K. A., Kamat, P. V.: The influence of solution matrix on the photocatalytic degradation of TNT slurries. Wat. Res., 31(6) (1997) 1439-1437
20. Arslan I., Balcioglu, I. A. Tuhkanen, T.: Advanced oxidation of synthetic dyehouse effluent by O_3, H_2O_2/O_3, and H_2O_2/UV processes. Environ. Technol. 20(9) (1999) 921-932
21. Hirvonen, A., Tuhkanen, T, Kalliokoski, P.: Removal of chlorinated ethylenes in contaminated ground water by hydrogen peroxide mediated oxidation processes. Environ. Technol. 17 (1996) 263-272
22. Nicole, I., De Laat, J., Dore, M., Duguet, J. P., Bonnel, C.: Utilisation du rayonnement ultraviolet dans le traitement de eaux: Measure du flux photonique par actinometrie chimique au peroxyde d'hydrogene. Wat. Res., 24 (1990) 157-168
23. Walling, C.:Fenton's reagent revisited. Acc. Chem. Res., 8 (1975) 125-131
24. Okamoto, K., Yamamoto, Y., Tanaka, H., Tanaka, M., Itaya, A.: Kinetics of heterogeneous photocatalytic degradation of chlorophenol over anatase TiO_2 powder. Bull. Chem. Soc. Jpn., 58 (1985) 2023-2028
25. Augugliaro, V., Palmisano, L., Sclafani, A., Minero, C., Pelizetti, E.: Photocatalytic degradation of phenol in aqueous titanium dioxide dispersions. Toxicol. Env. Chem., 16 (1988) 89-109

Recovery and Reuse in the Textile Industry – A Case Study at a Wool and Blends Finishing Mill

D. Orhon*, S. Sözen*, I. Kabdasli*, F. Germirli Babuna*, Ö. Karahan*, G. Insel*, H. Dulkadiroglu*, S. Dogruel*, N. Kiran**, A. Baban** and N. Kemerdere Kaya**

*Istanbul Technical University, Civil Engineering Faculty, Environmental Engineering Department, 80626 Maslak, Istanbul, Turkey
**Tübytak Marmara Research Center, Energy Systems and Environmental Research Institute, PO Box: 21, 41470, Gebze, Kocaeli, Turkey,
Dorhon@srv.ins.itu.edu.tr

Abstract

The paper provides experimental justification for a scientific approach to optimizing wastewater recovery and reuse potential in the textile industry. A wool and blends finishing mill, with a high wastewater generation rate, was selected as a case study. A detailed survey was carried out on a selected process to generate a comprehensive pollution profile, which identified a reusable fraction of more than 50%. Chemical settling and oxidation tests were performed to improve the quality of this fraction. COD fractionation and respirometric assessment of biological treatability were evaluated for the wastewater remaining after segregation of the recovered portion, with specific emphasis on the technological implications for appropriate biological treatment.

Introduction

Recovery and reuse of wastewater is of major importance in the textile industry due to the high water consumption. The large amount of wastewater generated by textile mills has a high organic pollutant load in addition to a strong color, a major problem for the receiving water bodies. One of the important in-plant control strategies for the textile industry is to select raw materials and auxiliaries that result in a lower pollutant load in the wastewater without changing the quality of the product. Another strategy is to recover the wastewater for reuse as process water in production. The first strategy has been investigated in detail for most of the basic processes in practice, whereas the latter still needs detailed case studies coupled with careful methodology and process surveys.

It was reported in the literature that slightly loaded rinsing baths from textile dyeing could be reused in many cases without being treated [1]. On the other hand, application of different treatment alternatives such as coagulation-flocculation, chemical oxidation, advanced treatment methods, etc. or their combinations could be required for the reuse of wastewaters [1,2,3,4].

This study is an investigation of wastewater recovery and reuse in a typical wool finishing textile mill. The goal was to select the reusable streams using a scientific approach to highlight the effects of segregating the reusable fraction on the treatability of the overall wastewater. In this case study, a wool-polyester blend fabric dyeing process was chosen for the application of the adopted methodology. The wastewater in each bath was characterized in detail, and then reuse opportunities were selected based on the literature. The experimental study investigated the overall wastewater, the reusable fraction, and the remaining wastewater fraction after segregation of the reusable streams. Overall and remaining wastewater streams were characterized for potential biological treatment, and chemical treatability studies were applied to the reusable fraction of wastewater to fulfill the reuse criteria. The study also specifically addressed the strength and treatability of the wastewater generated after the recovery of wastewater.

Description of the textile mill

The wool finishing textile mill in this case study is located in Istanbul, Turkey and manufactures wool, polyester, linen and wool-polyester blends of yarn and fabric. The plant is representative of the wool finishing sub-category with its high amount of production of around 3000 tons/year and a corresponding water consumption of 3200-3500 m^3/day. The manufacturing scheme of the plant involves four main process lines which may be identified as: (1) high temperature (HT) line, where fiber, yarn and tops are dyed, (2) tops dyeing line, (3) fabric dyeing line and (4) fabric finishing line. The wastewater originating from the fabric finishing line, which has a water consumption of approximately 2300 m^3/day, is separately collected and treated, so this line was not examined in the present investigation. The annual amount and distribution of the product types are reported in Table 1. Dyeing processes mostly involve batch operations, using seven different groups of dyes, namely: cationic (light), chromium (dark), reactive (light), disperse, 1:1 metal-complex (dark), 1:2 metal-complex (dark and light), 1:2 metal-complex and disperse mixture (dark).

Table 1. The distribution of the product types

Type of product	Annual Amount of Product Types (kg/year)				
	Wool	Wool/Polyester	Polyester	Miscellaneous	Total
Yarn	517 364	731 758	8 600	697 927	1 955 649
Fabric	383 873	350 416	-	109 451	843 740
Tops	112 615	-	112 615	-	225 230
Total	**1 013 852**	**1 082 174**	**121 215**	**807 378**	**3 024 619**
(%)	**33.5**	**35.8**	**4.0**	**26.7**	**100**

The dyehouse generates the largest portion of the total wastewater. Wastewaters originating from dyeing and subsequent rinsing operations typically contain residual dye, dye by-products, auxiliaries and surfactants. The dyes used can be categorized as follows; dark color dyes 62.5%, medium color dyes 30%, and light color dyes 7.5%, annually.

As can be seen from production figures given in Table 1, dyed wool and wool-polyester blends, yarn and fabric products account for 65.6% of the total production, also corresponding to 71.3% (67.7% dark color; 3.6% light color) of total dye consumption. Therefore, operations concerning wool and wool-polyester blends yarn and fabric were subjected to further investigation. Annual production and dye consumption data for wool and wool-polyester blends yarn and fabric are listed in Table 2.

As shown in Table 2, wool-polyester yarn dyeing accounts for most (24%) of the production at this textile mill, followed by wool yarn and wool-polyester fabric, with 15.4%, and 11.4%, respectively. Meaningful interpretation of these values is only possible when coupled with relevant data on the wastewater generation rate and the corresponding pollution load, conveniently expressed in terms of COD. The process profiles showing unit wastewater flows and COD loads (organic loads) are outlined in Table 3.

Table 2. Dye consumption and production data of wool and blends dyeing processes

Process	Raw Material	Type of Dye	Dye Color	Dye Consumption (%)	Annual Production (kg/year)	(%)
A	Wool Yarn	Chromium	Blackish	33.4	467 004	15.4
B	Wool Yarn	1:2 MC*	Light Silver	3.6	50 363	1.7
C	Wool Fabric	1:1 MC*	Dark Blue	5.0	129 500	4.3
D	Wool Fabric	1:2 MC*	Dark Blue	9.9	254 370	8.4
E	Wool/Polyester Yarn	1:2 MC*-D	Dark Blue	10.0	731 758	24.2
F	Wool/Polyester Fabric	1:2 MC* D	Blackish	9.5	350 416	11.6
Total				**71.4**	**1 983 411**	**65.6**

*MC-D: Metal-complex and disperse

Table 3. Process profile and organic loads of wool and blends dyeing processes

Process	Product per Batch	Wastewater Generated			COD		
	kg/batch	l/batch	l/kg product	m^3/day	mg/l	kg/day	%
A	840	20 000	24	41	1080	44.3	5.5
B	600	22 860	38	7	365	2.6	0.3
C	120	11 700	98	47	220	10.3	1.3
D	120	27 685	231	217	650	140.8	17.4
E	120	18 100	151	407	750	305.3	37.8
F	120	18 100	151	195	1560	304.2	37.7
Total				**914**		**807.4**	**100**

The total daily amount of wastewater generated from wool and blends dyeing processes was estimated from the average annual figures. Table 3 shows in fact that wool-polyester yarn (process E) and wool-polyester fabric (process F) dyeing, when evaluated in terms of production, unit wastewater generation and strength, appear to be the major polluting processes as they each contribute 38% to the total COD load of the wool and blends dyeing discharges. Since both dyeing processes have similar recipes but different water consumption levels, *wool-polyester dyeing* (process F) with concentrated wastewaters was selected for further detailed examination. A detailed survey of the segregated wastewater streams of process F with a high pollutant strength, is also significant, especially in terms of reuse applications. Characteristics of the source-based wastewater from selected processes are shown in Table 4.

Table 4. Wastewater characteristics of selected processes

Process	Bath	Flowrate l	pH	T °C	Conductivity µS/cm	TDS mg/l	TSS mg/l	COD mg/l
D	Dyeing	2110	4.2	80	3890	1790	25	1735
	1st Rinse	2400	5.8	30	1320	840	20	430
	2nd Rinse	2400	6.6	30	815	590	20	180
	3rd Rinse	18375	7.0	35	750	400	15	130
	4th Rinse	2400	7.0	35	735	450	15	40
	Composite	**27685**	**5.7**	**25**	**1550**	**1020**	**30**	**650**
A	Dyeing	6000	4.1		9620	6750	6750	3100
	1st Rinse	7000	3.9		2540	1600	1600	450
	2nd Rinse	7000	5.8		537	335	335	30
	Composite	**20000**	**4.1**		**428**	**3500**	**3500**	**1080**
B	Dyeing	5280	4.6	40	1214	2110	2360	995
	1st Rinse	6000	6.5	36	550	735	1115	90
	2nd Rinse	11580	7.1	35	532	1170	1210	45
	Composite	**22860**	**6.2**	**35**	**770**			**365**

Materials and methods

Source-based wastewater samples were collected from each batch of the dyeing process. A detailed wastewater characterization was conducted on these segregated effluents in order to obtain the pollution profile and to decide on the reusable streams. All the analyses were performed according to Standard Methods [5] except COD, which followed the ISO 6060 [6] method. The soluble fractions were defined by filtration through cellulose acetate membrane filters of pore size 0.45 µm. Color measurements were conducted on filtered samples with a Hach DR/2 model spectrophotometer at 455 nm wavelength while absorbance measurements were carried out at three different wavelengths, namely, 436, 525, and 620 nm. The conventional characterization studies were also conducted on three composite wastewater samples, which were prepared as mixtures of related segregated effluents proportional to their flow rates in the process. The composite wastewaters formed were (1) raw wastewater, (2) reusable wastewater, and (3) remaining wastewater after separation of the streams suitable for recovery and reuse.

Biological treatability studies involved the respirometric analysis of raw wastewater and remaining wastewater. For the determination of readily biodegradable substrate as proposed by Ekama et al. [7], oxygen uptake rate (OUR) measurements were conducted with a Manotherm® Respirometer RA-1000. A 2 l reactor was operated under batch conditions with an F/M ratio of 0.7 g COD/g VSS. The biomass was previously acclimated to both wastewater samples in a fill and draw reactor at a sludge age of 15 days. For these experiments the same amounts of initial biomass were used. In the experiments the samples were adjusted to a pH of 7-7.5 with phosphate buffer and the tests were carried out at room temperature.

Chemical treatability experiments were conducted in order to examine additional reuse alternatives. Chemical precipitation experiments were performed with lab-scale jar-test apparatus adjusted to provide 5 minutes flash mixing, 30 minutes flocculation and 30 minutes settling. For experiments run with sodium bentonite, on the other hand, a jar-test cycle of three consecutive 3 minutes flash mixing and 3 minutes settling was adopted. $FeCl_3H_2O$, alum ($Al_2(SO_4)_3 18H_2O$) and sodium bentonite were used as flocculants together with a suitable anionic polyelectrolyte. pH was adjusted with $Ca(OH)_2$, when necessary. Ozonation experiments were conducted on the reusable wastewater for the removal of color and COD, to meet the reuse criteria. Ozone was produced by the ozone generator PCI GL1 and the experiments were carried out under 103.45 kPa pressure, with an ozone flux of 15.4 mg/min, at room temperature in a 1.5 l semi-batch reactor, having an effective depth of 23 cm, with 1 l sample volume. Ozone was supplied from the bottom of the reactor through a sintered glass plate diffuser and the off gas was trapped in a series of two gas washing bottles containing 2 % (v/v) KI solution connected to the reactor.

Wool-Polyester fabric dyeing

The selected process was a Forosyn® dyeing recipe applied to 55% wool-45% polyester blend fabric. The dye was a mixture of 1:2 metal-complex and disperse dyestuffs. The dyeing equipment used was a batch type jet-dyeing machine where the material to be dyed moved and the bath itself was stationary. The process had a liquor ratio of 1/20 (weight of fabric/volume of water). As can be seen from the process flowchart given in Figure 1, the investigated process consisted of one dye bath followed by five rinsings.

The dye is a mixture of Forosyn® Navy Hen, Forosyn® Blue SE, Forosyn® Red SE, and Forosyn® Yellow SE. The auxiliaries used consist of wool preserver, pH adjustment agents, dye carriers, dispersant, and crease proofing agents.

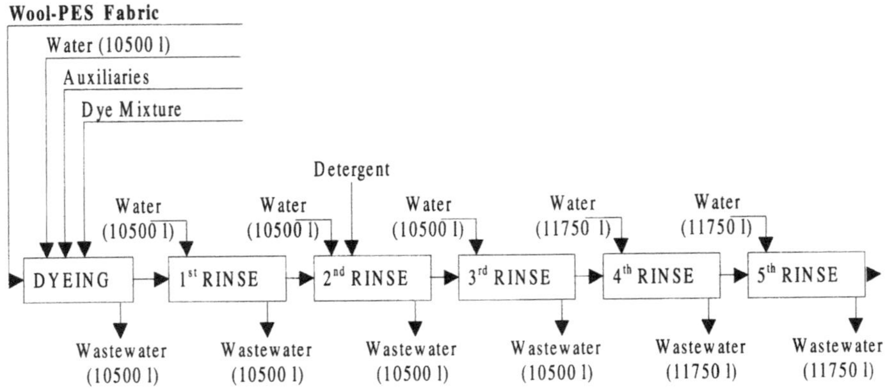

Fig. 1. Process flowchart of forosyn® dyeing

Reuse criteria obtained from two different sources are given in Table 5. One set of criteria is rather broad [3], while the other is rather strict [8]. Characteristics of the source-based wastewater from Forosyn® dyeing are shown in Table 6. The total COD of the composite sample was in agreement with the corresponding figure given in Table 3, providing support for the previous general survey. It must also be noted that close total and soluble COD values were measured for segregated streams. As can be seen from Table 6, wastewaters originating from 4th and 5th rinses, which represented 36% of the total wastewater generated in the process, could be reused without being subjected to a treatment. Another reuse alternative was to add 3rd rinse wastewaters after chemical treatment to the reusable portion, and that increased the reusable fraction to 52%. Because of this, chemical treatability studies were conducted on the wastewaters of 3rd rinse.

Table 5. Reuse criteria for textile dyeing wastewaters

Parameters		Li and Zhao [3]	Hoehn [8]
pH		6.5-8.0	6.5-7.5
Total COD	mg/l	0-160	< 50
TSS	mg/l	0-50	< 500
TDS	mg/l	100-1000	
Total Hardness	mg $CaCO_3$/l	0-100	90
Chloride (mg/l)	mg/l	100-300	< 150
Chromium	mg/l		0.1
Ferrous	mg/l	0-0.3	0.1
Manganese	mg/l	<0.05	0.05
Conductivity	µS/cm	800-2200	
Alkalinity	mg $CaCO_3$/l	50-200	

Table 6. Characteristics of segregated streams

Parameters	Segregated Streams						Composite
	Dyeing	1st Rinse	2nd Rinse	3rd Rinse	4th Rinse	5th Rinse	
Total COD (mg/l)	5675	1390	1385	370	120	65	1445
Soluble COD (mg/l)	5210	1385	1160	340	105	60	1320
TKN (mg/l)	350	65	23	7	1.7	0.3	73
NH_3-N (mg/l)	231	53	18	6.7	1.1	0.3	50.4
pH	6.1	6.8	7.2	7.4	7.5	7.6	7.0
TSS (mg/l)	30	30	15	<10	<10	<10	<10
TDS (mg/l)				390			
Sulfur (mg/l)	256	70	30	5.6	-	-	33
Hardness (mg $CaCO_3$/l)							41
Chromium (mg/l)	0.59	0.18	0.28	0.07			0.23
Cobalt (mg/l)	0.62	<0.1	<0.1	<0.1			0.16
Ferrous (mg/l)	0.35	0.17	0.07	0.06			0.19
Manganese (mg/l)	0.11	0.02	<0.01	<0.01			0.11
Conductivity (μS/cm)	3030	1070	695	495	470	460	910
Alkalinity (mg $CaCO_3$/l)	280	150	125	125	125	125	150
Color (Pt-Co)	4500	1075	990	250	45	15	1155
Absorbance 436 nm	1.457	0.374	0.337	0.089	0.022	0.007	0.360
Absorbance 525 nm	1.151	0.274	0.292	0.073	0.016	0.005	0.279
Absorbance 620 nm	1.345	0.339	0.394	0.100	0.018	0.004	0.332

Treatability Studies

The results of the chemical treatability tests are given in Table 7. Experiments performed with sodium bentonite showed over 80% color removal with a dosage over 300 mg/l. A COD removal of 40% was obtained with dosages over 500 mg/l. On the other hand, an alum dosage of 50 mg/l was adequate for COD and color removals of 50% and 80%, respectively. Although manufacturers discourage the use of iron salts, as they might stain the textile product and possibly cause problems in the dyeing process, $FeCl_3$ treatment was also investigated. It was found that efficient COD and color removals were attained with 100 mg/l of $FeCl_3$.

In order to meet the reuse criteria, COD and color removal efficiencies of over 16% and over 80%, respectively, must be achieved for the wastewater originating from the 3rd rinse. A sodium bentonite dosage of 500 mg/l and an alum dosage of 50 mg/l were found to be suitable to meet these criteria. Because of the success of chemically treating the 3rd rinse, the reuse stream became a combination of the treated 3rd rinse +4th rinse +5th rinse. The characteristics of the combined stream are given in Table 8. It must be noted that total hardness, TSS and total chromium concentrations in this combined reuse stream were under detection limits.

Table 7. Results of chemical treatment

Chemical Type	Dosage (mg/l)	pH	Total COD mg/l	Color Pt-Co	Removal Efficiency (%) COD	Color
Sodium Bentonite	200	7.28	315	80	15	68
	300	7.38	285	50	23	80
	400	7.34	265	50	28	80
	500	7.47	225	40	39	84
	600	7.64	210	35	43	86
	700	8.02	210	35	43	86
Alum	30	3.88	340	160	8	36
	50	6.61	260	40	30	84
	100	6.08	250	30	32	88
	200	6.74*	250	30	32	88
	300	6.48*	240	25	35	90
$FeCl_3$	50	6.61	280	150	24	40
	100	6.02	220	30	41	88
	150	6.45*	230	20	38	92
	250	6.93*	240	30	35	88

*with lime

Ozonation was also applied to wastewaters generated from 3rd rinse. With 155 mg of utilized ozone after 30 minutes contact time, 80% color removal could be achieved, but COD removal was negligible, as described in the literature [9, 10, 11].

Biological treatability studies consisted of oxygen uptake rate (OUR) measurements on raw wastewater and the wastewater stream remaining after segregation of the reusable 4th and 5th rinsings. OUR measurements were evaluated first for the assessment of the readily biodegradable substrate in wastewater, S_{S1}, and second for the interpretation of the kinetic structure of both wastewaters. The OUR profiles obtained in the experiments are shown in Figure 2. The high rate in the first phase relates to the utilization of the readily biodegradable substrate initially present in the wastewater. This initial rate stays constant over a period where S_S is high enough to sustain maximum growth rate. With the consumption of S_S, the OUR drops to a second plateau correlated with the readily biodegradable substrate generated by hydrolysis. For the assessment of S_{S1}, the heterotrophic yield coefficient has been accepted as 0.67 g COD/g COD [4].

Table 8. Characteristics of reuse stream after chemical treatment

Chemical	Dosage mg/l	COD mg/l	Color Pt-Co	Conductivity µS/cm	TDS mg/l	Alkalinity mg $CaCO_3$/l
Sodium Bentonite	500	132	33	445	325	130
Alum	50	143	33	440	320	110

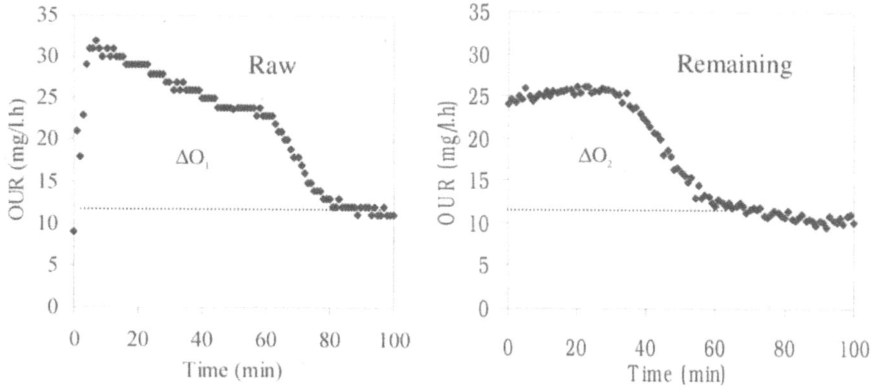

Fig. 2. OUR profiles for raw and remaining wastewaters

As shown in Table 9, the S_{S1}/C_{T1} ratio of 0.15 for the examined wastewater was in accordance with data given in the literature, which was in the range of 0.14 to 0.23 [12, 13]. The readily biodegradable substrates available in the raw and remaining wastewaters were approximately the same despite the fact that the remaining portion represented a much stronger wastewater with an initial COD of 2805 mg/l, 27% higher than the 2205 mg/l associated with the raw wastewater sample. This observation may be taken as an index of different reactions and factors negatively affecting biological degradability of the wastewater remaining after reuse.

Another point of interest is the perfect numerical agreement of the OUR profiles obtained for both samples as the experimental conditions were set the same as previously explained. Similar OUR profiles reflected compatible values for the kinetic coefficients describing microbial growth and hydrolysis. The only difference was that the remaining portion was likely to include a much higher slowly biodegradable fraction, thus requiring a higher treatment volume. From the experimental results, the basic problem appeared to be the level of residual COD, which might pose a serious problem for a more concentrated wastewater stream in meeting effluent limitations. This aspect deserves further experimental evaluation.

Table 9. COD fractions of wastewaters

Wastewater	C_{T1} (mg/l)	S_{T1} (mg/l)	S_{S1} (mg/l)	S_{S1}/C_{T1}	S_{S1}/S_{T1}
Raw	2205	1320	340	0.15	0.25
Remaining	2805	2020	345	0.12	0.17

Conclusions

A scientific approach was developed, and justified experimentally, for optimizing wastewater recovery and reuse potential in the textile industry. A wool and blends finishing mill, with a high wastewater generation rate, was selected as a case study. A full pollution profile, which included comprehensive information on the production scheme, applicable unit wastewater generation data and pollution characteristics of individual wastewater streams, was evaluated as a prerequisite for an acceptable evaluation. A detailed survey carried out on *wool and polyester fabric dyeing* discharges, selected as a representative process for the study, showed a significant fluctuation in the quality of different wastewater streams. The COD varied from 5675 mg/l in the dyebath to 65 mg/l in the last rinse discharge.

The recoverable wastewater fractions were selected in a way to minimize additional treatment before reuse. The percentage of the total discharge that was suitable for reuse without treatment amounted to 36%. Chemical treatment of an additional discharge stream was effective in increasing the reusable ratio to 52%. Ozonation did not prove beneficial to upgrade the wastewater quality, either in terms of COD or color removal. The study underlined the significant need to improve the quality criteria for reuse, applicable to the textile industry.

Segregation for reuse was observed to generate a stronger wastewater, without significantly affecting biodegradation kinetics applicable to the utilization and hydrolysis of different COD fractions. Reuse was evaluated to increase the relative amount of slowly biodegradable COD, necessitating a larger treatment volume. The issue of residual organics requires further study and deserves significant consideration in meeting the effluent COD limitations after reuse.

Acknowledgment

This study was conducted as part of the sponsored research activities of The Environmental Biotechnology Center of The Scientific and Research Council of Turkey. It was also jointly supported by the Volkswagen Stiftung Fund and The Research and Development Fund of Istanbul Technical University.

References

1. Sewekow, U.: Ullmann's Encyclopedia of Industrial Chemistry, Vol. A26 (1995) Chapter 14, VCH Verlagsgesellschaft, Germany
2. Perkins, W. S., Walsh, W. K., Reed, I. E., Namboodri, C. G. A.: Demonstration of reuse of spent dyebath water following color removal with ozone. Textile Chemist and Colorist, 28 (1995). (1) 31-37
3. Li, X.Z., Zhao, Y.G.: Advanced treatment of dyeing wastewater for reuse. Wat. Sci. Tech., 39 (1999) (10-11) 245-255
4. Orhon, D., Germirli Babuna, F., Kabdasli, I., Insel, G., Karahan, Ö., Dulkadiroglu, H., Dogruel, S., Sevimli, F., Yediler, A.: A scientific approach to wastewater recovery and reuse in the textile industry. 1st World Water Congress of the International Water Association (IWA), 3-7 July (2000) Paris, France (accepted for publication)

5. Standard Methods for the Examination of Water and Wastewater. American Public Health Association/American Water Works Association/Water Environment Federation (1998) Washington D.C., USA
6. ISO Water Quality – Determination of the chemical oxygen demand, Ref. No. ISO 6060-1986 (1986)
7. Ekama, G. A., Dold, P. L., Marais, G. v. R.: Procedures for determining influent COD fractions and the maximum specific growth rate of heterotrophs in activated sludge systems. Wat. Res., 11 (1986) 1049-1057
8. Hoehn, W.: Textile Wastewater – Methods to minimize and reuse. Textilveredlung, Reuse Standards-Thies-Handbuch fuer den Garnfaerber (1998)
9. Schönberger, H., Kaps, U.: Reduktion der Abwasserbehandlung in der Textilindustrie, Forschungsbericht 102 06 511 UBA-FB (1994) pp. 93-143
10. Kuo, W.G.: Decolorizing dye wastewater with Fenton's reagent, Wat. Res. 26 (1992) (7), 881-886
11. Solozhenko, E., Soboleva, N.M., Goncharuk, V.V.: Decolorization of azo dye solutions by Fenton's oxidation, Wat. Res., 29 (9) (1995) 2206-2210
12. Germirli Babuna, F., Orhon, D., Ubay Çokgör, E., Insel, G., Yaprakli, B.: Modeling of activated sludge for textile wastewaters. Wat. Sci. Tech., 38 (4-5) (1998) 9-17
13. Germirli Babuna, F., Soyhan, B., Eremektar, G., Orhon, D.: Evaluation of treatability for two textile mill effluents. Wat. Sci. Tech., 40 (1) (1999) 145-152

Different Treatment Methods for Effluent from a Pulp Mill and their Influence on Fish Health and Propagation

T. Engström and U. Gytel

Kemira Kemi AB, Box 902, S-25109 Helsingborg, Sweden
Ulla.gytel@kemira.com

Abstract

Effluent from a TCF bleach plant was biologically and/or chemically treated in order to remove organic material (COD). Ecotoxicological studies with wastewater were carried out on trout and zebra fish. Untreated effluent adversely affected spawning and hatching of zebra fish larvae. Although different types of organic matter were removed with different treatment methods there was no major difference in the impact of treated effluent on fish health and propagation. Biological treatment apparently influenced the uptake of sterols in the bile of exposed trout.

Introduction

Fresh water is one of the essential raw materials in the manufacturing of pulp and paper. The water systems in modern mills are to a high degree closed, and a major part of the water is recycled. Recycling of water has reduced the discharges of different compounds. A Kraft mill now uses 20-80 m^3 water per tonne of bleached pulp compared to 250 m^3 twenty years ago.

The main discharge from the pulp industry into water is dissolved organic matter originating from wood. It, therefore, differs greatly from municipal wastewater and certain food processing effluents where as much as 40-70% of the COD-load is associated with the suspended solids [1].

On environmental grounds the forestry industry in Scandinavia has switched from chlorine to chlorine dioxide (ECF or Elemental Chlorine Free) and hydrogen peroxide (TCF or Totally Chlorine Free) for bleaching. Discharges of AOX (chlorinated organic matter) have been substantially reduced. Introduction of new bleaching methods has decreased the molecular weight (MW) of organic materials in the effluent compared to older bleaching sequences. The higher MW fractions were believed to be biological inactive and non-toxic [2] but some ecotoxicological effects of biologically treated effluent have recently been reported.

The pulp and paper industry has for a long time used mechanical, biological and chemical treatment methods for wastewater. Effluents from these industries are low in nutrients, so phosphate and nitrate must be added before biological purification. Chemical treatment is mainly used for removal of the phosphate surplus and suspended matter after a biological treatment. It can however also be used as a single treatment step to eliminate any "non-biodegradable" high molecular weight material.

Materials and methods

Total effluent from a TCF Kraft pulp mill (softwood, bleaching sequence Q(EOP)Q(PO)) was treated on pilot plant scale by biological as well as chemical processes. The biological treatment was a 2-step biofilm/activated sludge process and the chemical treatment precipitation with an aluminium salt (AVR). Composite samples from a test period of 4 days were analysed and stored at 4°C until used.

Fractionation

Fractionation of samples was carried out in a stainless steel stirred cell (volume 300 ml, 50 rpm, membrane area $40 \cdot 10^{-4} m^2$) at an operating pressure of 5 bar. Before fractionation the samples were filtered on glass microfibre filters by a special procedure in order to remove fibres. A small portion (10-15 ml) of the sample was evenly distributed on the surface of an 11 cm glass filter on a büchner funnel. A new filter paper was used for the next portion of the sample. This procedure minimises the removal of colloidal material by the fibre pad. Details of membranes and filters used are presented in Table 1.

Table 1. Characteristics of membranes and filters used in experiments

Supplier	Type	Specifications Pore size µm Molecular weight cut-off [kDa]	Pure water flux $lm^{-2}h^{-1}$	Surface charge at pH 8 mV
Whatman	GF/A	1.6 µm		
Millipore	PLHK	100 kDa	1320	-17.3
Millipore	PLTK	30 kDa	390	-16.4
Millipore	PLGC	10 kDa	65	-7.5
Millipore	PLBC	3 kDa	22	-9.2
Millipore	PLAC	1 kDa	15	-11.6

Ecotoxicological studies:

Juvenile rainbow trout (*Oncorhyncus mykiss*) were, after 1 week of acclimation, exposed to dilute pulp mill effluent for 8 weeks in aerated 670-l tanks at ambient temperature (starting at +10°C, ending at +2°C). Fresh water from Lake Fryken was continuously pumped to the tanks at a flow rate of about 2.7 l/min and the pulp mill effluent was added in amounts corresponding to dilution factors of 1:400 and

1:2000, respectively. Every test group, including two controls without wastewater, consisted of 32 fish [3]. This procedure was used by "Environmental Research Group-MFG" in a similar study on effluent from an ECF mill. Before and after the test period analyses of morphometry, haematology and liver function were carried out. The contents of fatty acids, sterols and resin acids in the bile were subsequently determined.

The zebra fish (*Brachydanio rerio*) embryo/larvae system was used to examine subacute effects of untreated and treated effluent on fish reproduction. The test procedure is a further development of the standard test for acute toxicity (SS 028193) commonly used at IVL [4]. After being exposed to sublethal concentrations of untreated and treated effluent for 6 weeks spawning, deformation and survival of larvae were studied. The reversibility of any effects was studied after a 6-week postexposure period in the absence of wastewater.

Results

As shown in Figure 1 and Table 2, removals of organic material (COD) by chemical treatment, biological treatment and a combination of both were 47%, 71% and 93%, respectively. Biological treatment eliminated >95% of the (easily degradable) BOD. Chemical treatment removed, as can be seen from the increase in the BOD/COD ratios, mainly "non- biodegradable" material, phenolic and coloured compounds (lignin, lignans, etc.) and phosphate. About 50-60% of the chelating agent (EDTA) was removed by chemical precipitation.

Table 2. Chemical analysis of treated effluent

		Untreated	Chemical treatment	Biological treatment	Bio + Chem treatment
COD	mg/l	1728	959	497	135
BOD	mg/l	770	530	5	<3.0
TOC	mg/l	610	380	170	51
BOD/COD ratio		0.44	0.55	0.01	<0.02
Lignin	mg/l	157	85	148	66
Carbohydrates	mg/l	53	29	n.d.	n.d.
Fatty acids	μg/l	2100	133	40	26
Resin acids	μg/l	630	46	10	3
Sterols	μg/l	71	8	4	5
H_2O_2	mg/l	76.5	2	2	3.1
EDTA	mg/l	70	34	55	22
Colour	mg/l Pt	180.0	59	260	29
P_{tot}	mg/l	1.1	0.12	0.78	0.12

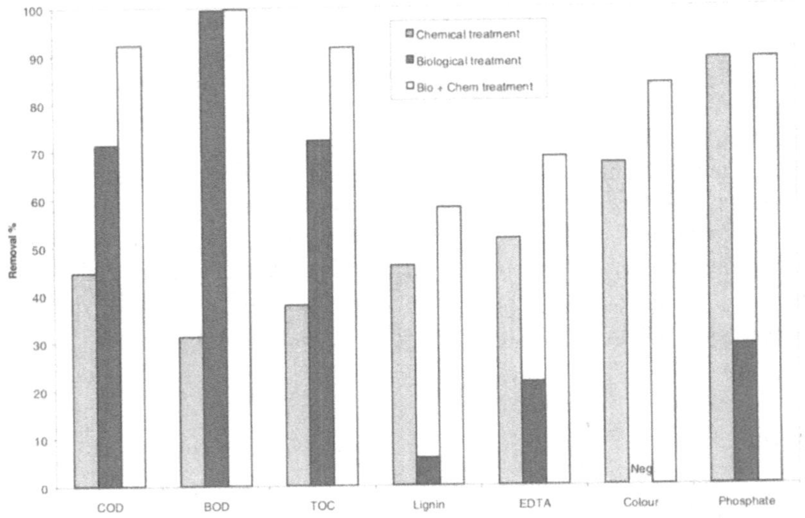

Fig. 1. Removal of different compounds from a TCF mill effluent by biological and chemical treatment methods (Neg = increase)

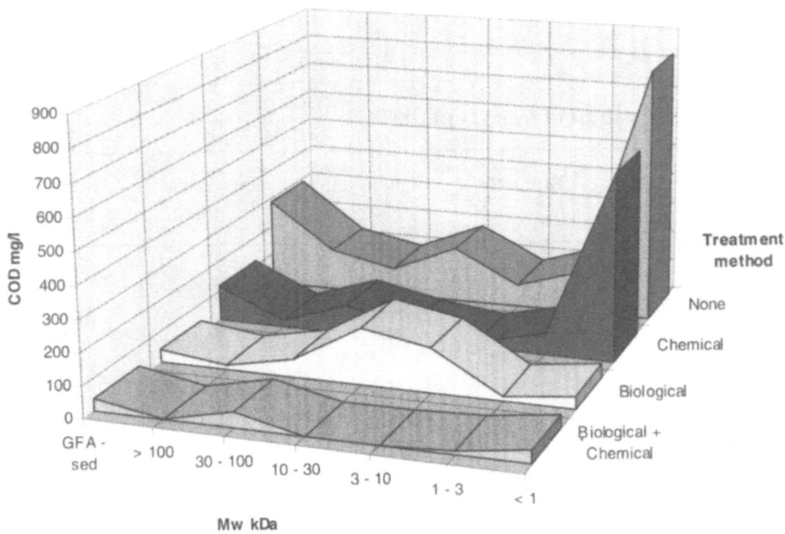

Fig. 2. Fractionated wastewater samples from a TCF pulp mill

Fractionations

There is a considerable difference among the various treatment procedures on the removal of COD. Figure 2 shows that biological treatment removes most of the easily degradable low molecular weight (<3 kDa) material whereas chemical precipitation primarily removes the medium weight (approx. 20 kDa) material. The biological treatment also reduces the mean molecular weight of the organic compounds. The decomposition of high MW compounds can be seen from the shifting of the middle part of the curve towards lower MW (fewer 30-100 kDa, more 3-10 kDa). A combination of the two treatment methods removes virtually all organic material.

Ecotoxicological studies

Small haematological changes were observed for all effluent-exposed trout. When the guidelines for the evaluation of fish tests were applied to the results, none of the effluents tested were found to have any significant impact on fish health.

As shown in Table 3, significant effects must be observed for one primary or (at least) 3 secondary responses within a group of prioritised physiological functions (Morphometry, Liver function, Immune system and Pathology/ Haematology) [5,6].

Table 3. Summary of ecotoxicological studies on trout. Significant effects ($p<0.05$) compared to control groups are marked + if the value increased as a result of exposure and – if it decreased. Responses in **bold** are primary responses.

	Treatment Method	Untreated		Chemical treatment		Biological treatment		Bio+Chem treatment	
	Dilution factor	400	2000	400	2000	400	2000	400	2000
Function	Response								
Morphometry	**Growth**								
	CF								
	LSI								
	Glycogen (liver)				+				
Liver functions	LSI								
	Glycogen				+				
	EROD								
	Lipids		–			+		+	
	Histology								
Immune system	Lymphocytes								
	Granulocytes						–		
	Trombocytes								
Pathology, Haematology	**Diseases**								
	Haematocrit	+							
	Haemoglobin								
	RBC	+			–				
	Immature RCB			–			–	–	

CF=condition factor, LSI=liver somatic index, RBC=red blood cells

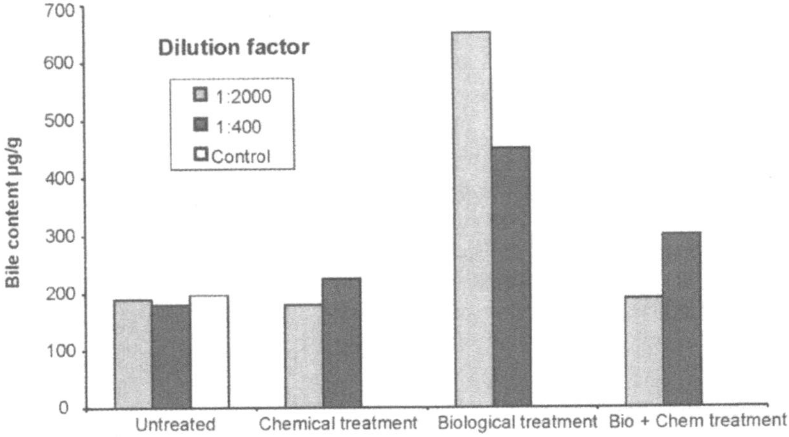

Fig. 3. Sterols in bile from effluent exposed trout

Increased concentrations of resin acids are found in bile from trout exposed to untreated wastewater. The amounts of fatty and resin acids are correlated to the effluent concentrations but for sterols no such correlation exists. Although the concentrations in the effluents are similar, there are considerably fewer amounts of sterols in bile from trout exposed to chemically treated effluent, with or without a biological step, compared to a single step biological treatment. Trout exposed to the lowest concentration of biologically treated effluent (dilution factor 1:2000) have the highest sterol content, as shown in Figure 3.

Untreated effluent significantly influenced the hatching (frequency/time) of zebra fish and survival of larvae (NOEC 30%, 40% and 22%, respectively). No spawning was observed at wastewater concentrations exceeding 2.5%. When spawning occurred there was no adverse effect on hatching and survival of larvae.

On exposure to untreated and biologically treated effluent an increased incidence of deformation of larvae was observed. Treated effluent caused no significant changes in the frequency of spawning, hatching, and survival of larvae. All observed changes were fully reversible.

Table 4. Summary of reproduction tests

Treatment method	Untreated	Chemical treatment	Biological treatment	Bio + Chem treatment
LOEC* (%)	10	82	n.d.	n.d.
NOEC** (%)	2.5	>10	>40	>40
Recovery NOEC (%)	2.5	>10	>40	>40

*LOEC = Lowest observable effect concentration
**NOEC = No observable effect concentration

Discussion and Conclusions

Although the composition of a biologically treated effluent from a TCF pulp mill is very different from a chemically treated effluent, the impact on fish health and propagation is almost identical. All studied treatment procedures were very effective in removing compounds that affect the reproduction of zebra fish. Ecotoxicological studies indicate that untreated as well as treated effluent have no unacceptable environmental impact" on fish health.

The amount of fatty and resin acids in the bile of exposed trout is closely correlated to the concentration of the effluents. For sterols and biologically treated effluent this is not the case. These results may indicate that the digestibility of sterolic compounds is changed during the treatment. Further studies are required to confirm this hypothesis.

References

1. Hahn, H.H., Klute, R. (eds.): Pretreatment in Chemical Water and Wastewater Treatment, Springer-Verlag (1988)
2. Södergren, A (ed.): Bleached pulp mill effluents, Swedish Environmental Protection Agency, Report 4047 (1993)
3. Grahn, O., Grotell, C.: Summary of ecotoxicological studies on fish exposed to untreated, chemical and biological treated pulp mill effluent, Miljöförskargruppen F99/038 (in Swedish) (1998)
4. Neilson, A.: Incorporation of a Subacute Test with Zebra Fish into a Hierarchical System for Evaluating the Effect of Toxicants in the Aquatic Environment, Ecotoxicology and Environmental Safety 20 (1990) 82-87
5. Swedish Environmental Protection Agency, Report 4695, Environmental impact of pulp and paper mill effluents. Guidelines for future evaluation of environmental risks (in Swedish)
6. Larsson, Å.: Choosing biological functions and test factors for fish tests. University of Gothenburg (1999) (in Swedish)

Sludge Treatment

Acid Extraction of Heavy Metals from Bio-Waste and Bio-Solids

M. Schaefer*, H.H. Hahn** and E. Hoffmann**

*Korea Institute of Science and Technology – Europe, Saarbrücken, Germany
schaefer@kist-europe.de
**Institute for Aquatic Environmental Engineering, University of Karlsruhe, Germany
hermann.hahn@bau-verm.uni-karlsruhe.de

Abstract

The opportunity to market secondary fertilizers such as bio-wastes and sewage sludges, for example, depends significantly on the pollution load of the final products (e.g. composts, soils) and on the acceptance of people to reuse these products. To ensure a high quality, different kinds of regulations and seals of approval (e.g. Bio-Waste Ordinance) have been passed or introduced in Germany and elsewhere. At present about 90% of the waste treatment operators in Germany can meet these standards, but an increasing amount of separately collected bio-waste will lead to a reduction in raw substrate quality. To support a high recycling rate, payable technologies have to be developed to reduce the pollutant load, especially for highly loaded fractions or charges. Regardless, all efforts have to be undertaken to reduce the amount of pollutants in order to minimise the necessary treatment.
This paper focuses on the possibilities for reducing the heavy metal loads of bio-waste, primary, and excess sludge. The leaching experiments were performed in lab scale reactors, varying the dosage of chemicals or acids. The internally-produced and externally-dosed acids and chemicals were citric acid, acetic acid, hydrochloric acid and a CaCl solution.

Introduction

The recycling of municipal organic wastes resulting from separate collection systems as well as sewage sludges, which are not or only slightly affected by industries, is expanding and becoming more important not only in Germany but all over the world. Experiences gathered in the field of bio-solids reuse during the last few years show that the recyclability of sewage sludge depends significantly on the product quality, which is related to the nutrient content and pollution load, or the ratio of nutrients to pollutants (for example, heavy metals). This means that potential users will accept only harmless bio-solids and bio-wastes, making these wastes competitive in the difficult recycling market. The idea of closing the material cycle implies, however, that as much as technologically possible should be reused from the cost/benefit standpoint. Therefore increasing amounts of bio-solids and bio-wastes have to be (pre-)treated to reduce the organic and inorganic pollutant load prior to reuse.

In order to reduce the harmful, polluted fractions of bio-solids and bio-wastes, the following three treatment technologies can be applied:

- Mechanical pre-treatment and separation of highly polluted material
- Biological degradation
- Physical-chemical treatment to separate fractions by changing their raw substance properties (e.g. solubility, electrophoretic mobility)

Of these alternatives the last one, physical-chemical treatment, was investigated in detail. The objective of these experiments was to study the influence of different chemical boundary conditions and dosages of internal (produced during the anaerobic process) and external (added) acids and chemicals on the leaching potential of heavy metals. The investigated materials included municipal bio-waste, primary sludge, and excess sludge.

Materials and methods

The experiments were carried out using stirred beakers with a volume of 1 l and a 3 step percolation reactor system of 60 l with a hydraulic loading of about 0.25 l H_2O/kg DM (Figure 1).

The raw materials were treated by dosing different amounts of citric acid, acetic acid, hydrochloric acid, or a CaCl solution. The waste residence time was about 24 h in the stirred beakers, and in the range of 2 up to 6 days in the percolation reactors depending on the variation of temperature throughout the year.

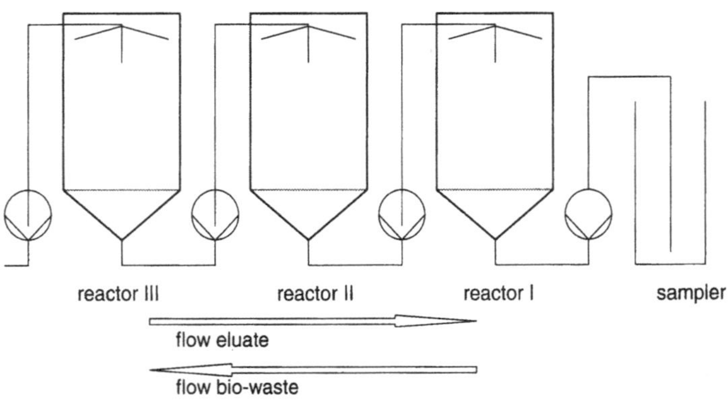

Fig.1. Flow schematic of the percolation reactors

The input material itself (bio-waste and also sewage sludge) was taken directly from the treatment plants with a catchment area of about 125 000 inhabitants. As a pre-treatment step for the bio-waste a two rotor device was used to separate the particles into two size fractions: those less than 1 cm and those greater than 1 cm. The temperature in both reactor types was kept in the range of 30 - 35 °C.

Sample treatment and analysis (DM, VDM, heavy metals by AAS) were done according to the German Standard Methods (DEV).

Results and discussion

There have been several reports in the literature [1, 2, 3, and others] on the leachability of heavy metals when different mineral and organic acids were added to sewage sludge, waste and dredged soil. The reports show that the leachability varies depending on the specific heavy metal, its concentration, the pH level and the redox potential. For example, lead is only barely leachable at a pH of 0.5, while zinc or cadmium can already be eluted at a significantly higher pH.

Table 1. Fraction, transport form, and bio-availability of heavy metals [4]

Fractions	Transport Form
Particulate	
Biol. bound	Bound
Complexable	and
Adsorptive bound	Settleable
Dissolved	Dissolved
Dialyzable	

Bio-Availability	
Not Available	Abiotic amount
Biol. Matrix	Bio-available and Biol. effective
Hardly Available	
Easily Available	

The leachability is not only influenced by the aforementioned more general parameters. In addition the chemical or biological matrix itself affects the leaching result. Table 1 describes the different fractions, their possible transport form, and their biological effects. The dialyzable fraction can be separated and then analysed, using dialysis tubing with a pore diameter of 2.4 nm, while the dissolved fraction can be collected by filtration using a pore size of 0.45 µm. The other fractions are bound to particulates and settleable. These fractions can also be classified as: (i) biologically available and (ii) biologically not available. The biologically available fraction can be further subdivided into three fractions, which differ in the strength of the chemical bonds: (i) the biologically bound fraction is only available after the organic matter is degraded; (ii) the second fraction can be eluted by complexing agents like EDTA; (iii) the third fraction can be described as adsorbable or desorbable and can be eluted with a CaCl solution [4].

The biological/chemical matrix itself depends on the source of the heavy metals and the origin of the sludge or the bio-waste. In the case of sewage sludge, however, most of the investigations have focussed only on digested sludge, but do not refer to the different kinds of sludge as primary, surplus and tertiary sludge.

Primary sludge contains a high amount of material, which is discharged from the urban surfaces into the combined sewer systems. This inorganic material, especially the smaller particle fraction, is highly loaded with heavy metals but it is biologically not or only slightly available. On the other hand, surplus sludge contains mainly the portion of the heavy metals that was transported in dissolved form to the wastewater treatment plant thus passing the primary sedimentation tank. The chemical/biological matrix of the heavy metals in the surplus sludge can be described as complexable or desorbable. Accordingly, these heavy metals are still (easily) biologically available and effective. After the digestion of sewage sludge, especially the easily biologically available metals are transformed into other matrices, which are significantly less bio-available. This could be validated by some leaching experiments with sewage sludge and hydrochloric acid at a pH of 1.5 (see Table 2). A study on the heavy metal balance through a wastewater treatment plant [5] also confirmed this trend. While the amount of dissolved heavy metals was rather low (Cu: ~27 %, Pb: ~16 %) in the influent of the treatment plant, the dissolved fraction increased significantly at the influent of the activated sludge treatment (Cu: ~64 %,Pb: ~50 %).

Although there are no detailed similar studies concerning the leachability of heavy metals from bio-wastes, the leachability is probably analogous. The inorganic content of bio-wastes amounts to about 30 % [7], which comes from atmospheric deposition and sweeping, for example. The heavy metal concentration of this inorganic fraction is just as high, as is known from the previously described primary sludge fractions, and its chemical/biological matrix is less bio-available. This means that this inorganic fraction should be separated in an independent treatment step (drum sieve or flotation) comparable to the primary sedimentation tanks in wastewater treatment plants. A feasible treatment technology to remove this inorganic fraction from bio-waste is described in [8].

The heavy metals bound to the organic fraction are bio-available. The amounts of desorbable and complexable fractions at a pH level of 4 (normal pH level in a separate biological hydrolysis step) are shown in Figure 2. It can be seen that zinc and nickel are very easy to elute and possess a high desorbable amount. On the other hand, copper and lead are only slightly leachable, and they possess a rather small amount of a desorbable matrix. But even lead can be eluted up to 10% under these equilibrium conditions between the dissolved and the bound phase.

Table 2. Amount of leachable heavy metals in different kinds of sewage sludge (HCl was used to adjust the pH to 1.5) [6]

	Primary sludge	Excess sludge	Digested sludge
Pb	10-30%	50-70%	< 10%
Cu	5-25%	30-80%	< 10%
Cd	15-40%	60-70%	< 20%

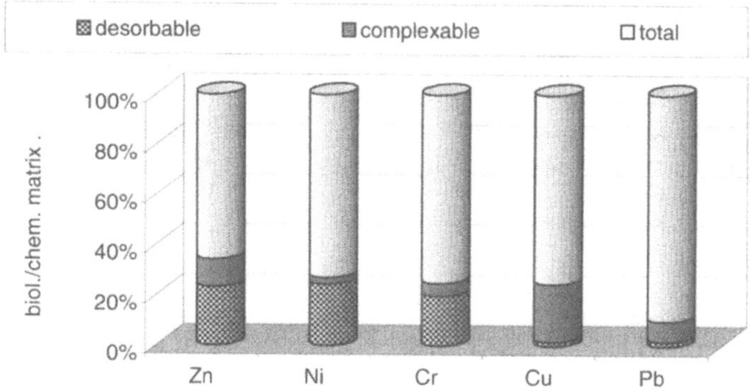

Fig. 2. Biological/chemical matrices of heavy metals in bio-waste under equilibrium conditions

In order to investigate the elution potential of the heavy metals under continuous, non-equilibrium conditions, bio-waste was tested in the previously described 60 l percolation reactors, using either internal acetic acid produced in the biological hydrolysis step or internally-produced acetic acid with an addition of citric acid, as described in [9].

Figure 3 shows the pH development and the corresponding heavy metal concentrations in the effluent of the percolation reactor. With no addition of citric acid, the pH increased from 3 to 4 during the first few hours up to 5 to 6 after 50 hrs. This change in pH may or may not affect the heavy metal concentration in the effluent, depending on the solubility of the heavy metal. For example, lead is a barely soluble heavy metal. Figure 3 shows that as pH increased, the lead effluent concentration remained low over time. In contrast, zinc is a more readily soluble heavy metal. In the pH range below 4, its concentration was high, but it decreased as the pH increased.

The addition of small amounts of citric acid (up to 1 % of the eluate flow) leads to a significant decrease in pH, which has a significant influence on the dissolved heavy metal concentration. The concentration of the dissolved lead could be increased 4 to 6 times, and the discharge of zinc could be doubled by adding citric acid (Figure 3).

The reduction percentages of heavy metals in the raw bio-waste reflect the solubilities of the heavy metals. Table 4 shows that the readily soluble elements zinc and cadmium could be eluted up to around 40 % with only the biological self-production capacity of acetic acid. On the other hand, the input load reduction for the barely soluble elements lead and copper was only about 5 %. These efficiencies, however, could be increased significantly by adding a small amount of citric acid. Only copper did not increase significantly in reduction efficiency. Schlaak et al. [9] reported similar results for copper under comparable conditions. The reasons for this are not known. One explanation might be the lower input load, which might have reduced the diffusion velocity.

It was of further interest to determine if the concentration of citric acid affected the amount of heavy metals leached out in the effluent. Figure 4 shows that there is no sense to use higher acid concentrations as 1 % of the eluate, because higher doses of citric acid did not leach out more heavy metal (Pb). These results contradict to those of Schlaak et al. [9], who found a significant dependency of leachable heavy metal load on the citric acid dosage. The reason for the difference in results may be related to the mixture of internally-produced acetic acid and externally-dosed citric acid (as the main complexing agent) used in this study.

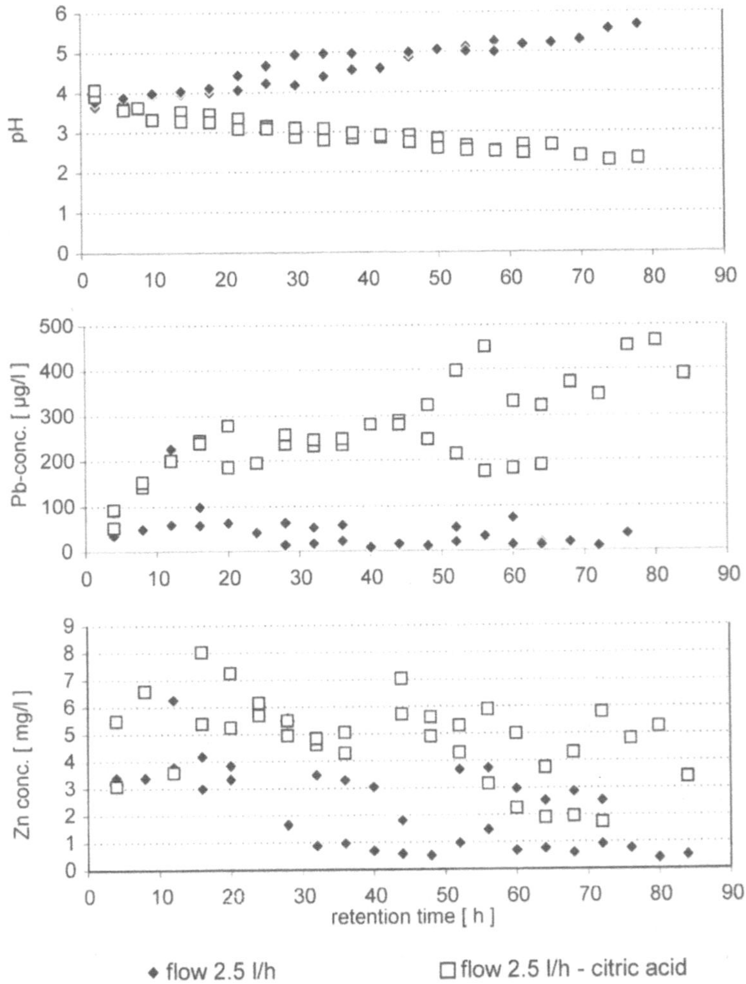

Fig. 3: PH and heavy metal concentrations in the effluent of the percolation reactors [10]

Table 3: Reduction efficiency with and without the addition of citric acid as an additional complexing agent

Heavy metal	Reduction efficiency (%) Without citric acid (pH range: 4 - 6)	With citric acid (pH range: 2 - 4)
Pb	5	18
Zn	40	53
Cu	5	5
Cd	43	48

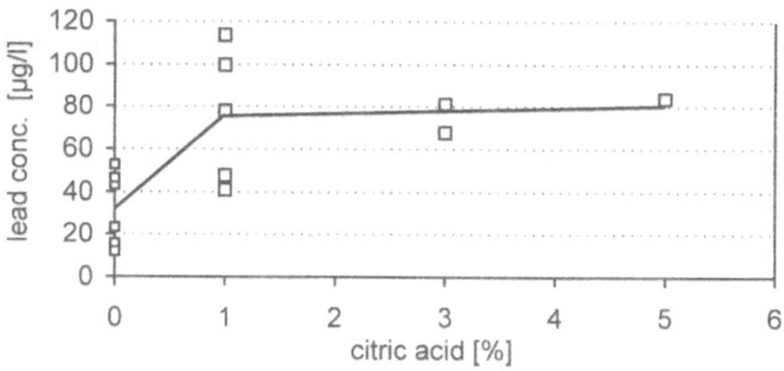

Fig. 4. Lead concentration in the effluent as a function of the amount of citric acid added

Summary and Conclusions

The results of the present study, as well as those of related investigations [9, 6, 11] suggest that, if heavy metal load reduction is the primary objective, a two-step anaerobic treatment, with a separate hydrolysis step, is superior to a single-step treatment. This conclusion applies to bio-wastes and to sewage sludges.

The production of internal organic acids in the biological hydrolysis step under limit conditions (pH up to 2) makes it possible to elute heavy metals without an additional chemical treatment step and without adding more chemicals. The elution efficiencies (the separated heavy metal discharge), however, did not totally achieve the expected levels in terms of increasing the bio-waste quality by significantly lowering the heavy metal loads. One explanation for this might be the relatively low specific loads of the investigated bio-waste. Zinc, for example, was in the range of 80 mg/kg DM to 230 mg/kg, while lead ranged from 10 mg/kg to 40 mg/kg. If the heavy metal load would have been higher (Zn > 300 mg/kg, Pb > 100 mg/kg; the maximum allowable heavy metal load according to the German Bio-waste Ordinance [12]), then the leaching efficiency would probably have been higher too.

The addition of a small amount of citric acid (up to 1 % of the eluate flow) as an additional complexing agent increases the heavy metal reduction efficiency. There is no need to recycle the citric acid, because the required dosage is rather low, and especially because it is biodegradable. It can be reused in the subsequent methanisation step for energy production.
Ongoing and future investigations must focus on increasing the elution efficiencies by varying the operational conditions such as, for example, the input concentration and contact times. Furthermore, there is an urgent need to examine in more detail the source of the raw material and the bonds between the heavy metals and the organic matter (particulate, biologically-bound, complexable, desorbable or dissolved).

Acknowledgement

We are very grateful for funding provided by the Deutsche Bundesstiftung Umwelt, Osnabrück, Germany. They have been part of the project "Behandlung von schadstoffbelasteten organischen Abfällen, insbesondere Biomüll – Ausschleusung und Entfrachtung organischer und anorganischer Schadstoffe" (AZ. 08819).

References

1. Müller, G.: Schwermetallelimination im Hinblick auf die Wiederverwertung von Klärschlamm in der Landwirtschaft. In: Klärschlamm – Ressource oder kostenintensiver Abfall? (Hahn, H.H. and Trauth, R., eds.) Schriftenreihe des Inst. f. Siedlungswasserwirtschaft, Universität Karlsruhe, Volume 71 (1994) pp. 81-90.
2. Oliver, B.G., Carey, J.H.: Acid solubilization of sewage sludge and ash constituents for possible recovery. Water Research, 10, (1976) 1077-1081
3. Strasser, H., Brunner, H., Schinner, F.: Leaching of iron and toxic heavy metals from anaerobically digested sewage sludge. Journal of Industrial Microbiology, 14 (1995) 281-287
4. Schulze, G., Gunkel, G.: Verteilung und Umsetzung von Schwermetallen in der biologischen Stufe einer kommunalen Kläranlage. Vom Wasser, 70 (1988) 209-220
5. Schäfer, M., Hoffmann, E.: Reststoffmanagement auf kommunalen Kläranlagen. Teil 1 - Stoffbilanzen. gwf Wasser Abwasser, 9 (1998) 586-591
6. Daniel, M.: Methodische Optimierung der Schwermetallausschleusung. Unpublished thesis (Diplomarbeit) at the Institut für Siedlungswasserwirtschaft (1996) Universität Karlsruhe (TH)
7. Kranert, M., Hartmann, A., Grau, S.: Determination of sand content in digestate. Proceedings of ORBIT (1999) , Weimar, Part II, pp. 313-317
8. Schäfer, M., Hahn, H.H., Hoffmann, E., Hilligardt, D.: Nassmechanische Ausschleusung von Schadstoffen aus Bioabfall. Müll & Abfall, 8, (2000) (in print)
9. Schlaak, M., Bödecker, H., Siefert, E., de Vries, J.: Elution von Schwermetallen aus Biomüll und Kompost durch schwache organische Säuren. Müll und Abfall, 12 (1994) 816-827

10. Sauter, J.: Betriebsoptimierung eines mehrstufigen quasikontinuierlichen Perkolationshydrolysereaktors. Unpublished thesis (Diplomarbeit) at the Institut für Siedlungswasserwirtschaft (1999) Universität Karlsruhe (TH).
11. Langhans, G.: Bio-chemical reduction of heavy metals in organic wastes by advanced digestion process. Anaerobic Conversions for Environmental Protection, REUR Technological Series 51 (1997) FAO Regional Office for Europe, Gent
12. Anonymous: Bioabfallverordnung des Bundesministers für Umwelt, Naturschutz und Reaktorsicherheit (1998)

Utilisation of Fractions of Digester Sludge after Thermal Hydrolysis

G. Eliasson*, E. Tykesson**, J. la C. Jansen**and B. Hansen***

* Malmö Water & Sewage Works, S-205 80 Malmö, Sweden
gabriella.eliasson@malmo.se
** Dept of Water and Env. Eng, Lund Institute of Technology, P.O. Box 118, 221 00 Lund, Sweden
***Kemira Kemi AB, Kemwater, P.O. Box 902, S-251 09 Helsingborg, Sweden

Abstract

Acidic thermal hydrolysis dissolves a large portion of the organic and inorganic particles in digested sludge. After dissolving, different fractions can be separated from the solution by chemical and physical methods. Several possibilities for fractionation exist governed by the expected use of the different fractions. One fraction, the hydrolysate, contains the dissolved organic substances and precipitation agents (iron). These are considered valuable in case of carbon shortage for denitrification or even biological phosphorus removal. The organic substances can promote growth of bio-P bacteria and iron can stabilise the process.

At Öresundsverket Waste Water Treatment Plant (WWTP) in Helsingborg, Sweden thermal hydrolysis has been in full-scale operation for several years. After separation of particles and phosphorus the remaining dissolved organic substances and precipitant return to influent waste water at the WWTP. Since the plant has changed the method of phosphorus removal from two trains with simultaneous precipitation and two with biological phosphorus removal into biological phosphorus removal in all four trains, the combined effect of precipitation agents and dissolved organic substances on the process is of special importance. Laboratory experiments revealed that the organic substances have significant potential for biological phosphorus removal and that the iron is valuable as precipitant. The impact of the combined effect was evaluated.

At Sjölunda in Malmö, Sweden, the potential of thermal sludge hydrolysis was considered but here the dissolved organic substances are intended for production of biogas in existing digesters. For more than one year, laboratory experiments were carried out in two lab scale digesters to evaluate the gas production based on the soluble organic fraction from thermal hydrolyses of digester sludge. The experiments have demonstrated that the process is feasible and that the hydrolysate can be co-digested with normal primary sludge as well as digested alone. Separation of the soluble organic fraction for digestion is of special interest in Sweden since biogas could be a valuable alternative as fuel for heavy vehicles.

The KREPRO process

KREPRO, Kemwater Recycling Process, is a thermal, acidic process for hydrolysis of sludge. The process is able to treat both raw and digested sludge in a continuous process as shown in Figure 1.

Initially sludge of 5% dry substances (DS) is acidified with sulphuric acid to a pH below 2 and most of the inorganic salts dissolves. The next step is to heat the sludge in heat exchangers before it enters a reactor, where additional heat is added so that the temperature increases to 140°C and the pressure is raised to 3.5 bar. In the reactor, where the average retention time is one hour, 40% of the organic matter and previously non-dissolved organic salts dissolve. After the reactor the sludge is cooled in the heat exchangers and, at the same time, the inlet sludge is heated up.

The sludge now contains dissolved phosphorus, ferrous and COD but also organic particles. These are now separated in a centrifuge where an organic solids fraction is produced. The organic fraction has a DS of 45-50% and an energy content equal to wood chips [1], but with ash content higher than that of wood.

To the liquid phase a ferric salt and some alkali are added. Ferric phosphate precipitates and is afterwards separated in a second centrifuge. The ferric phosphate has very low heavy metal content and high phosphorus content and is suitable for agricultural use.

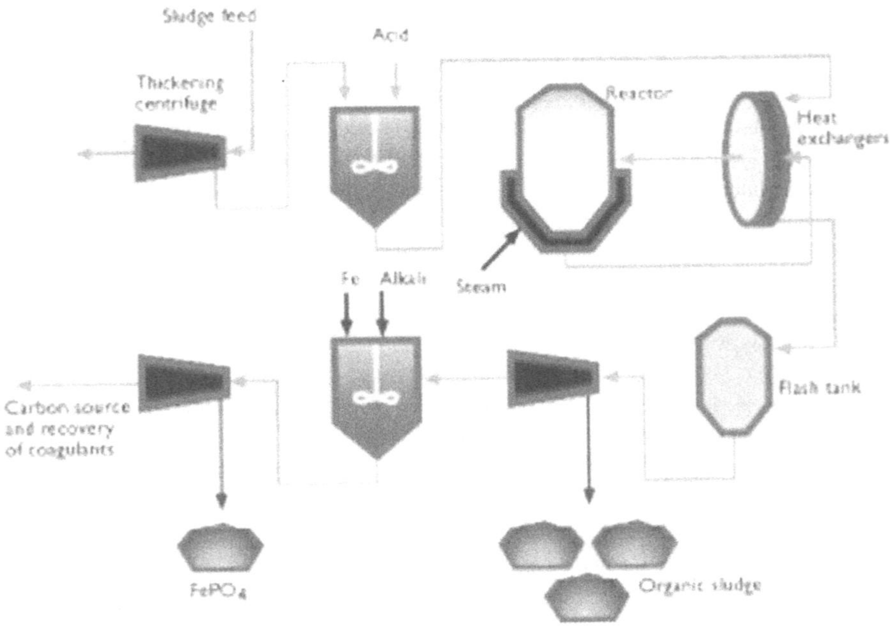

Fig. 1. Principle of the KREPRO process

The reject water from the ferric phosphate centrifuge contains high concentrations of dissolved COD, ferrous and nitrogen. The COD can be used for improved nitrogen removal, biogas production or as a carbon source for biological phosphorus removal. The ferrous can be used for phosphorus precipitation. The nitrogen is an internal load of nitrogen at the waste water plant unless the reject water is used for biogas production and ammonia stripping is used for removal of ammonia from the digester supernatant.

The KREPRO plant at Öresundsverket in Helsingborg

A full-scale pilot plant has been in operation for several years at Öresundsverket [2]. The plant was constructed in order to test and optimise the processes with raw as well as digested sludge. Later the plant was rebuilt in order to process most of the digested sludge from the plant. A general description of Öresundsverket can be found in Jönsson et al. [3]. The waste water treatment plant operated with four trains in parallel. For long periods two lines were operated with simultaneous precipitation and two with biological phosphorus removal. A two media sand filter acted as a safeguard in case of poor performance of the biological phosphorus removal. Lately the lines with precipitation were converted to biological phosphorus removal.

The reject water was characterised with normal chemical analyses, and Table 1 gives the typical composition. Great variation exists depending on the sludge and the operational conditions. At Öresundsverket the reject water is directed to the inlet of the plant and is used for denitrification and chemical precipitation. Table 1 also gives the typical dry weather concentration of the waste water together with the increase from the supply of hydrolysate.

It is seen from the table that the hydrolysate is high in COD, which can be utilised for biogas, and that the contribution of especially COD and iron may have an influence on trains with biological phosphorus removal at Öresundsverket.

Table 1. Typical composition of the hydrolysate, inflow concentrations to Öresundsverket and the extra contribution from hydrolysate

	Hydrolysate	**Inflow**	**Contribution from hydrolysate**
COD (mg/l)	16 000	390	60
Volatile Fatty Acids (mg/l)	1 100	20	4
Suspended Solids (mg/l)	5 500		
Total Nitrogen (mg/l)	1 900		
Total Phosphorus (mg/l)	250		
Ammonia (mg/l)	800		
Iron (mg/l)	2 400	3	11
pH	3		

Recycling of precipitant

The iron in the hydrolysate is mainly present as ferrous iron, a very poor coagulant compared to ferric. However ferrous iron can also precipitate phosphates and can therefore be useful for phosphorus removal at the waste water treatment plant. The ferrous in the hydrolysate can be compared with copperas, ferrous sulphate, a product used in many waste water treatment plants for phosphorus removal.

Prior full-scale trials at Öresundsverket showed that when KREPRO was running and precipitant recycled, no additional copperas was needed. Because these trials were not very well documented, a laboratory test was used to evaluate the potential of ferrous in the hydrolysate for phosphorus precipitation.

The content of the hydrolysate produced from Sjölunda WWTP is shown in Table 2. It is seen that the hydrolysate not only brings iron into the waste water but also some phosphates as well as COD.

Table 2. The composition of hydrolysate used in phosphorus precipitation trials

pH	Total P (mg/l)	Total Fe (mg/l)	COD (mg/l)
2.05	330	6 500	17 000

Two trials were conducted, one conventional flocculation jar test with 10 seconds of intensive mixing, 10 minutes of flocculation and finally 10 minutes of sedimentation before samples were taken. The second trial was similar to the first but after the iron (hydrolysate) was dosed the waste water was aerated for one hour before the flocculation. The dosage varied between 0 and 30 mg Fe/l. The results were compared with waste water without any addition of hydrolysate but otherwise treated the same way, with or without aeration (dosage 0 mg Fe/l). The treated waste water was analysed with regard to pH, total phosphorus, orthophosphate, total and dissolved iron and total COD. The phosphorus removal results are shown in Figure 2.

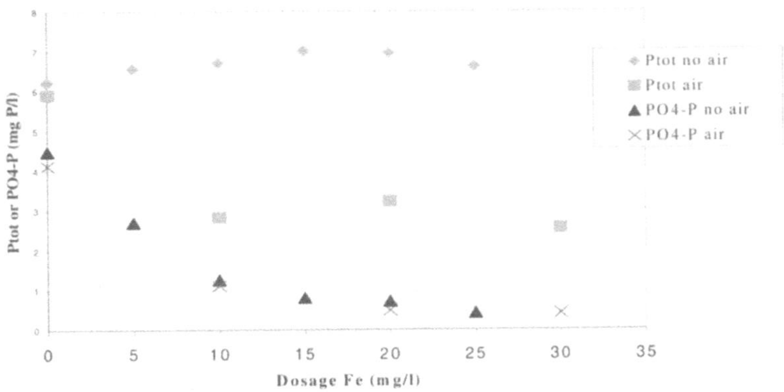

Fig. 2. Residual total P and orthophosphate as a function of the dosage of iron

The orthophosphate was removed both in the aerated and non-aerated samples. The difficulty is to remove the resulting ferrous phosphates since they hardly form large flocs. This can be done by aerating the water when the hydrolysate is added. Most of the ferrous is then oxidised to ferric and is able to form flocs. The flocs were small but did anyhow improve the reduction rate of total phosphorus.

Other results from the trial showed that the turbidity removal increased slightly when the water was aerated. Residual total iron was high when the water was not aerated and if it was, iron was quite high at the higher dosages of hydrolysate. Notable was that the increase of residual COD was very low even if as much as 80 mg/l were added by the hydrolysate. When the water was aerated, all the COD values were lower than in the non-aerated trial.

Hydrolysate as carbon source for biological phosphorus removal

The thermal hydrolysate has a high concentration of carbon and iron. Laboratory experiments were performed in order to investigate the capacity of the hydrolysate as an additional carbon source for biological phosphorus removal, and also to investigate the impact of the iron.

The process of biological phosphorus removal in activated sludge is often explained as a result of changing anaerobic and aerobic conditions. When the sludge is treated anaerobically, without oxygen or nitrate, stored polyphosphates within the cells are released into the water at the same time as easily degradable carbon is taken up. During aerobic conditions phosphorus is taken up again while stored organic matter is utilised for energy and growth.

An excess of an easily degradable carbon source, volatile fatty acids (VFA), during the anaerobic phase is of great importance for the process. If the VFA content in the waste water is not sufficient, biological phosphorus removal can be improved by internal hydrolysis or by addition of an external carbon source.

A laboratory experiment was performed with three different reactors of one litre each. The experiment was performed in two phases. The reactors were filled with activated sludge from one of the trains with biological phosphorus removal at Öresundsverket. Phosphorus was added at a concentration of 15 mg PO_4-P/l in order to enable maximum phosphorus uptake in the first aerobic phase, which lasted one hour. The aeration was then stopped and nitrogen gas was added to the reactors to create anaerobic conditions for the activated sludge. At the same time carbon sources were added to the reactors. Sodium acetate was added to the first reactor at an amount of 300 mg COD/l. For reactor two and three the carbon source consisted of hydrolysate, with different content of ferrous iron. The dose corresponded to about 300 mg COD/l. The hydrolysate had a high concentration of iron, but a lower content was obtained by increasing the pH and then sedimenting the metal hydroxides. During the anaerobic period phosphorus was supposed to be released from the bacteria, resulting in a higher phosphate concentration in the solution. The hydrolysate used for the test was more dilute than the typical hydrolysate described in Table 1. The increase of the concentrations of COD and iron after addition is shown in Table 3.

Table 3. Increase of COD and iron by addition of carbon sources

	COD (mg/l)	Fe (mg/l)
Acetate	300	0
Hydrolysate, iron reduced	300	15
Hydrolysate	300	73

Throughout the experimental period samples from the activated sludge were removed, directly filtered, and analysed. The phosphate concentrations in filtered samples from the experiment are shown in Figure 3.

During the aeration period the three reactors acted similarly. The bacteria stored phosphate and the concentrations decreased in the reactors. At time 1:00 the conditions became anaerobic and carbon sources were added. The phosphorus concentrations increased, which means that the bacteria released phosphate. The reactor with acetate had the highest release of phosphate. At time 1:00 there was a significant and sudden drop before the release started in the reactor with hydrolysate, which is a result of precipitation. The drop was less significant when hydrolysate with reduced iron content was added.

The maximum phosphorus release, measured as the difference between maximum and minimum phosphate concentration in the three reactors, is compared in Figure 4.

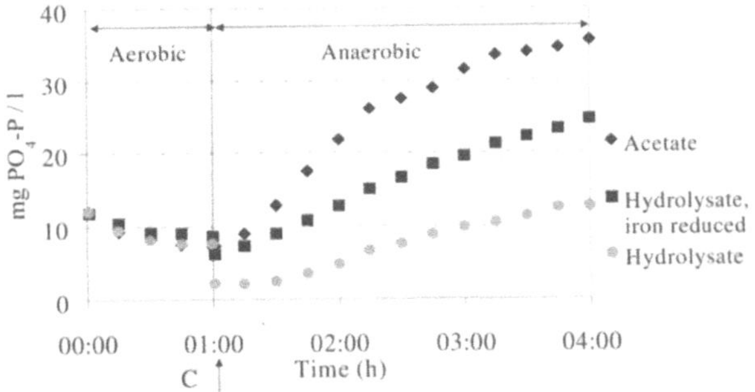

Fig. 3. The P-release in a batch experiment using three different carbon sources

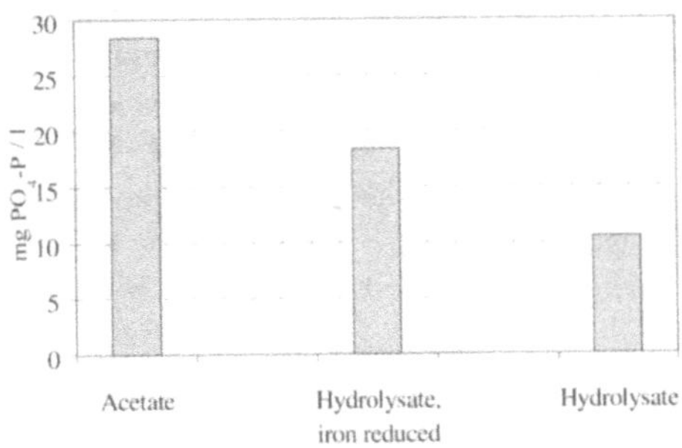

Fig. 4. Maximum P-release with different carbon sources

The difference in maximum phosphorus release might be due to a smaller fraction of degradable carbon in the hydrolysate than in the acetate. VFA is only a small part of the total COD in the hydrolysate, as seen in Table 1. Even if the VFA addition is more than enough for complete phosphorus release in the acetate reactor, the VFA in the hydrolysate reactors might not be sufficient. The iron content may also have an influence on the phosphorus release measured in the experiment. The release might be the same in all reactors, but the resulting phosphorus concentrations could actually differ because of different iron precipitation.

A complementary experiment was made in order to study if ferrous has an effect on the phosphorus release or if it just reduces already released phosphorus by precipitation. An experiment similar to the one with hydrolysate was carried out. Acetate was added as carbon source to two different reactors at the beginning of the anaerobic period, and ferrous sulphate was added at the same time to one of the reactors. The amount of ferrous was chosen to get the same ratio of ferrous to COD as in the hydrolysate. The results show, as expected, that the maximum phosphorus release, measured as the amount of dissolved phosphate at the end of the anaerobic period, is lower in the reactor with iron. The carbon source taken up during the phosphorus release was measured as total organic carbon (TOC). Samples from the two different reactors show similar TOC reductions, which indicates that the phosphorus release in reality is the same. The different maximum release values measured are just the result of precipitation and the release of phosphorus is not directly influenced by iron.

The reduced phosphorus release with addition of hydrolysate is therefore expected to be partly attributed to precipitation of phosphorus and partly to the complex composition of the hydrolysate where only a minor portion of the COD is directly available for biological phosphorus removal.

Hydrolysate as a carbon source for biogas production

At Sjölunda WWTP the dissolved carbon in the reject water after thermal hydrolysis of dewatered sludge was examined as a possible carbon source for biogas production. On a laboratory scale, the hydrolysate was digested from November 1998 to November 1999 to determine if the hydrolysate:

- was degradable
- was stable in a biogas process
- produced biogas.

The laboratory experiments were performed with two continuously stirred digesters heated to the mesophilic range. Produced gas was collected in a gas storage tank and was measured daily. One of the digesters was fed with raw thickened sludge and used as a reference and the other one was fed with a mixture of raw sludge and hydrolysate.

The load of the reactors was maintained at an organics load of about 1 kg VSS/m^3reactor and day and at a retention time of 20 days.

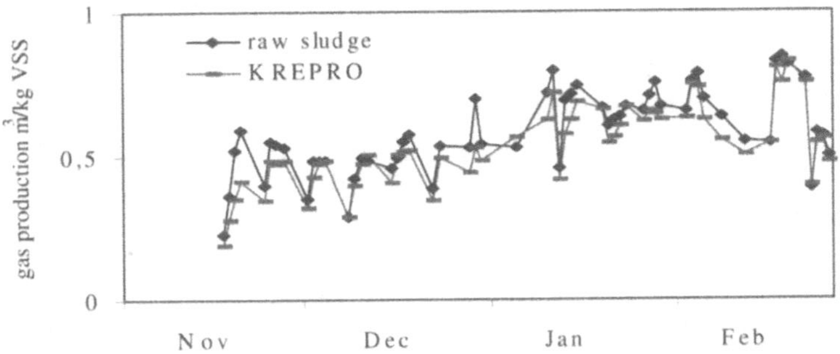

Fig. 5. Gas production in the laboratory digesters

At first the hydrolysate was co-digested with raw thickened sludge from Sjölunda WWTP. The hydrolysate made up 20% of the VSS, corresponding to reliable full-scale figures if KREPRO was in operation and the hydrolysate was co-digested in the existing digesters.

The digesters were operated for three months in parallel in order to ensure stable microbiological conditions. Daily variations in gas production could be seen in both digesters, and this was due to variations in dry substances (DS) in the incoming raw sludge.

As seen in Figure 5 the gas production was similar in the reactors, with a slightly reduced production in the reactor fed with hydrolysate, indicating that the process was stable and no disturbances occurred.

In subsequent laboratory experiments, the ratio of hydrolysate was increased in order to investigate if any inhibition of the digestion took place. The increase was performed over a period of six months until 100% hydrolysate could be digested. Figure 6 shows the gas production in the period ending with digestion of hydrolysate alone.

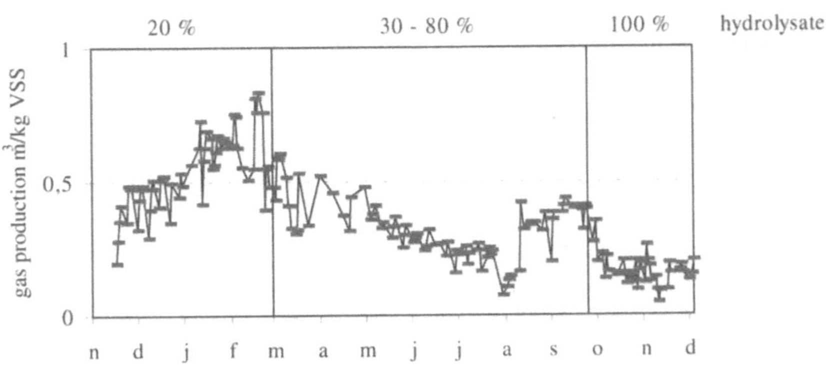

Fig. 6. Gas production over 13 months in a laboratory digester fed with different ratios of hydrolysate and raw sludge

Digestion of hydrolysate alone resulted in a gas production on average of about 0.15 m^3 gas/kg $VSS_{supplied}$. This amount corresponds to about one third of the gas production in the reference reactor. The reduced gas production as the proportion of hydrolysate was increased from 30% to 80% is a result of a lower gas-producing potential of the VSS from hydrolysate than that from raw sludge, and is not a result of operational problems with digestion of hydrolysate. Even over the three-month period with digestion of 100% hydrolysate, the daily operation showed no signs of disturbances in the process.

Conclusions

Acidic, thermal hydrolysate of digested sludge contains high concentrations of COD and ferrous iron. The hydrolysate can be utilized for chemical precipitation, but separation of the flocs by sedimentation is difficult. Precipitation and floc formation can be improved, however, by dosing under aerated conditions.

Phosphorus release tests with sludge from Öresundsverket showed a much smaller P-release with hydrolysate than with acetate, but the release was significantly increased when the iron content of the hydrolysate was reduced.

The same uptake of organic matter was obtained with acetate and with acetate together with ferrous sulphate even if the phosphorus release was significantly reduced when iron was included.

The potential of hydrolysate for biological phosphorus removal is greater than evaluated directly from a phosphorus release experiment.

Hydrolysate can be co-digested with raw sludge in a stable process without disturbances even with 80% hydrolysate and 20% raw sludge. The hydrolysate can even be digested alone.

The gas production from hydrolysate is significantly smaller than that from raw sludge when the same load, in terms of VSS, is supplied.

References

1. Cassidy, S.: Recovery of valuable products from municipal wastewater sludge. In Chemical Water and Wastewater Treatment V (Hahn, H.H., Hoffmann, E. and Ødegaard, H., eds.) Springer-Verlag, Heidelberg, (1998) pp. 325-340
2. Hansen, B., Karlsson, I., Cassidy, S. and Pettersson, L.: Operational experiences from a sludge recovery plant. Presented at IAWQ conference on sludge management for the 21st century: A value-added renewable resource, Perth Australia (1999)
3. Jönsson, L.E, la Cour Jansen, J., Magnusson, P.: Long Term Dosage of Aluminium in a Full Scale to Improve Activated Sludge Settleability. In Chemical Water and Wastewater Treatment IV (Hahn, H.H., Hoffmann, E. and Ødegaard, H., eds.) Springer-Verlag, Heidelberg, (1996) pp. 263-273

The Influence of Free Water Content on Sewage Sludge Dewatering

J. Kopp and N. Dichtl

Institute of Sanitary Engineering, Postal Box 3329, Technical University of Braunschweig, D-38106 Braunschweig, Germany, Tel. +49 531 391 (7942), Fax (7947), J.Kopp@tu-bs.de

Abstract

The paper describes a thermogravimetric measurement of the free water content to predict the dewaterability of sewage sludges. Dewaterability can be described by the total solids concentration of the sludge cake and the polymer demand for conditioning. The total solids concentration depends on how water is distributed in the sludge cake. Four different types of water can be distinguished mainly by their distribution in the cake and by the intensity of their physical bonding to the solids. These types are free water, which is not bound to the particles; interstitial water, which is bound by capillary forces between the sludge flocs; surface water, which is bound by adhesive forces; and intracellular water. Only the free water component can be removed during mechanical dewatering. By measuring this free water content in the sludge, it is possible to make an exact prediction of full-scale dewatering results. Polymer conditioning does not influence free water content, but it increases the velocity of the sludge water release. The percentage of excess sludge in the total sludge mass is a factor that affects the dewaterability of sewage sludge. No definite relationship could be found between volatile solids content and dewaterability. The thermogravimetric method thus shows potential in determining the full-scale dewatering result of sewage sludges.

Introduction

The product of mechanical dewatering can be described in terms of the total solids concentration of the sludge cake and the amount of conditioning agents. Of the four types of residual water that can be identified in the sludge, the free water fraction is particularly important, since only this fraction can be separated during mechanical dewatering. The addition of conditioning agents, such as cationic polymers, causes the sludge to coagulate, thus accelerating the release of water from the sludge. The polymer-demand is basically determined by the anionic surface charge of the sludge particles. To predict the dewaterability of a sewage sludge it is thus necessary to measure the percentage of the free water fraction in the sludge and to determine the polymer-demand. This paper describes a thermogravimetric method to measure the free water content in sewage sludge.

The four types of water in sewage sludge are mainly distinguished by their distribution, and by the intensity of their physical bonds with the solids. The bonding forces counteract the steam pressure of the water molecules, thus decreasing the steam pressure [1]; in other words, the higher the bonding forces, the lower the steam pressure. To clarify this point, the bonding forces can be understood as an attraction between the sludge particles and the adsorbed water molecules. In a sewage sludge suspension four different types of water can be distinguished according to their physical bonds with the sludge particles. These are:

- free water, which is not bound to the particles
- interstitial water, which is bound by capillary forces between the sludge flocs
- surface water, which is bound by adhesive forces and
- intracellular water

The free water content represents the largest fraction in sewage sludge. The water moves freely between the individual sludge particles, is not adsorbed by them, is not bound to them, and is not influenced by capillary forces. This fraction can be separated mechanically, for example, by centrifugal forces or filtration. The interstitial water is found in the interstices of the sludge particles and microorganisms in the sludge floc. This water is bound physically by active capillary forces. The surface water covers the entire surface of the sludge particles in several layers of water molecules and is bound by adsorptive and adhesive forces. The surface water is physically bound to the particles and cannot move freely. Water that is chemically bound in exopolymers is also considered to be part of the surface water. The intracellular water includes the water inside cells and also the water of hydration. Intracellular water can only be determined together with the surface water and is often called bound water content [2]. Bound water makes up the smallest fraction of residual water, has the strongest physical-chemical bonding to the particles and can only be removed thermally. It must be noted that the term "bound water" is not clearly defined and differs according to the method used for measurement [3].

Materials and methods

Thermogravimetric and dilatometric laboratory tests [4,5] were used to measure the fractions of residual water. At the Technical University of Braunschweig these methods were adjusted and calibrated, so that it was possible to determine directly the maximum suspended solids content in the sludge cake after mechanical dewatering in centrifuges and filter-presses [6].

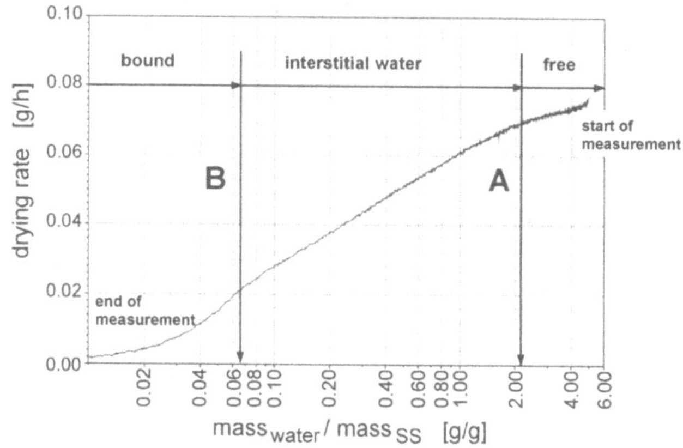

Fig. 1. Semi-log drying curve of a digested sewage sludge

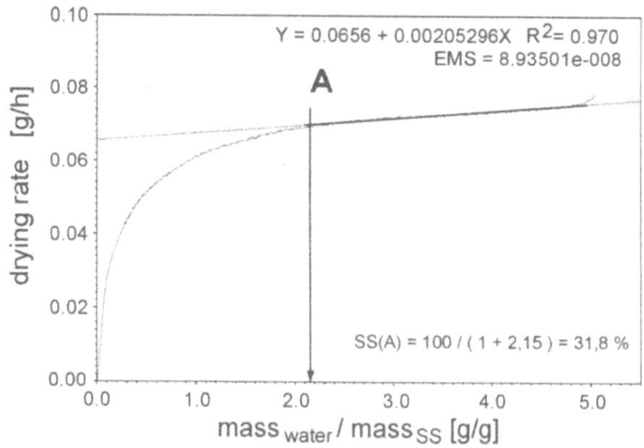

Fig. 2. Drying curve of a dDigested sewage sludge using an arithmetic scale

For thermogravimetric measurements the sludge sample was dried under constant conditions (air flow, temperature). The different water fractions could be identified by plotting the drying rate as a function of the moisture content ($mass_{water}/mass_{SS}$) of the sample. Prior to drying, the sample was concentrated in a laboratory beaker centrifuge at 1000 x g for 30 min. The drying procedure must be carried out very slowly, because otherwise the different fractions of water cannot be distinguished using the drying curve because of the high energy input.

Figure 1 shows an example of a drying curve of a digested sewage sludge sample. Chronologically, the drying curve starts at the top right-hand corner with a high moisture content ($mass_{water}/mass_{SS}$) and ends at the bottom left, when all the water has been removed from the sample. Two critical points A and B can be seen on the drying curve. Point A marks the end of the free water content and point B the end of the interstitial water.

The most interesting point, before the background of mechanical dewatering, is to determine exactly the free water content, namely, point A of the drying curve.

Point A cannot be readily identified in the semi-log plot shown in Figure 1, but this point can be determined when the drying curve is plotted using an arithmetic scale (Figure 2). As long as there is free water in the sludge sample, the drying rate is linear. On this part of the curve, the drying rate is described by the weight loss of the sludge for each unit of time. At point A the drying rate decreases because of the capillary bonding of the interstitial water to the sludge. The calculated tangent no longer describes the curve. At point A, the solids content of the sludge (SS_A) can be derived from the moisture content of the sample.

An interesting observation could be made by monitoring the drying process continuously with a video camera. First the water level in the sludge sample went down until all the free water evaporated. Then the sludge sample began to shrink. This observation can be explained by the fact that after all the free water has evaporated, the interstitial water begins to evaporate from the capillary tubes. Once the capillary tubes are no longer filled with water, they start to deform during the subsequent drying process.

The amount of surface and intracellular water can be determined summarily as bound water content using dilatometric measurements [7]. The measuring principle is based on the fact that the bonding forces are so strong, that bound water does not freeze at -25°C. The difference between the volume expansion of the frozen water content and the total water content of the sample determines the bound water content. If one assumes that the difference between the free water content and the bound water content is equivalent to the interstitial water content, then it is possible to determine the relative water content of the four fractions. Laboratory tests were carried out with digested sludge from various wastewater treatment plants.

The highly cationic active polymer Zetag-87 (Allied Colloids) was used for the conditioning of the sewage sludge. Conditioning agent demand was determined in a series of experiments. An optimum polymer dosage is reached when no electrostatic repulsive forces affect the sludge particles anymore and the value of the zeta potential in the centrate lies between -3 and ±0 mV [8]. The zeta potential was measured electrophoretically (Malvern Zetamaster) in the centrate water. Particle size distribution, capillary flow time, and the surface charge of the sludge were additional dewatering parameters that were determined.

Results and discussion

Effect of temperature and air flow when determining the water fractions

The measurement depends on two conditions, air flow and temperature, so that a calibration of the drying instrument is necessary before an exact evaluation of the free water content can take place. Figure 3 shows the effect of these two conditions on the moisture content (A). The relationship between temperature and moisture

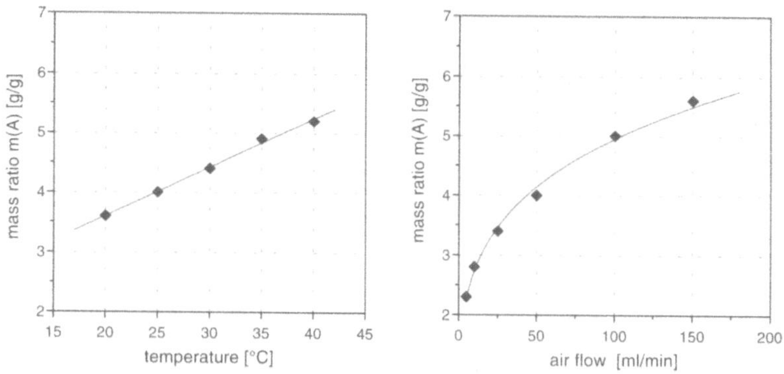

Fig. 3. Drying Curve as a function of laboratory conditions

content is linear. The relationship between air flow and moisture content, however, is not linear, with the moisture content leveling off at high air flows. As these measurements are very time consuming *E. coli* bacteria were used for the calibration, because these samples can be reproduced with a constant quality.

Calibration of the drying instrument

Silica particles were chosen for the calibration of the drying instrument as it is inevitable for the exact determination of the free water content. The advantage to using these particles is that all variables are known for the calculation of interstitial water content and bound water content according to the model of Batel [9] and Schubert/Pietsch [10]. The contact angle of silica comes to 60-70°, at natural bulking it has a porosity of 40%, and a coordination number of 8 can be expected. 0.1 µm was assumed in order to determine the strength of the adsorptive layer. Figure 4 shows the results of the model calculation and the measured values for five particle sizes. It can be seen that the values agree very well with the more exact model calculation of Schubert/Pietsch.

Prediction of full scale dewatering results

Table 1 gives sample data on the water fractions, polymer-demand, solids content (SS), and volatile suspended solids (VSS) of four sewage sludges. Due to the present state of the art in science and technology, mechanical dewatering processes can only remove the free water content (w_{free}) from sludge. This water content was calculated with the drying instrument. The solids content in the sludge cake, which would exist after removal of the free water content, is derived as the parameter (SS_A). The solids content at point A (SS_A) is therefore the maximum solids content

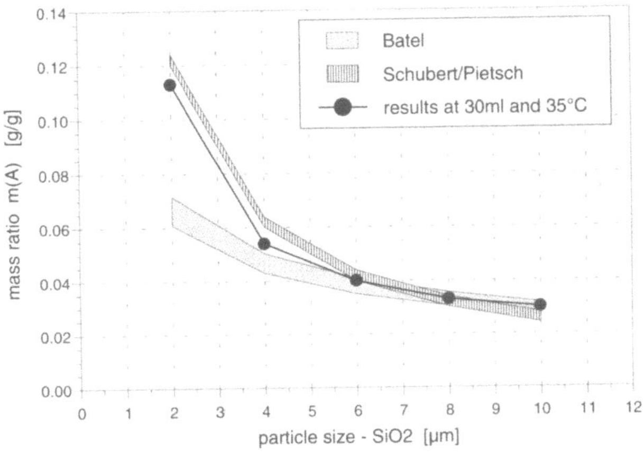

Fig. 4. Calibration of the drying instrument using standard particles

in the sludge cake, which can be achieved with mechanical dewatering. The content of interstitial water (w_{inter}) in the end determines the dewaterability, because this water content remains in the sludge cake after the dewatering process. For sewage sludges the interstitial water content amounts to approximately 4-13%. For the mechanical dewatering process, the bound water content (w_{bound}) measured with dilatometric tests is of minor importance. On the one hand this water content always remains in the sludge cake, but, on the other hand, it is very small. The values lie at about 1% SS.

Figure 5 compares the SS_A of digested sewage sludges calculated using the drying instrument (in other words, the SS after separating the free water content) with the dewatering results obtained with full-scale centrifuges and filter-presses. The accuracy of the thermogravimetric measurement lies at about ±1% SS, thus giving an exact prediction of the solids content after dewatering.

Table 1. Water fractions in municipal sewage sludge

Sample	SS [%]	VSS [%]	Poly* [g/kg]	SS [%]	w ** [%]	W ** [%]	W ** [%]
Excess sludge 1	2.46	68	10.0	20.0	0.8	11.6	87.6
Excess sludge 2	1.53	61	2.5	26.3	1.0	7.7	91.3
Primary sludge	3.77	64	4.5	37.0	0.5	4.8	94.7
Raw sludge	3.72	63	6.5	27.8	0.6	7.5	91.9
Digested sludge 1	4.28	55	7.5	25.0	0.7	8.6	90.7
Digested sludge 2	4.59	52	5.8	31.8	0.9	5.7	93.4

*= Polymer-demand for Zetag-87 ** The percentages are related to 3% SS of the sludge suspensions

Factors that affect the water fractions

It is of great interest to determine what factors influence the water fractions. One of these factors is the dosage and type of polymer, which was examined for various digested sewage sludges. Figure 6 shows the solids content SS(A) of three digested sludges when different polymer dosages were applied. For these tests high under- and over-flocculation was used, as well as an optimum polymer dosage for the conditioning. Contrary to the expectation that the free water content should be the highest for an optimum polymer dosage, the free water content remained constant for all dosages. This effect was confirmed during tests with other sludges and polymer products (results not shown). A possible explanation for this result is that the interstitial water is bound in the flocs, and during conditioning flocs are combined in larger agglomerates, thus increasing the velocity of the sludge water release. In the final analysis the dewatering result achieved in full-scale centrifuges is independent of the polymer dosage and the polymer product. The retention time of the sludge in the centrifuge, however, is so short that optimum dewatering results can only be reached when the velocity of water release from the sludge is optimal.

Another important factor that affects sewage sludge dewaterability is the percentage of excess sludge in the total sludge mass [11]. Figure 7 shows the solids content at point A of the drying curve (SS_A) for various sludges consisting of excess and primary sludge, using raw as well as digested sludge mixtures. Usually primary sludges are easy to dewater. The amount of free water is high because little water is bound by capillary forces as interstitial water. In contrast, excess sludges are difficult to dewater, since the percentage of free water is low. As the percentage of excess sludge in the total sludge mass increases, the amount of free water decreases, thus decreasing the quality of the dewatered product.

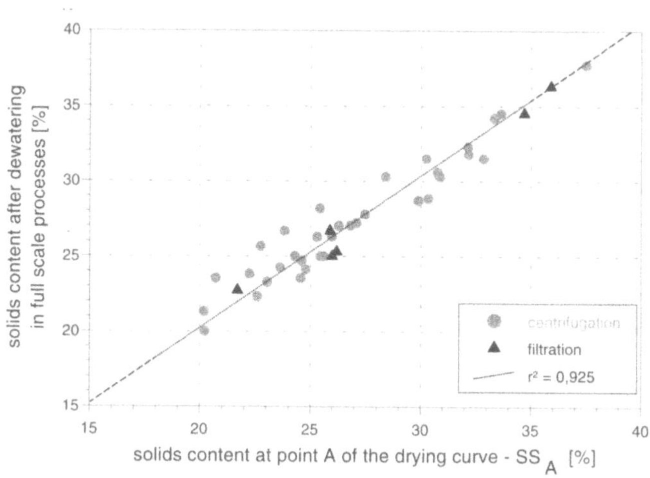

Fig. 5. Correlation between SS_A - solids content at point A of the drying curve and the solids content after dewatering in full-scale centrifuges

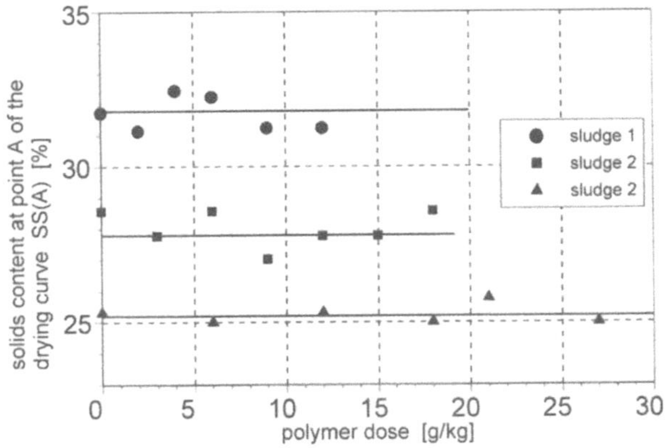

Fig. 6. Effect of polymer dose on the drying curve

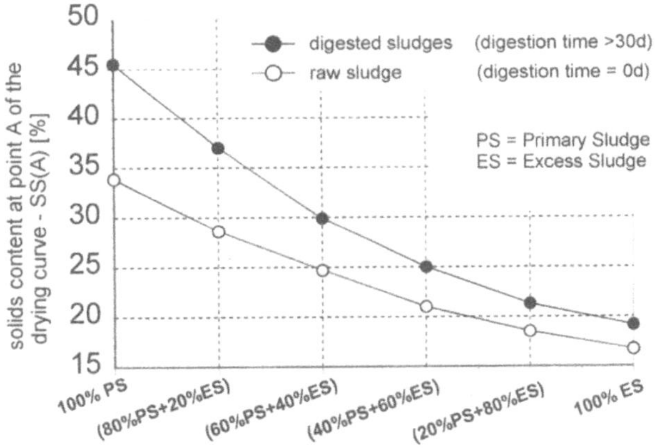

Fig. 7. Effect of the percentage of excess sludge in the total sludge mass on the solids content that can be achieved by dewatering (SS_A)

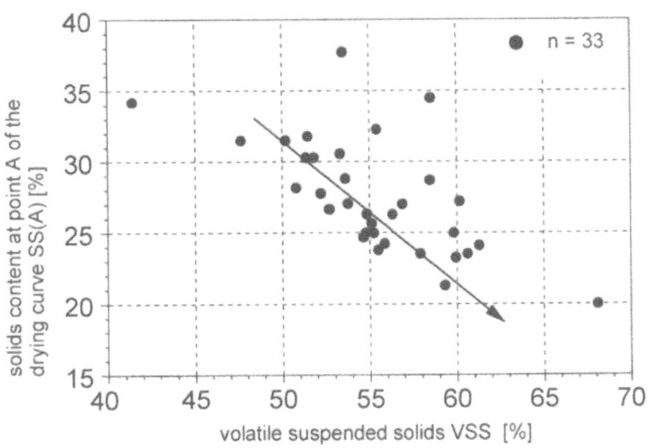

Fig. 8. Effect of volatile suspended solids on the solids content that can be achieved by dewatering (SS_A)

The volatile solids content represents the organic part of a sewage sludge and is the parameter most often used to characterise the dewaterability of sludge. The higher the amount of organic compounds in a sludge, the lower the density of the sludge particles ($\rho_{organic} \approx 1.0$ g/cm³; $\rho_{inorganic} \approx 2.3$ g/cm³). In addition, more water is bound, since the contact angle of organic particles is smaller than that of inorganic particles, thus binding more water by means of capillary forces. That means that sludges with a volatile suspended solids content (VSS) are in general more difficult to dewater. Figure 8 illustrates how the solids content after separation of the free water content (SS_A) is affected by the volatile solids content. One detects a general trend whereby the dewatering result improves with decreasing volatile solids content. The measured values are so scattered, however, that the dewatering result of a sludge cannot be determined from the volatile suspended solids content. Similar problems can be observed for other characterisation parameters, such as the amount of colloids.

Summary

Four types of residual water fractions can be distinguished in sewage sludge. The free water fraction is not bound to the sludge particles and can be separated by mechanical dewatering processes. The other three fractions (interstitial, surface, and intracellular water) remain in the sludge cake after the dewatering process.

Two methods were used to determine the proportion of these water fractions in a sludge sample. The thermogravimetric method, adjusted and calibrated at the Technical University of Braunschweig, was used to measure the free water fraction, which is the only fraction that can be separated from the sludge. Being able to measure this fraction makes it is possible to predict the maximum dewaterability of a sewage sludge for full-scale centrifuges and filter-presses. The second method, the dilatometric method, was used to measure the so-called bound water fraction (surface and intracellular water).

Both methods were carried out after determining the polymer demand of a sludge in laboratory experiments. This was done by increasing the polymer dosage until the zeta potential in the centrate approached the isoelectric point.

Factors that affect the dewaterability of sewage sludges were also studied. It was found that as the percentage of excess sludge in the total sludge mass increased, the amount of free water was reduced, resulting in lower sludge dewaterability. Polymer conditioning had no effect on the free water content, and no definite relationship could be found between volatile solids content and dewaterability.

References

1. Lück, W.: Feuchtigkeit: Grundlagen - Messen - Regeln, Verlag R. Oldenbourg, München (1964) Wien
2. Smith, J. K. and Vessilind, P. A.: Dilatometric Measurement of Bound Water in Wastewater Sludge, Wat. Res. 29(12) (1995) 2621-2626
3. Vessilind, P.A.: What is Bound Water and how does Bound Water affect Sludge Dewaterability ? Wat. Sci. Tech. 36(11) (1997) 87-91
4. Jones, E.V., Gortner, R. A.: Free and Bound Water in Elastic and Non-elastic Gels, J. Phys. Chem. 37 (1932) 387
5. Smollen, M.: Categories of Moisture and Dewatering Characteristics of Sewage Sludge, Proceedings of the 4th World Filtration Congress, Ostend, Belgium, 22-25 April (1987) 1435-1441
6. Kopp, J., Dichtl, N.: Prediction of full-scale dewatering of sewage sludge by thermogravimetric measurement of the water distribution, Proceedings of the 8th World Filtration Congress, 3-7 April 2000, Brigthon UK,Vol 1 (2000), pp. 161-164
7. Smith, J. K., Vesilind, P.A.: Dilatometric measurement of bound water in wastewater sludge, Water Research 29(12) (1995) 2621-2626
8. Kopp, J., Dichtl. N.: Influence of Surface Charge and Exopolysaccharides on the Conditioning Characteristics of Sewage Sludge, In: Chemical Water and Wastewater Treatment V, (Hahn, H.H., Hoffmann, E., Ødegaard, H., Eds.) Springer Verlag (1998) Berlin, pp 285-296
9. Batel, W.: Menge und Verhalten der Zwischenraumflüssigkeit in körnigen Stoffen, Chemie Ingenieur Technik 33(8) (1961) 541-546
10. Schubert, H., Pietsch, W.: Rumpf, H.: Haftkraft, Kapillardruck, Flüssigkeitsvolumen und Grenzwinkel einer Flüssigkeitsbrücke zwischen zwei Kugeln, Chemie Ingenieur Technik 39 (1967) 885-893
11. Nellenschulte, T., Kayser, R.: Change of particle structure of sewage sludges during mechanical and biological processes with regard to the dewatering result, Proceedings of the 4th International Conference on "The role of particle characterisation in separation processes", 28-30 October (1996) Jerusalem Israel, pp 270-286

Resources Reuse

Phosphorous Recycling from Pre-Coagulated Wastewater Sludge

Y. Watanabe*, T. Tadano**, T. Hasegawa***, Y. Shimanuki****
and H. Ødegaard*****

*Graduate School of Engineering, Hokkaido University, Sapporo 060-8628, Japan
yoshiw@eng.hokudai.ac.jp
**Graduate School of Agriculture, Hokkaido University, Sapporo 060-8589, Japan
***Suido Kiko Kaisha, Ltd., Sakuragaoka, Setagaya-ku, Tokyo 156, Japan
****Asahi Yukizai Industry, Ltd., Nakanose-cho 2-5955, Nobeoka 882-8688, Japan
*****Faculty of Civil and Environmental Engineering, Norwegian University of Science and Technology, Trondheim, N-7491 Norway

Abstract

The amount of rock phosphorous remained in the world is limited. Therefore, it is important to develop the technology and construct the social system for use of recycled phosphorous. Municipal wastewater contains 5 to 10 mg/l of phosphorous on suspended and soluble forms. This paper deals with separation of phosphorous from municipal wastewater by pre-coagulation, and use of plant functions to secrete acid phosphatase and organic acids that lead to release of soluble inorganic phosphate from the pre-coagulated sludge. Acid phosphatase and organic acids such as citric acid will release soluble inorganic phosphate from organic phosphates and hardly soluble inorganic phosphates, respectively. Production of pre-coagulated sludge took place in a solid liquid separation unit consisting of the jet mixed separator and inclined tube settlers. The unit removed more than 90 % of total phosphorous from municipal wastewater with the addition of poly aluminum chloride of 5 mgAl/l. A production method to depress the polymerization of the polysilicato iron, a new coagulant that seems to be preferable to aluminum coagulants in the pre-coagulation scheme, is also described. Phosphorous compounds contained in the pre-coagulated sludge were transformed to soluble inorganic phosphates by acid phosphatase and organic acids. The excess sludge worked as a phosphate fertilizer for sugar beet, resulting in about 5 times the dry weight compared that obtained in the phosphate-deficient condition when acid phosphatase was added in the root compartment.

Introduction

Phosphorous is the eleventh most common element on earth, essential to all living organisms. In nature phosphorous always occurs combined with oxygen and other elements, forming phosphates. Around 80 % of the phosphates produced by the world's industry today are used in fertilizers, with a further 5 % being used to supplement animal feeds. These phosphates are manufactured from phosphate-containing rock mined from deposits in several countries. Around 140 million tons of phosphate rocks are extracted each year across the world. In 1989, the Phosphorous Resources Institute of Japan estimated that the phosphate rock in the world will remain for 50 years if we spend as much money recycling it as we do now, and it will remain for 100 years if we spend twice as much. Therefore, we must develop a system for recycling phosphates. The quantities of phosphorous present in municipal wastewater are significant. EC regulations and local environmental objectives are making phosphorous removal from wastewaters increasingly widespread. These two factors mean that phosphorous recovery for recycling could become economically viable in the future.

Crites and Tchobanoglous [1] presented the expected phosphorous load factor from individual residences as 3.28 g P/capita/day (1.2 kg P/capita/year). It corresponds to about 20 % of P used as fertilizers in Japan (6 kg P/capita/year). Japan imports 350,000 tons of rock phosphorous as P_2O_5 every year. We estimate that the municipal wastewater sludge produced in Japan last year, contained about 100,000 tons of P_2O_5. Therefore, we recognize that municipal wastewater is an important phosphorous resource.

The nutritional status of plants has a great influence on the root secretion. When plants grow under nutrient stress conditions, such as phosphorous deficiency, iron deficiency and aluminum excess, plant roots secrete functional compounds in order to avoid these conditions [2,3,4]. As far as a phosphorous deficiency is concerned, two organic compounds, acid phosphatase and organic acids are very important. Acid phosphatase and organic acids may release soluble inorganic phosphate from organic and hardly soluble inorganic phosphate compounds, respectively. These plant functions are useful for the development of a method for the recycling use of phosphorous contained in municipal wastewater.

Based on the background mentioned above, this paper deals with the separation of phosphorous from municipal wastewater by pre-coagulation, and subsequent use of plant functions to secrete acid phosphatase and organic acids resulting in the release of soluble inorganic phosphate from the pre-coagulated sludge.

Pre-Coagulation of municipal wastewater for phosphorous recovery

Flocculation/Sedimentation Unit

The jet mixed separator (JMS), incorporated with inclined tube settlers, was used as a unit for pre-coagulation of municipal wastewater. Figure 1 shows a schematic drawing of the treatment unit, which has been operated in the Soseigawa municipal wastewater treatment plant in Sapporo City.

The JMS is a solid-liquid separator with 4 porous plates inserted vertically in the channel perpendicular to the flow. Simultaneous flocculation and sedimentation occur in the JMS part of the unit and residual flocs are removed in the part of inclined tube settlers. Watanabe et al. [5,6,7] have published several papers concerning the theory and performance of JMS.

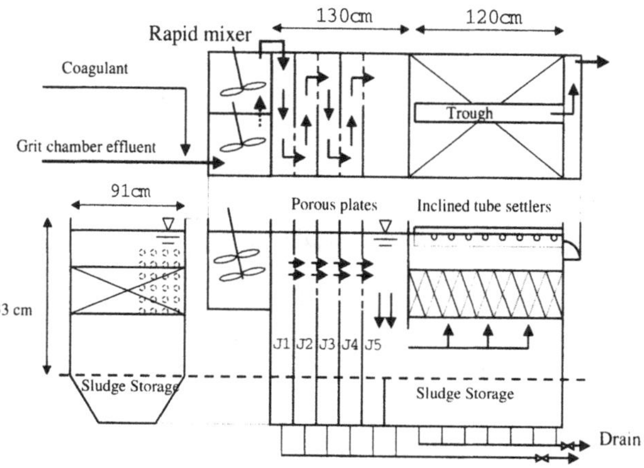

Fig. 1. JMS combined with inclined tube settlers at the Soseigawa municipal wastewater treatment plant

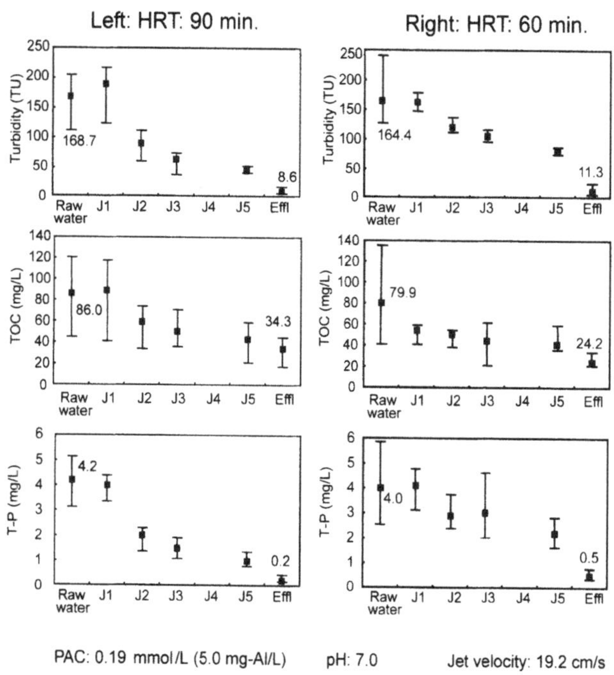

Fig. 2. Performance of JMS with tube settlers (J1 – J5: sampling points)

In this experimental study, the total hydraulic retention time in the unit was 60 min. or 90 min. The jet velocity through the porous plates was fixed at 19.2 cm/s. Effluent from the grit chamber of Soseigawa municipal wastewater treatment plant was pumped to the experimental unit and coagulated by poly aluminum chloride (PAC) before it was fed to the JMS. Turbidity, TOC and total phosphorous concentration were measured at each stage of JMS and effluent through inclined tube settlers.

Figure 2 shows the experimental results. Comparing with the previous experimental data using a JMS without inclined tube settlers [6,7], inclined tube settlers significantly reduced the turbidity, TOC and total phosphorous concentration. Settled sludge was automatically discharged at a designated time interval.

The new coagulant - Polysilicato Iron (PSI)

When phosphate contained in the sludge is to be recycled, iron coagulants are preferable to aluminum coagulants. The long term ecological and health effects of aluminum are not currently known. However, it is bio-accumulative and is suspected of being a health risk

For these reasons, Hasegawa et al. [8,9] developed a new inorganic polymer coagulant, called polysilicato iron (PSI), which has strong bridging properties derived from the polysilicate acid particles. Table 1 summarizes the characteristics of PSI.

Table 1. Characteristics of polysilicato-iron

Composition	Fe_2O_3: 3% SiO_2: 2 – 11%
Molar ratio of Si to Fe	1:5 to 1:1
Molecular weight	200,000 – 500,000 (Dalton)
Optimum pH range for coagulation:	
Suspended particles	5.0 – 8.5
Soluble organics	4.5 – 7.0

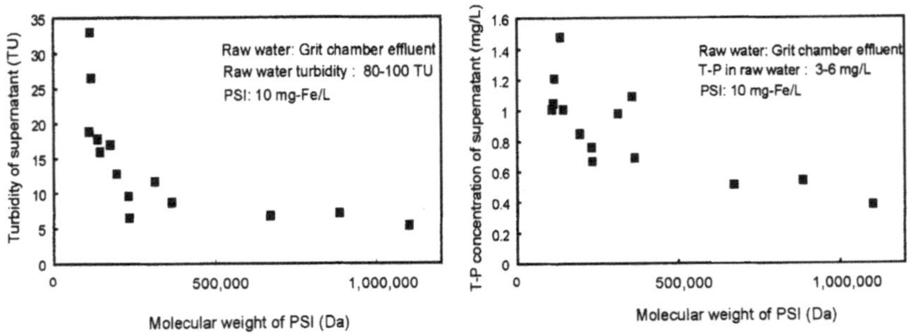

Fig. 3. Effect of PSI molecular weight on removal of turbidity and total phosphorus.

Figure 3 shows the effect of molecular weight of PSI-3 (molar ratio: Si/Fe = 3) on the removal of turbidity and total phosphorous obtained in jar tests. The optimum molecular weight of PSI seems to be around 500,000 Da. PSI has been successfully applied to the coagulation and separation of turbid and organic colored particles, and blue-green algae, Microcystis sp. and Stephanodiscus sp. in laboratory scale experiment. PSI was also applied for pre-coagulation of municipal wastewater using the JMS [9]. In this case, the average turbidity and total phosphorous concentration in the JMS effluent were 27 turbidity units and 0.6 mg P/l, respectively. However, PSI has been produced in a beaker scale unit and cannot be stored for long periods because of its easy gelation.

We have now developed a large-scale unit which can continuously produce the PSI, and also a production method to keep the coagulant stable for long periods without gelation. The method is based on a combination of removal of Na^+ and addition of ethanol. Table 2 shows the effect of Na^+ removal and ethanol addition on the gelation time of the PSI-3.

Table 2. Gelation time of PSI at various production conditions

Case	Si Concentration (%)	Fe Concentration (%)	Removal Efficiency of Na^+ (%)	Content of Ethanol (%)	Gelation time (hr)
1	1.1	0.7	0	0	1000
2	1.1	0.7	60	0	2030
3	1.1	0.7	60	10	>2700
4	1.1	0.7	90	0	>1700
5	1.1	0.7	90	10	>1700

A possible explanation for the positive effect resulting from the proposed method may be as follows: elimination of attractive force between Na^+ and silica, and reduction of the nucleophilic substitution reaction activity of silica by ethanol. The most effective production method may be case 3 where the removal efficiency of Na^+ is 60 % and ethanol content in PSI solution is 10 %. Figure 4 shows the depression effects of polymerization by the proposed production method. The y-axis, limiting viscosity is an index of a polymer's molecular weight, and the relationship between the limiting viscosity and molecular weight of PSI was made in a previous paper [9]. The significance of having a longer time of storage but a relatively constant limiting viscosity is the stability of PSI. If PSI is stable, it can be stored for a long time in the water or wastewater treatment plants before use. A longer time storage can reduce the transportation cost of PSI from the production site to the plants. The coagulation ability of the stored PSI-3 for 3 months was evaluated by jar tests using polluted pond water.

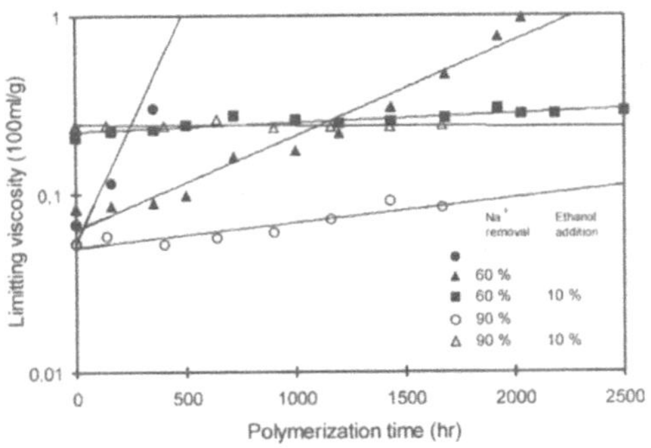

Fig. 4. Depression effects of polymerization by removal of Sodium and addition of Ethanol with PSI at 30 ° C

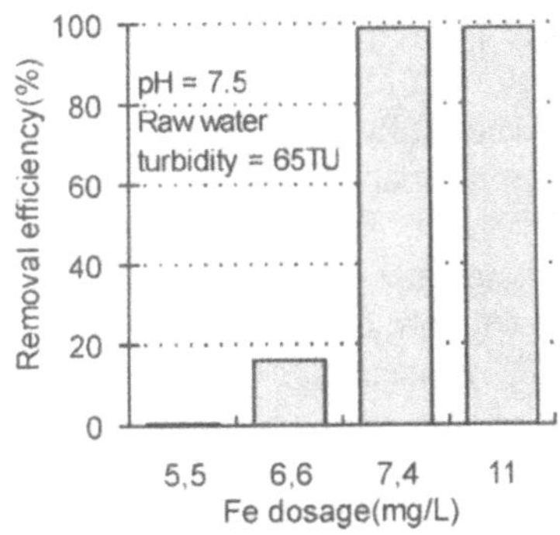

Fig. 5. Effect of different concentrations of PSI-3 on turbidity removal efficiency

The pH in the jar tests was fixed at 7.5, and PSI-3 dosage was changed at 5.5, 6.6 , 7.4 and 11 mg Fe /l . Figure 5 shows the removal efficiency of turbidity. Even though pH was rather high compared to the optimum pH of 5 to 6, a very high removal efficiency was obtained with the PSI addition of 7.4 mg Fe /l. This results are almost the same as the previous experiment using the fresh PSI [9]. Therefore, the stored PSI may have almost the same coagulation ability as fresh PSI.

Recycling use of pre-coagulated sludge as phosphate fertilizer

Functions of plants to secrete acid phosphatase and organic acids

Plants have functions to secrete acid phosphatase and organic acids from roots when they grow in soils with a low phosphate supply. Acid phosphatase secreted from roots may release inorganic soluble phosphate from organic phosphate compounds existing in the rhizosphere and the roots may absorb the released phosphate efficiently. On the other hand, organic acids secreted from roots such as citric acid, malic acid and oxalic acid may form chelate compounds with iron, aluminum and calcium contained in hardly soluble iron phosphate, aluminum phosphate and calcium phosphate existing in the rhizosphere, to release soluble inorganic phosphate.

Functions of nine species of crop plants to secrete acid phosphatase from roots were compared by growing these plants under phosphate-sufficient and deficient conditions, as shown in Figure 6 [2]. Wasaki et al.[11] discovered that there are two types of cDNA encoding the acid phosphatase, named as LASAP1 and LASAP2 . LASAP1 and LASAP2 are considered to encode the acid phosphatase localized in the cell membrane and wall, and secreted from the roots to rhizosphere, respectively. The secreting function was low under the phosphate-sufficient condition, but it increased under the phosphate-deficient condition.

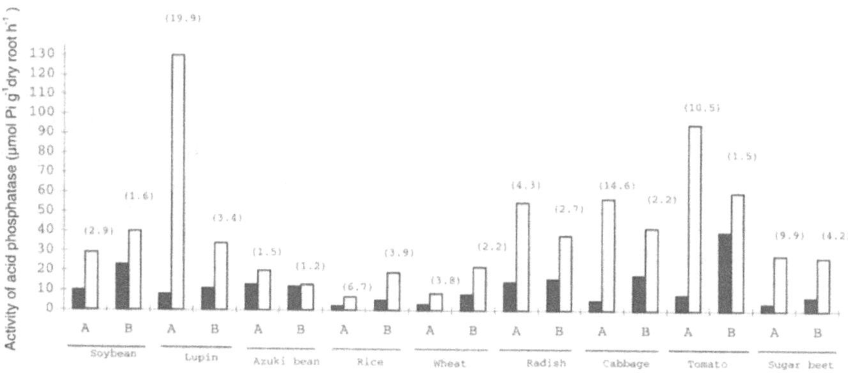

■ control treatment

❑ low P treatment ( ) increased rate of the activity of acid phosphatase in the low P treatment (increased rate x times) A: secreted acid phosphatase from the root. B: Fixed acid phosphatase in the cell membrane and cell wall

Fig. 6. Effect of phosphorus deficiency on the activity of acid phosphatase in various crop plants. (Fixed acid phosphatase was eluted by 100 mM NaCl on a root dry weight basis) white: low P-treatment; black: control treatment

As seen in Figure 6, there is a large difference in the secreting function of acid phosphatase among plant species. It is very high in beans such as lupin and soybean, and low in rice, wheat and sugar beet, which are important crop plants for food production. The increased rate under the phosphate-deficient condition compared with the phosphate-sufficient condition was also different among species, as seen in Figure 6. Not only the secreting function, but also the increasing rate was especially high in lupin. Acid phosphatase activity in the cell wall also increased under the phosphate-deficient condition in all plants.

The phosphate releasing function was investigated with a rizobox in which rhizosphere soils were separated with thin vinyl sheets [3]. The acid phosphatase activity was highest at the root compartment and decreased with the increase of distance from the root compartment. The activity was observed within 2.5 mm from the root surface. It was found that only one acid phosphatase isozyme was secreted, this protein was purified and molecular weight was determined. It was estimated that the molecular weight of this protein was 140,000 Da, which was a homo-dimer with a molecular weight of 72,000 Da for the subunit [4]. Enzymatic characteristics of this enzyme are shown in Table 3.

It has been shown that organic acids such as citric acid are also secreted from lupin roots [3]. Thus, distribution of organic acids secreted from lupin roots in the rhizosphere and the phosphate releasing function were examined. It was found that citric acid , malic acid and oxalic acid were secreted. Among the organic acids, citric acid was secreted in the highest amount, and the concentration decreased with distance from the root compartment similar to acid phosphatase.

The authors have proposed a concept for the recycling use of phosphate, in which pre-coagulated wastewater sludge is combined with the functions of plants to secrete acid phosphatase and organic acids, as described in Figure 7. Firstly, the appropriate plants must be selected for the recycling process. We selected Lupin. These plants should be able to secrete large amounts of acid phosphatase in response to phosphate deficiency. Genetic engineering can be used to produce transgenic plants that have the acid phosphatase gene. Wasaki, et al. [10] have already made the genetic analysis of acid phosphatase, and succeeded to transfer its gene to the tobacco plant [11].

Table 3. Enzymatic characteristics of acid phosphatase secreted from lupin roots

Molecular Weight	140 kDa (by Bio-Gel P-200) 72 kDa (by SDS-PAGE)
Optimum pH	4.3
Isoelectric point	4.7
pH Stability	4.0 – 9.0
Temperature Stability	≤ 70 ° C
Km value (Substrate:*p*-nitrophenylphosphate)	0.027mM
Glycosylation	Glycoprotein

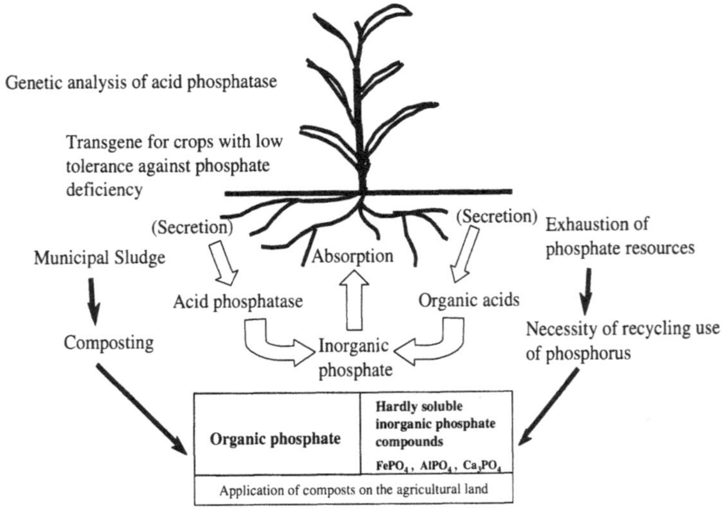

Fig. 7. Use of acid phosphatase and organic acids secreted by plants to recycle phosphate from municipal sludge

Phosphate release from coagulated sludge by acid phosphatase and organic acids

The total phosphorous in the coagulated sludge can be generally divided into four fractions as shown in Figure 8. The STS (Schmidt –Thannhause-Schneider) method has been used to measure the concentration of various types of phosphorous compounds contained in the sludge [12]. Figure 9 shows two measured examples of the phosphorous forms contained in pre-coagulated wastewater sludge drained from the JMS [13]. The composition of phosphorous forms may depend on the characteristics of sewerage system including the flow patterns through collection system and length of sewer.

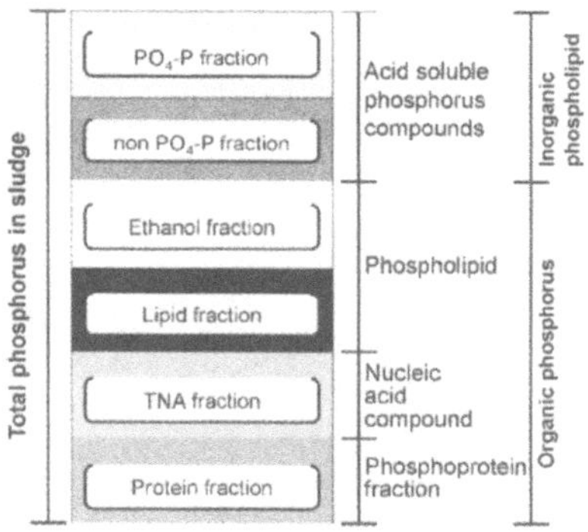

Fig. 8. Composition of phosphorus compounds in municipal wastewater sludge based on the STS method

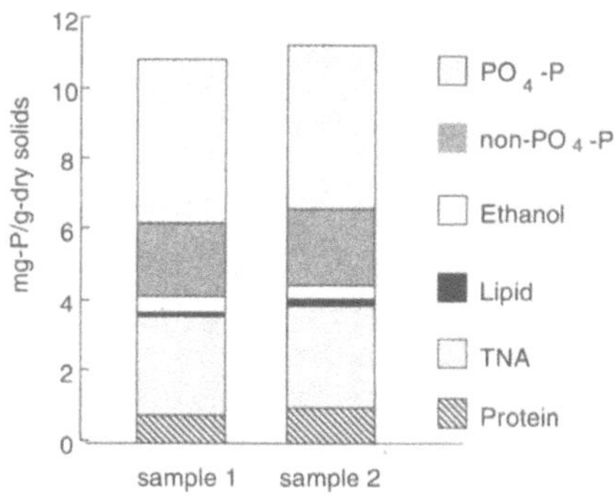

Fig. 9. Composition of phosphorus compounds in pre-coagulated sludge drained from the JMS

The phosphorous recovery efficiency from pre-coagulated sludge by $FeCl_3$, $Fe_2(SO_4)_3$ or Poly aluminum chloride (PAC) was measured by adding the acid phosphatase secreted from lupin [13]. The pre-coagulated sludge was produced by jar tests using the grit chamber effluent in Soseigawa sewage treatment plant. Coagulant dosage was fixed at 10 mg Fe/l or 10 mg Al/l. The pre-coagulated sludge was dried at 70 °C, then 1 g of the dried sludge was added into a test tube with 30 ml deionized water. 0.03 unit of pure acid phosphatase and 7.5 ml of 0.2 N acetate buffer solution were added into the test tube to keep the pH at 4.3. After this, the test tube was continuously shaken for 24 hours at 29 °C, then 15 ml of 0.2 N NaOH was added into the test tube to stop the enzyme reaction. The supernatant after centrifuged at 10,000 rpm for 10 min. was collected and concentration of total phosphorous, inorganic phosphate (PO_4 -P and non PO_4 -P) and organic phosphate was measured. And residual concentration of various phosphorous compounds remained in the sludge was also measured. Figure 10 shows the composition change of the phosphorous compounds by the reaction of acid phosphatase. Almost 50 % of the organic phosphorous were released from the sludge to the supernatant. Because of a low pH condition, inorganic non PO_4-P was also released.

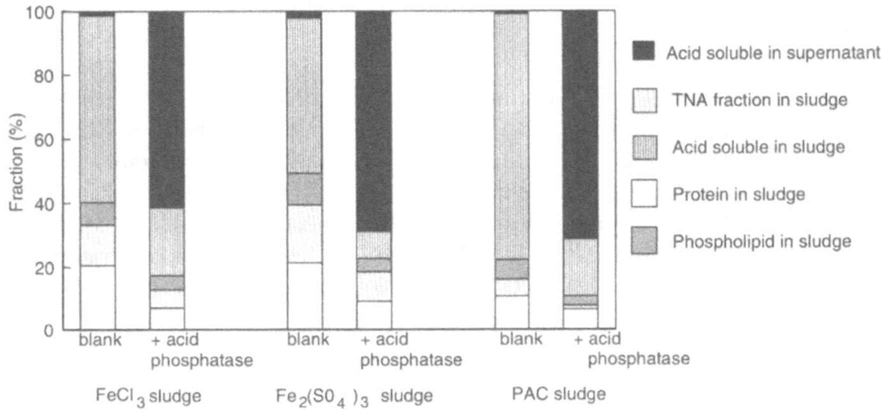

Fig. 10. Phosphorus forms in sludge and in supernatant before and after treatment with acid phosphatase

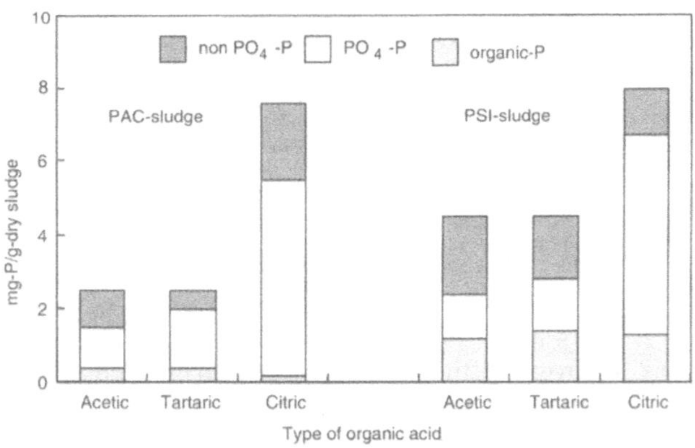

Fig. 11. Effect of organic acids on dissolution of coagulated phosphorus

0.05 g of dried pre-coagulated sludge by PAC or PSI were added to a test tube with deionized water and 1 m mol of organic acid (acetic acid, tartaric acid or citric acid) was added. The pre-coagulated sludge by PSI was made by jar test with its addition of 10 mg Fe/l. After continuous shaking for 24 hours at 20 °C, solution in the test tube was filtered through 0.45 •m membrane. Then the concentration of total phosphorous, inorganic phosphorous and organic phosphorous was measured. Figure 11 shows the results of phosphorous release by three organic acids. The amount of secretion of citric acid is the highest one and it is also the most active one to release soluble phosphate from the pre-coagulated sludge.

Excess sludge from an existing municipal wastewater treatment plant was used as a phosphorous fertilizer to grow sugar beet in pot tests, which has a low secretion ability of acid phosphatase. Figure 12 shows the test results concerning the effects of adding acid phosphatase and citric acid secreted by lupin roots on absorption and growth of sugar beets [14]. From the results shown in Figure 12, it can be seen, firstly, that the presence of phosphate fertilizer in the soil had a positive effect on both P absorption and growth. Secondly, the sludge from the wastewater treatment plant had about the same effect as phosphate fertilizer on P absorption and growth. Thirdly, the highest P absorption and growth were achieved when the soil was supplemented with acid phosphatase alone. Finally, more P was absorbed in the shoots than in the roots.

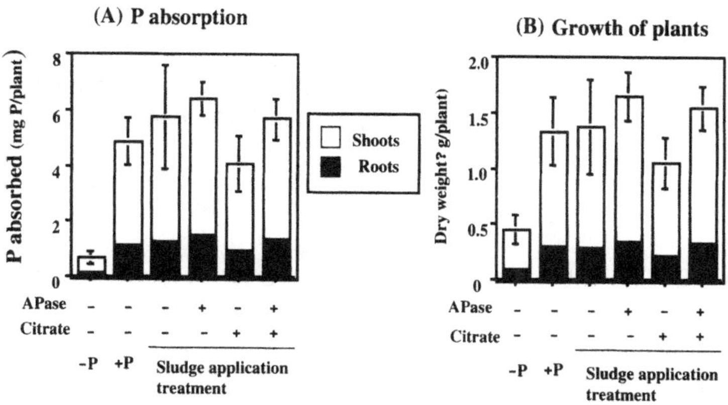

Fig. 12. Effect of supplementary acid phosphatase (APase) and citric acid on the phosphorus absorption and growth of sugar beets(– P and + P refer to the soil without and with phosphate fertilizer, respectively).

Conclusions

Phosphorous is essential to all living organisms. However, the amount of rock phosphorous remained in the world is limited and recycling schemes have to be developed.

In this paper, a concept of the recycling use of phosphorous has been proposed. The concept combines recovery of phosphorous from municipal wastewater by the pre-coagulation, and use of functions of plants to secrete organic compounds which release soluble inorganic phosphate from pre-coagulated sludge. The conclusions from this study can be summarized as follows:

1. The jet mixed separator incorporated with inclined tube settlers worked very well as a unit for the phosphorous recovery by pre-coagulation of municipal wastewater.
2. A production method to depress the polymerization of the polysilicato iron, a new coagulant applicable to the pre-coagulation, was also presented. Removing Na^+ and adding ethanol can significantly depress the gelation of polysilicato iron.

3. Acid phosphatase and citric acid secreted from Lupin roots efficiently release soluble inorganic phosphate from the organic phosphate and hardly soluble inorganic phosphate compounds, respectively contained in the pre-coagulated sludge.
4. These results suggest that phosphorous contained in the pre-coagulated sludge produced from municipal wastewater with low heavy metal contents may become an important phosphate fertilizer.

Acknowledgements

This research was partially supported by CREST (Core Research for Evolutional Science and Technology) of the Japan Science and Technology Cooperation.

References

1. Crites, R.W., Tchobanoglous, G.: Small and decentralized wastewater management systems, McGraw-Hill Series in Water Resources and Environmental Engineering, Boston, WBC/McGraw-Hill (1998) 180 – 181
2. Tadano,T., Sakai, H.: Secretion of acid phosphatase by the roots of several crops species under phosphorous- deficient conditions, Soil Sci. Plant Nutr., 37 (1991) 129 – 140
3. Li, M., Shinano, T., Tadano, T.: Distribution of exudates of lupin roots in the rhizosphere under phosphorous – deficient conditions, Soil Sci. plant Nutr., 43(1997) 237-24
4. Tadano, T., Komatsu, K.: Utilization of organic phosphorous in soil by plant roots, Proc. of the 15th World Congress of Soil Science (Eds; J.D.B. Etchevers, A.S. Aguilar, R.E. Nunez, G.G. Alcanter, P.G. Sanchez), Acapulco, Mexico 9 (1994) 521-522
5. Watanabe, Y., Fukui, M., Miyanoshita, T.: Theory and performance of a jet mixed separator, Journal of Water SRT-Aqua, 39(6) (1990) 387-395
6. Watanabe, Y., Kanemoto, U., Takeda, K., Ohno, H.: Removal of soluble and particulate organic material in municipal wastewater by chemical flocculation and biofilm processes, Wat. Sci. Tech., 27(11) (1993) 201-209
7. Watanabe, Y., Kasahara, S. , Iwasaki, Y.: Enhanced flocculation/sedimentation process by a jet mixed separator, Wat. Sci. Tech.,37(10) (1998) 55-67
8. Hasegawa, T., Hashimoto, K., Tambo, N.: Characteristics of metal-polysilicate coagulants, Wat. Sci. Tech., 23 (1991) 1713-1722
9. Watanabe, Y., Hashimoto, K. Hasegawa, T., Kameda, S., Suzuki, H.: Application of polysilicato-iron coagulant to coagulation of algae and municipal wastewater. In: Chemical Water and Wastewater Treatment V, Proc. of the 8th Gothenburg Symposium (Hahn, H.H., Hoffmann, E., Ødegaard, H. , eds.) Springer, Berlin (1998) pp. 3-13
10. Wasaki, J., Omura, M., Osaki, H., Matsui, H., Shinano,T., Tadano,T.: Structure of a cDNA for an acid phosphatase from phosphate deficient lupin (*Lupinus albus* L.) roots, Soil Sci. Plant Nutr., 45 (1999) 439-449
11. Dateki, H.: Transformation of secretory acid phosphatase gene from lupin into tobacco plants, Master's thesis, Graduate School of Agriculture (2000) Hokkaido University

12. Mino,T., Matsuo,T., Kawakami, T: Studies on phosphorous composition and phosphorous metabolism in activated sludge-analysis of phosphorous composition in activated sludge by STS method, Journal of Japan Sewage Works Association 20 (228) (1983),28-36(in Japanese)
13. Prasityousil, J., Watanabe, Y. Kameda, S., Tadano,T.: Recovery of phosphorous from municipal wastewater by coagulation and dissolution of the coagulated phosphorous by enzyme and organic acids secreted from plant roots, Environmental Engineering Research, Japan Society of Civil Engineers, 36 (1999) 121-128 (in Japanese)
14. Yamamura, T.: Recycling use of municipal sludge by application of acid phosphatase and organic acids secreted from plant roots, Master's thesis, Graduate School of Agriculture, Hokkaido University (2000)

Recovery and Re-Use of Aluminium Coagulants from Coagulation Sludge by Liquid-Ion Exchange

J.-Q. Jiang

Department of Civil Engineering, University of Surrey, Guildford, Surrey GU2 7XH, England
j.jiang@surrey.ac.uk

Abstract

The residual Al/Fe content in clarifier sludge resulting from coagulant use is a valuable resource. Chemical conditioning is carried out in previous years to recover the Al/Fe coagulants. However, the major problem associated with the use of the recovered coagulants in potable water treatment is the contamination of the recovered coagulants by heavy metals, colour-causing natural organic matter and manganese.

This paper describes the research conducted to evaluate the applicability of liquid-ion exchange to Al ion recovery and the coagulation performance of the recovered coagulants for water treatment. A range of factors affecting the liquid-ion exchange efficiency have been studied including the type and concentrations of extracting agents, mixing strength and retention time and the type and concentrations of stripping agents. A jar test experimental procedure was used to evaluate the coagulation performance of the recovered coagulants in comparison with the commercial aluminium sulphate (AS). The results have shown that the liquid-ion exchange process will recover more than 95% of the Al from influent sludge and the recovered AS can be of the same quality and concentration as commercial AS, and can perform as well as the commercial AS for potable water treatment.

Introduction

Removal of suspended solids, colloidal materials and colour-causing natural organic matter is primarily determined by the effectiveness of coagulation process. A coagulant (e.g., alum) is added into water to destabilize the colloidal particles and to form larger aggregates, onto which dissolved organic matter is adsorbed, thereby aiding their removal in subsequent sedimentation and filtration stages. As a result of this operation, a huge amount of sludge is produced, which now has to be disposed of to a registered landfill site due to the introduction of the Environmental Protection Act. Landfill tax was introduced in the UK in 1996 [1] and this has significantly increased in the costs of landfilling. With tighter regulations and disposal costs for sludge, many companies are looking for alternative, cost effective and environmentally friendly sludge treatment and disposal options. Coagulant recovery is one of such options.

The sludge produced from the use of alum is gelatinous in structure with a feathery, bulky nature. The alum sludge is difficult to dewater when disposed, it will remain in its semi-fluid state indefinitely unless its physical characteristics have been changed. Although the hydroxide sludge does settle readily, it is almost impossible to dewater it without prior treatment. On the other hand, the residual aluminium (Al) content in clarifier sludge resulting from coagulant use, is a valuable resource. Chemical conditioning is carried out in previous years [e.g. 2] to recover the metal coagulants. This is accomplished by adding acid to the thickened sludge at a concentration determined by the amount metal hydroxide in the sludge and the desired level of recovery. The dissolved Al/Fe is separated from the residual solids by a gravity separator and stored, and the residual sludge is disposed of by landfilling after pH adjustment. However, the major problem associated with reusing the recovered coagulants in potable water treatment is the contamination of the recovered coagulants by heavy metals, colour-causing natural organic matter and manganese.

Liquid–ion exchange (extraction) can be used to recover the metal coagulants because of the selectivity of the extractant used, which can separate the desired metals (e.g. Al in this study) from the other contaminants. In the process, a small quantity of an organic soluble chemical (the extractant) is dissolved in a second organic liquid (the diluent) to form an extractant solvent (organic phase). During extraction the extractant reacts chemically with the desired metal in the aqueous phase to form a metal-extractant complex that is soluble in the diluent. The ion exchange occurs at the interface of the two phases. The system is operated by agitating the solvent and water mixture to form a dispersion of small water droplets in the organic acid. The Al exchanges for the positive hydrogen ion, the hydrogen ion enters the aqueous phase, and the positively charged Al ion moves to the organic-water interface, where a rapid coalescence occurs, resulting in an Al-rich organic phase and an Al-free aqueous phase.

The second half of the liquid-ion exchange operation is called a stripping step, which is to recover the desired metal, e.g. Al, from the organic phase into the aqueous phase with high concentration, and the extractant solvent is recycled. In general, the extraction step performs the major purification, whereas the purity Al coagulant is recovered in the stripping step. In this secondly step, Al in the organic phase exchanges for protons in the acid-aqueous phase, producing aluminium sulphate (AS) and regenerated extractant for the extraction step. The stripping step is operated at an organic-to-acid phase ratio (volume ratio) that will concentrate the Al to the desired level.

This paper describes the research conducted to evaluate the applicability of liquid-ion exchange to Al ion recovery and the coagulation performance of the recovered coagulant for water treatment. The study investigated a range of factors affecting the liquid-ion exchange efficiency, including the type and concentrations of extracting agents, mixing strength and retention time and the type and concentration of stripping agents. A jar test experimental procedure was used to evaluate the coagulation performance of the recovered AS in comparison with the commercial AS.

Methods and experimental procedures

The synthetic Al solutions were used in this study as acidified Al containing sludge since this allowed for the availability of large quantities and for easy variation of the other parameters. A standard Al reagent (1000 mg/l, Merck Ltd, England) was used to prepare synthetic solutions with different concentrations of 200, 500, 750 and 1000 mg/l as Al. The solution pH was adjusted to 2, since this is the pH of the acidified sludge in most cases and at which, all the Al species will present as the dissolved ions [3]. The Al concentrations in the synthetic solutions and in the extractant phase were measured based on the standard chlorimetric methods (Standard Methods for the Examination of Water and Wastewater, AWWA, 1992) and using a double beam scanning UV-Vis spectrometer (Helios Alpha, Unicam Cambridge, England).

The extractants selected were in the type of alkyl phosphates, i.e., di-n-hexyl phosphoric acid (DHPA) and mono (2-ethylhexyl) phosphoric acid (MEHPA) (ABCR-Gelest Ltd, Manchester). For the entire tests, the extractants were diluted in kerosene (MERCK Ltd, England) with the desired ratio of the extractant to the diluent. The stripping agent was either sulphuric acid or hydrochloric acid which causes the Al to leave the organic phase and enter the stripping phase.

The liquid-ion exchange tests were conducted at bench scale. Several synthetic solutions (aqueous phase) with varying Al concentrations were added and thoroughly mixed with the extractant (organic phase) using a shaker (KS125, Merck Ltd, England) which provides variable shaking speed up to 1100 rpm. The phases were then allowed to be separated and the Al concentrations in the two phases were analyzed. Different quantities of the Al will have been transferred from aqueous to the organic phase, and this can be evaluated by the Al percentage recovery, which can be expressed as:

E% = (equilibrium Al concentration in organic phase)/(Al concentration in synthetic solution)

The E% is a function of such variables as pH, concentration of Al in the aqueous phase, and the concentrations of the extractants and diluents, and phase ratio.

The opposite of E% is the stripping percentage, S%, which measures how well a given stripping agent (e.g., sulphuric acid) can recover the Al from organic phase to aqueous phase.

The effects of the extraction/stripping conditions such as the shaking speed, the type and concentrations of the extractants and stripping agents and the phase ratios (defined as the volume ratio of the organic phase to aqueous phase) on the extraction/stripping performance were investigated.

The coagulation performance of the recovered AS in comparison with that of the commercial AS (MERCK Ltd, England) was evaluated. The raw water samples were collected from the lake on the University of Surrey campus and the water quality characteristics were shown in Table 1. A jar test apparatus (Flocculator 2000, Kemira KemWater) with six 1-L beakers were used. Jar test experiment was conducted using 800 ml raw water sample for each beaker and at various AS doses and a given coagulation pH. The fast and slow mixing speed and time were 400 rpm for 1 min, and 35 rpm for 20 mins, respectively. After 30 min sedimentation, the supernatants were withdrawn and filtered (filter paper grade 42,

Merck Ltd.). The water qualities of the filtered supernatants were analyzed based on the standard chlorimetric methods (Standard Methods for the Examination of Water and Wastewater (18th Ed.), AWWA, 1992) with a double beam scanning UV-Vis spectrometer (Helios Alpha, Unicam Cambridge, England).

Table 1. The water qualities of the lake on the University of Surrey campus

pH	7.8
Al (mg/l)	0.10
Mn (mg/l)	0.08
Cu (II) (mg/l)	0.55
Zn (II) (mg/l)	0.15
Turbidity, (NTU)	11.2
Colour (Vis_{420}), (m^{-1})	4.5
UV_{254} (m^{-1})	15.8
COD (mg O_2/l)	9.7

Results and Discussion

Extraction Process

The extracting organic phase was consisting of two parts, namely extractant and the diluent. Many parameters that are important in ion-exchange resin production are also important in extractant and diluent selection, and this can be summarised as:

- stability
- density difference to water
- low aqueous solubility
- number of active sites (i.e., extractant concentration)
- affinity of active site for the ion removed, and
- the degree of cross-linking of the polymerisation of the extractant.

In this experiment, kerosene has been selected as the solvent due to its stability, lower density and insolubility in water. In addition to these, use of kerosene was also due to its availability, comparatively low cost, and relative safety in handling.

The extractant to be selected for this study is primarily considered for its reactivity, affinity and selectivity for Al. Low aqueous solubility of the extractant is necessary to insure minimal loss of extractant and to prevent contamination of the recovered Al. Another important factor to be considered is the degree of polymerisation of the extractant. At high concentrations in kerosene, the extractants have a tendency to polymerise with each other, thereby reducing the number of sites available for Al gelation.

The extractants selected for this study were di-n-hexyl phosphoric acid (DHPA) and mono (2-ethylhexyl) phosphoric acid (MEHPA). They can react with cations and therefore, have potential for Al extraction. The functional group '=P(O) OH' is common for all members of this class of extractants. Owing to the presence of both electron acceptor and donor groups in the '=P(O) OH' grouping, acidic organophosphates typically undergo such various specific interactions as self-association and molecular complex formation with diluent or other solutes.

The effects of shaking speed and mixing time on the Al percentage recovery can be seen in Figure 1. The experiments were conducted with 100 mM MEHPA, a synthetic solution with pH of 2 and Al concentration of 500 mg/l as Al, and a phase ratio of 1:1 on a volume basis of organic to aqueous. It can be seen that a 95% of Al recovery can be achieved at shaking speed greater than 700 rpm with mixing time of approximately 50 min. All the subsequent experiments were therefore run at shaking speed of 700 rpm with mixing time of 60 min in order to ensure that equilibrium had been achieved.

Figure 2 shows the effect of the type and concentration of the extractants on the extraction efficiency which is evaluated by the Al recovery percentage for a given Al concentration in synthetic solution.

The experiments were conducted with a synthetic solution of 500 mg/l as Al, and a phase ratio of 1:1 on a volume basis of organic to aqueous. It can be seen that a 95% of Al recovery can be achieved with MEHPA at concentration of 100 mM, but with DHPA at concentration 200 mM. It was observed that certain amounts of third phase formation developed when DHPA was used as extractant. The third phase formed consisted of small amounts of precipitated alkyl phosphate salts, water, and kerosene and this was undesirable due to the operational difficulties and the loss of alkyl phosphate salt. Thus, the DHPA was not future used.

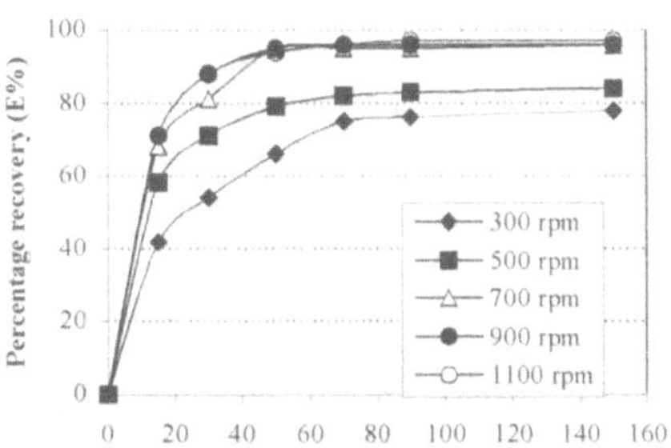

Fig. 1. The effect of mixing speed and time on the extraction efficiency

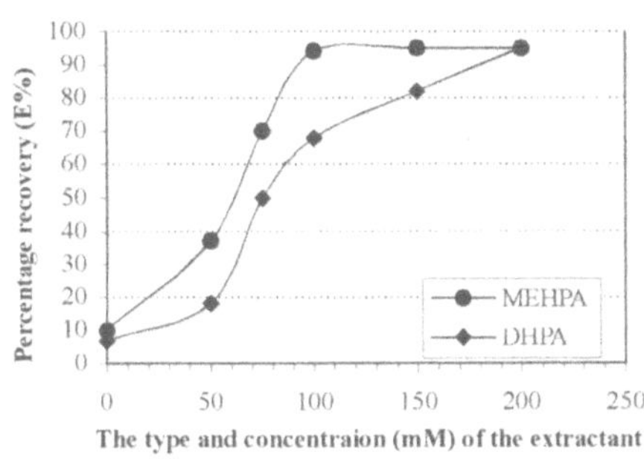

Fig. 2. The effect of the type and concentration of the extractant on the extraction efficiency

Figure 3 shows the concentration of MEHPA needed to extract more than 95 percent of the Al from the synthetic solution with various Al concentrations. The experimental conditions were given based on the previous study results, i.e., shaking speed of 700 rpm with mixing time of 60 min; solution pH of 2; and a phase ratio of 1:1. The results shown in Figure 3 indicated that the MEHPA concentrations should be increased with increasing solution Al concentrations if a 95% recovery percentage needs to be achieved.

The selectivity of the extraction for various metals that could be present in alum sludge was also studied. The metals chosen for study were copper (II), manganese (II), zinc (II), iron (II) and iron (III). The experiments were conducted with a feed solution at a concentration of 500 mg/l as Al and 10 mg/l of the selected metal, and the solution was in contact with 100 mM MEHPA at a phase ratio (organic to aqueous) of 1:1. The selected metal concentrations were considered to be close to that of the clarifier sludge.

The results can be seen in Figure 4. The selectivity ratios ranged from 6 (Al/Fe(III)) to more than 930 (Al/Zn(II)), indicating that MEHPA has a high selectivity for Al over potential metal contaminants under conditions encountered in alum sludge. The selective ratio is the ratio of the extraction coefficient of Al to the extraction coefficient of the selected metal, and the extraction coefficient is the ratio of the metal concentration in the organic phase to the metal concentration in the aqueous phase at equilibrium. With such high extraction selectivity for Al, the purity of the final recovered Al coagulants can be expected and be used for potable water treatment.

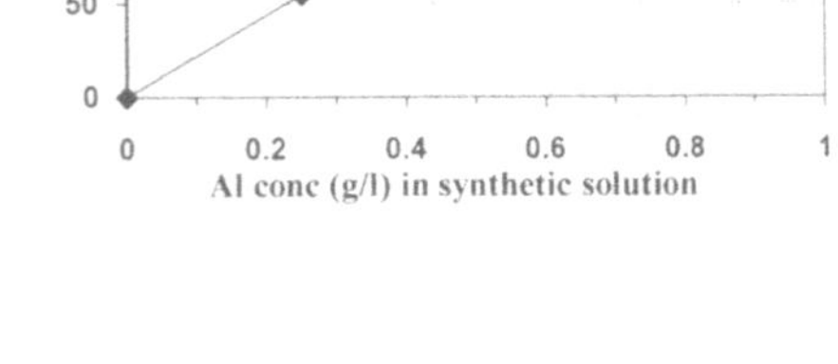

Fig. 3. The concentrations of MEHPA required to achieve 95% Al recovery

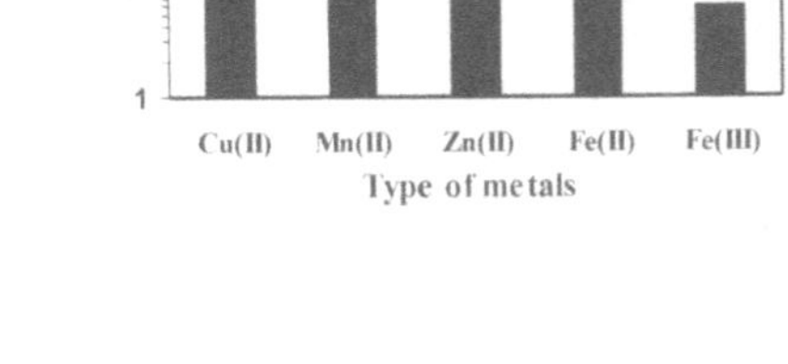

Fig. 4. The selectivity of the extraction for various metals (Test conditions: 100 ml of 500 mg/l Al and 10 mg/l selected metals, and 100 ml of 100 mM MEHPA).

Stripping process

The second half of the liquid-ion exchange operation is a stripping step, which is to recover the Al from the organic phase into the aqueous phase with a highly concentrated acid, and the extractant solvent is recycled. In this step, Al in the organic phase exchanges for protons in the acid-aqueous phase, producing AS and regenerated extractant for the extraction step. The stripping step is operated at an organic-to-acid volume ratio that will concentrate the Al to the desired level.

Stripping kinetic studies were conducted in the same manner as the extraction kinetic tests. The effects of various mixing speed and time on the stripping efficiency are shown in Figure 5. A 9 N H_2SO_4 was mixed with extractant containing 500 mg/l Al at the phase ratio (an organic-to-acid volume ratio) of 2:1. The maximum percentage of stripped Al was achieved at mixing speed of 800 rpm with a minimum time of 70 min. Increasing in the mixing speed did not increase the amount of Al stripped but fasten to achieve the equilibrium. All the rest-stripping experiments were conducted at a mixing speed of 800 rpm with mixing time of 80 min. The effects of the type and concentrations of acids used on the stripping efficiency can be seen in Figure 6. Both hydrochloric acid and sulphuric acid were evaluated. An extractant containing 500 mg/l as Al was mixed with a given acid for 80 min at 2:1 phase ratio and at various acid concentrations. The results show that 6 N HCl and 9 N H_2SO_4 were equally effective in Al stripping. Since sulphuric acid is relatively cheaper than hydrochloric acid, it was selected for all subsequent tests.

The further tests were conducted to examine the optimal phase ratio (an organic-to-acid volume ratio) since it is expected to affect the stripping efficiency and the running cost. An extractant containing 500 mg/l Al was contacted with sulphuric acid at concentration of 9 N and phase ratios of 2:1, 4:1 and 10:1. The stripping time was ranging from 50 min to 200 min and mixing speed was 800 rpm. Figure 7 shows that the greatest stripping efficiency was achieved at a phase ratio of 2:1.

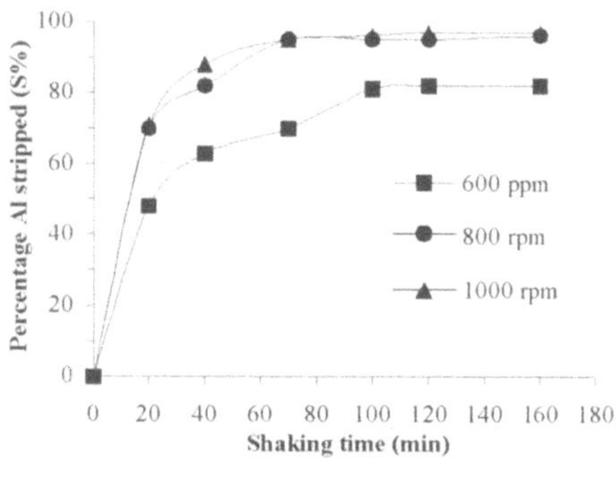

Fig. 5. The effect of the mixing speed and time on the stripping efficiency (Phase ratio of 2:1 and using 9 N H_2SO_4)

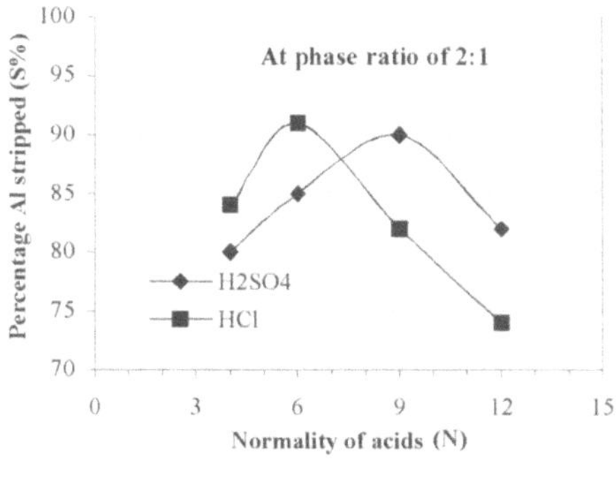

Fig. 6. The effect of the type and concentration of acids used on the stripping efficiency

It also can be seen that the phase ratio of 10:1 could be used in order to achieve Al stripping efficiency of greater than 90% if the stripping time is extended to more than 200 min.

The greater the phase ratio, the less amount of acid will be required and the higher Al concentration in the acid phase can be achieved. For example, if the organic phase contains 4 g/l Al^{3+}, and the phase ratio is adjusted to 4:1, the recovered Al concentration in the acid will be 16 g/l, whilst if the phase ratio is adjusted to 10:1, Al concentration in the acid will be 40 g/l Al^{3+}, if the stripping efficiency is 100%. Thus, from the cost-effective point of view, phase ratio of 10:1 could be selected but the stripping time should be longer than 250 min in order to achieve 95% Al recovery.

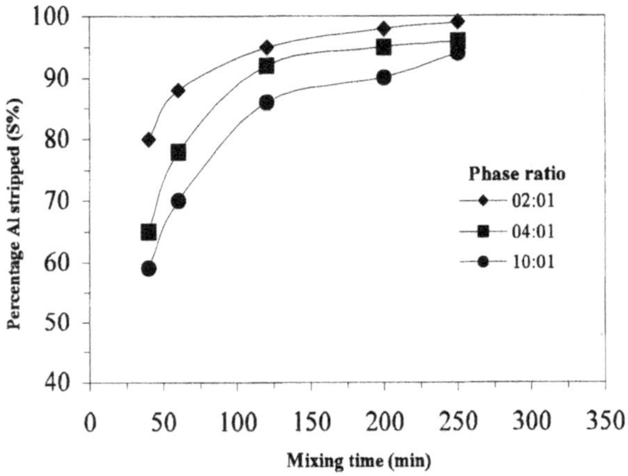

Fig. 7. The effect of the phase ratio and mixing time on the stripping efficiency (Acid used: 9 N H_2SO_4; mixing speed: 800 rpm)

Comparative coagulation performance of recovered AS

The coagulation performance of the recovered AS was evaluated in comparison with the commercial AS and the results are summarised in Figure 8. At a coagulation pH 6.5, which is proved as the optimal pH for AS coagulation in most cases, the recovered AS demonstrated its similar performance to the commercial AS in terms of the removal of turbidity, colour, UV-254 and COD.

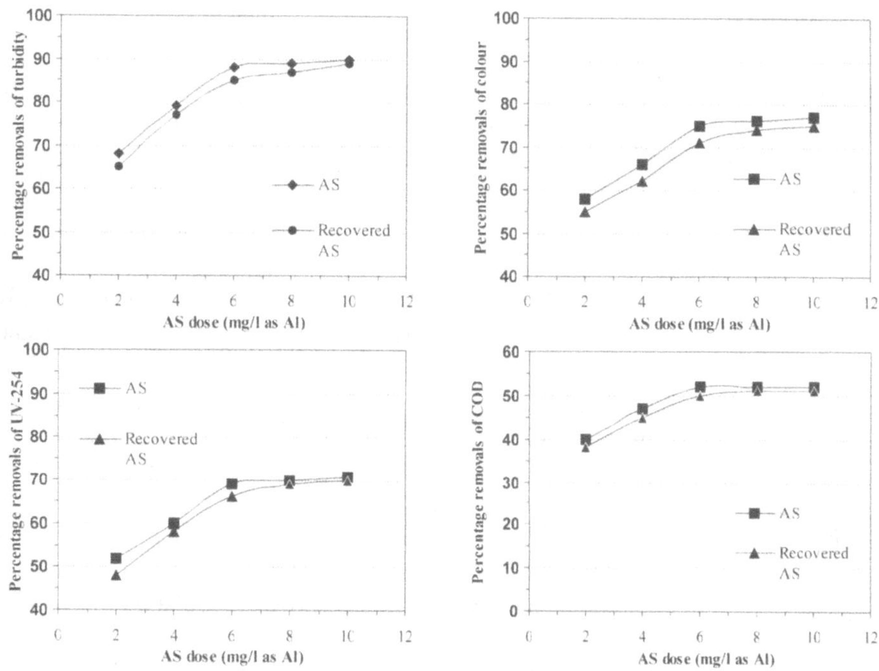

Fig. 8. The comparative performance of the commercial and recovered AS for water treatment at coagulation pH of 6.5

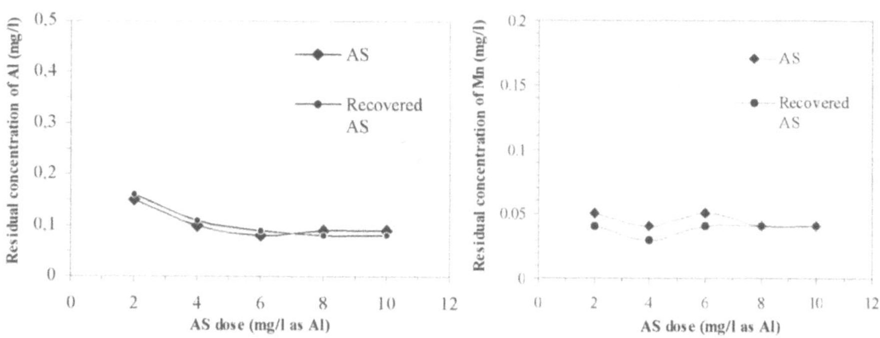

Fig. 9. Residual concentrations of Al and Mn in the coagulated water with commercial and recovered AS

From Figure 9, it can be seen that the residual concentrations of Al and Mn in the coagulated water were also close, no matter which type of the coagulants used, recovered AS or commercial AS.

In summarizing the results presented in Figures 8 and 9, a conclusion can be made that the recovered AS can perform as well as the commercial AS for potable water treatment.

Conclusions

This paper focused on examining the applicability of using liquid-ion exchange to recover Al ion from the coagulation sludges and the coagulation performance of the recovered AS. Laboratory-scale tests have shown that:

1. The liquid-ion exchange process will recover more than 95% of the Al from influent sludge.
2. The extracting agent, mono (2-ethylhexyl) phosphoric acid (MEHPA), is highly selective for Al over potential contaminants such as copper, manganese and zinc.
3. The optimum Al extracting conditions under this study were: shaking speed of 700 rpm with mixing time of 60 min. and an organic to aqueous phase ratio of 1:1. The MEHPA concentrations should be increased with increasing the feed Al concentrations if a 95% recovery percentage needs to be achieved.
4. The optimum stripping conditions in this study were: the stripping agent was 9 N H_2SO_4, the phase ratio (an organic-to-acid volume ratio) of 2:1, and the mixing speed of 800 rpm for a time of 70 min. However, the phase ratio of 10:1 could also be used but the stripping time should be longer than 250 min in order to achieve 95% Al recovery.
5. The recovered AS can be of the same quality and concentration as commercial AS, and can perform as well as the commercial AS for potable water treatment, in terms of the removal percentages of turbidity, colour, UV-abs, and of the reduction in the concentrations of COD and residual Al and Mn.

Acknowledgements

The author is gratefully to acknowledge the funding of this study by the Research Funding Group of University of Surrey.

References

1. Gronow, J: The role of landfills in the United Kingdom. Waste Management & Research, 17 (6), (1999) 409-412
2. Bishop, M.M., Cornwell, D.A., Rolan, A.T., Bailey, T.L.: Mechanical dewatering of alum solids and acidified solids - an evaluation. Journal American Water Works Association, 83 (9), (1991) 50-55
3. Base, C.F., Mesmer, R.E.: The Hydrolysis of Cations. John Wiley and Sons, New York (1976)

Chemical vs. Biological Treatment of Grey Water

S. A. Parsons, C. Bedel and B. Jefferson

School of Water Sciences, Cranfield University, Cranfield, Bedfordshire, MK43 OAL, UK
s.a.parsons@cranfield.ac.uk

Abstract

The major difficulty in treating grey water is the large variation in quality observed over short timescales. The use of cleaning agents has also been reported as having a major impact on treatment. Many treatment schemes proposed use mainly physical and biological processes and have problems adjusting to the shock loading of organic matter and chemicals.

A series of experiments were undertaken to investigate the use of chemical processes to treat grey water. The processes investigated were coagulation / flocculation to remove solids and TOC, and advanced oxidation using titanium dioxide and UV to remove TOC. Results were compared with results from biological processes such as biological activated filters (BAF) and membrane bioreactors.

Jar tests using ferric sulphate on both real and synthetic (weak and strong) grey water showed that at a dose of 30 mg/l ferric chloride, 90% of the solution's turbidity and 80% of the TOC could consistently be removed. This compares well with the 85% COD removal reported for grey water treated by biological activated filters. The effects of pH and alkalinity on removal were also investigated.

The use of titanium dioxide and UV as a treatment for removing organic carbon from real grey water was also investigated. The results show that the combined TiO_2/UV process could reduce organic carbon concentrations from 160 mg/l to 30 mg/l within 3 minutes, an 82% reduction. The rate of the oxidation reaction has been shown to be dependent on the concentration of TOC and the concentration of TiO_2.

Whilst most operational grey water recycling schemes in the UK use physical or biological processes, the present study has shown that both traditional and novel chemical processes can treat grey water to the required level whilst dealing with changes in influent quality.

Introduction

Over the coming decades, the management of water resources will become one of the most important issues across the industrialised world. Key drivers for this are region specific, but relate to climate changes, population increase and other demographic factors. Compounding these in the UK is the long term trend towards hotter summer temperatures and drier winters. In fact, annual rainfall in London is similar to that for Istanbul at around 610 mm p.a. and is less than locations such as Lisbon and Texas that are perceived to be dry.

Recycling and reuse can make a contribution to the sustainability of available water resources. Current preference in the UK is for internal water recycling whereby water is reused from sources local to demand such as industrial or urban reuse. Central to this approach is the concept of the utility of water, whereby water is used in such a way that quality is commensurate with its application. This then permits the exploitation of waters that are not suitable for all applications but may be available in large quantities. Urban water recycling is a particularly attractive option in the UK due to a relatively high domestic water consumption coupled with an intensive population. Urban water recycling makes use of one of three possible sources: (1) Rainwater, (2) Grey water, or (3) Black water. In particular, the reuse of grey waters is attracting interest in the UK and is also practiced in countries such as Japan and America.

Grey water arises from domestic washing operations and is generated by the use of soap or soap products for body washing. The macro and microscopic characteristics of grey water vary significantly due to factors such geographical location, demographics and level of occupancy (Table 1). Grey water can have a similar organic strength to domestic wastewater but is relatively low in suspended solids and turbidity, indicating that a greater proportion of the contaminants are dissolved. Reported mean COD values vary from 40 to 371 mg/l between sites, with similar variations arising at an individual site. This variation has been attributed to changes arising in the quantity and type of detergent products employed during washing. Residence time also affects the characteristics of grey water through natural degradation with the exact rate of change being related more to the original grey water composition and system temperature than to the hydrodynamic conditions of storage.

Currently a plethora of technologies which vary greatly both in complexity and performance are being developed or are installed in sites around the UK. Processes developed range from simple systems in single houses to very advanced treatment trains for large scale reuse. A number of generic types of technologies can be identified: (1) short residence time, (2) physical, and (3) biological. The most widely installed type of grey water recycling system is the single house system. A number of commercial systems are available but most operate on a similar principle. The grey water is passed through a coarse filtration stage and then pumped to a header tank where the water is contacted with a chemical disinfectant such as chlorine or bromine. The technological strategy in such cases is to reduce the residence time of the grey water to a minimum, supposedly negating the need for treatment as the water is said to remain "fresh". As such the water remains virtually unchanged in its organic and physical characteristics and, as long as the

Table 1. Grey water characteristics. (adapted from Jefferson et al. [4])

	Shower	Bath	Hand basin	HoR2[a]	HoR3[a]
BOD_5 (mg/l)	146 (55)	129 (57)	155 (49)	33	96 (103)
COD (mg/l)	420 (245)	367 (246)	587 (379)	40	168 (91)
TOC (mg/l)	65.3 (44.6)	59.8 (43)	99 (142)	-	49 (53)
Turbidity (NTU)	84.8 (70.5)	59.8 (43)	164 (171)	20	57 (138)
Total Coliforms (cfu/100ml)	6800 (9740)	6350 (9710)	9420 (10,100)	-	5,200,000 (360,000)
E. coli (cfu/100ml)	1490 (4940)	82.7 (120)	10 (8750)	-	-
PO_4^- (mg/l)	0.3 (0.1)	0.4 (0.4)	0.4 (0.3)	0.4	2.4 (0.7)
NH_3 (mg/l)	-	-	-	1.1	0.8 (0.7)
pH	7.52 (0.28)	7.57 (0.29)	7.32 (0.27)	-	7.7 (0.4)

a) hall of residence

disinfection system is working, the water is normally free of coliforms. However, the variable nature of the raw grey water imparts an unpredictable demand on the disinfectant as the organics themselves react with the disinfectant chemical, accounting for up to 99% of the total usage.

Physical systems such as depth filters and/or membranes have been developed as they produce a water quality much better than that from the single house systems described above. Typical removal efficiencies are 63% BOD_5 and 28% NTU for a depth filter, and 86% BOD_5 and 99% NTU for membrane systems. Both options reduce the organic load and more importantly its turbidity, although the microbiological removal can be variable. Membranes produce the highest level of removal as the membrane acts as a direct sieve whereby all particles and molecules above the pore size of the membrane are removed. However, problems with effluent reliability have been reported, which have been attributed to grey water degradation during long storage times producing smaller molecules that can be transported through the membrane.

Biological treatment is thus required to guarantee reliability in terms of the removal of dissolved organics. This is especially important in large schemes such as hotels and community-based recycling where organic levels in the treated water may cause re-growth in the distribution systems and loss of effectiveness in the disinfection system. Membrane bioreactors (MBRs) in particular, which combine the benefits of an activated sludge reactor and a membrane, have been used successfully around the world. Numerous currently operating schemes exist in America and Japan in office blocks and residential buildings. Typical removal efficiencies of 98% BOD_5 and turbidity have been reported. Moreover, the combination of the membrane and the biological reactor has improved microbiological rejection such that effluents are free from coliforms without the need for chemical disinfectants.

A major issue with using biological processes is their susceptibility to toxic shock loads. These concerns have concentrated mainly on occasional substance spiking (OSS), which refers to substances that may enter into a grey water source, but are not a regular component of it. A recent survey of the public has shown that bactericidal agents, such as bleach and bathroom cleaners, are the most likely, with mud, oil and food as secondary concerns. The impact of the specific substances has been assessed by measuring changes in the respiration rate of an active culture of biomass upon OSS addition (Figure 1). In most cases, an initial increase in respiration rate was observed followed by a rapid decrease. A critical point was assessed for each substance that referred to the concentration at which the normalised respiration fell below 1. In the case of bleach, the critical concentration was 1.4 $ml \cdot L^{-1}$ with massive inhibition occurring at concentrations as low as 3 $ml \cdot L^{-1}$. These concentrations are likely to occur in systems that serve small populations such as small offices and residential buildings. In these cases biological treatment may not be the most suitable technological solution to the treatment and subsequent reuse of grey water.

This aim of the work presented here was to investigate the potential of using physical-chemical processes for the treatment of grey water. Their relative performance in solids and organic carbon removal, and disinfection are reported and compared to biological processes.

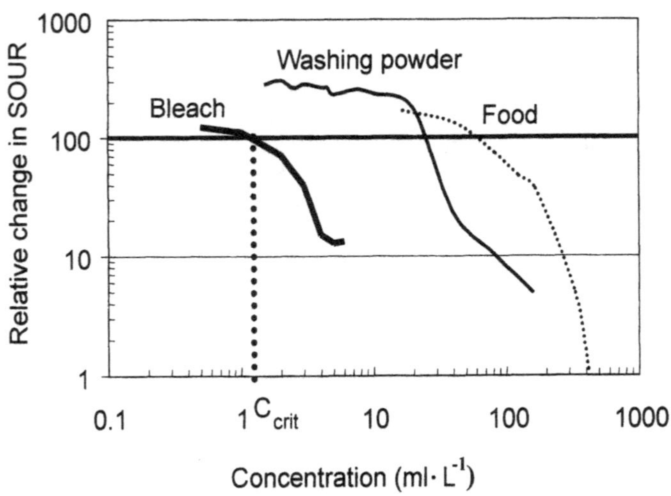

Fig. 1. Examples of the effects of occasional substance spiking on biological performance (SOUR: Spec.Oxygen Uptake Rate)

Materials and Methods

The experimental rig consisted of a recirculation loop containing a centrifugal pump, an UV tube rated at 35 W (Hanovia Limited, Slough) and a sampling station which permitted filling and sampling of the recirculation line. The flow rate was set to maintain turbulent flow conditions throughout the experiment such that the TiO_2 remained in solution. The synthetic grey water was made according to a recipe previously developed [3] and consisted of soap (62 g), shampoo (160 ml), oil (8 ml) and hair (2 g) in 1 m^3 of tap water. The real grey water was collected from a grey water testing facility housed at Cranfield University and consisted of the grey water produced from 16 single-person flats.

Two coagulants were studied: ferric sulphate (W grade, Huntsman Tioxide, Grimsby, UK) and aluminium sulphate (GPR grade, BDH, Poole, UK). Jar tests were carried out using a 6 x 1 litre jar tester (Camlab), and coagulants were simultaneously dosed using a 15 ml dosing bar (Camlab). Coagulants at 5 ml were dosed into 800 ml of grey water. pH was measured but remained as influent pH (between 7-8). Coagulant addition was followed by 1 min of rapid mixing at 230 rpm to ensure uniform coagulant dispersal. Rapid mixing was followed by 15 min of slow mixing at 30 rpm to encourage flocculation, followed by 30 min settling time. 200 ml of jar test effluent was decanted from each beaker and analysed for TOC and SS.

Results and discussion

Physical-chemical treatment of water and wastewater by coagulation is well known and reasonably well understood. Aluminum and iron salts have been widely used as coagulants to improve the removal of colloidal solids and organic carbon. A series of jar tests were undertaken to investigate the removal of turbidity and organic carbon from grey water coagulated with aluminum sulphate (alum) and ferric chloride (ferric) coagulants, and the results are shown in Figure 2. Both coagulants can remove over 90% of the grey water turbidity and 80% of its TOC at their respective optimum dose. Alum coagulation achieved the maximum percent removal of turbidity (97 % at 60 mg/l) compared to 88 % for ferric. Increasing coagulant dose to 100 mg/l did not affect the removal significantly. TOC removal was increased from 0 % for no coagulant dose to between 70 and 85 % over the coagulant dose range investigated.

PH is the most important parameter in optimising coagulant performance. The effect of pH on coagulant performance is evident for grey water, as shown in Figure 3. As the pH changes from pH 2.5 to pH 12 the coagulant either interferes with or enhances turbidity removal. Turbidity removal was found to be maximum in the sweep flocculation zone (pH 7 – 10).

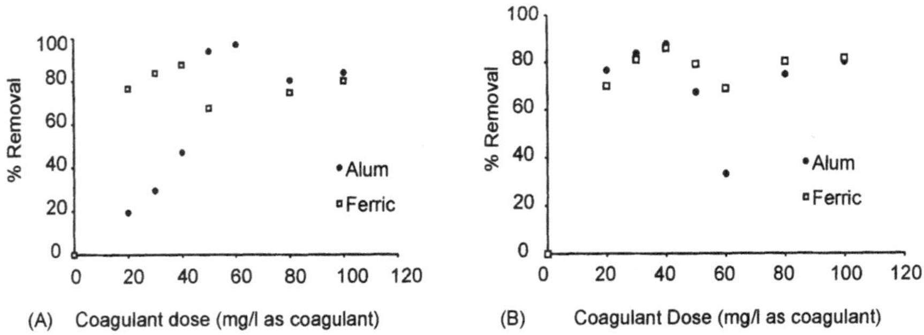

Fig. 2. Variation in (A) turbidity and (B) TOC of synthetic grey water as a function of alum and ferric chloride dose (initial pH = 7.4, TOC_0 = 85 mg/l, initial turbidity = 64 NTU)

Photocatalytic Oxidation

Photocatalytic oxidation (PCO) is an emerging technology that could be well suited for treating the recalcitrant organic compounds found in grey water. The photocatalytic process oxidizes organic reactants at a catalyst surface in the presence of ultraviolet (UV) light. Certain solid-phase semiconductors, such as TiO_2 have been shown to be excitable by near-UV light from the solar or a lamp-generated spectrum. In the presence of water and oxygen, the redox reaction produces hydroxyl radicals. In turn, the hydroxyl radicals can oxidize most organic pollutants, as they do in UV/hydrogen peroxide and UV/ozone treatment systems. Given enough exposure time, organic wastes will be oxidized into CO_2, water, and in the case of halogenated compounds, weak mineral acids.

The first stage of the PCO process is the adsorption of organic carbon onto the TiO_2 catalyst. Figure 4 shows the removal of organic carbon by adsorption and also by oxidation to CO_2. For this weak grey water (TOC_0 = 25 mg/l) 40 % of the

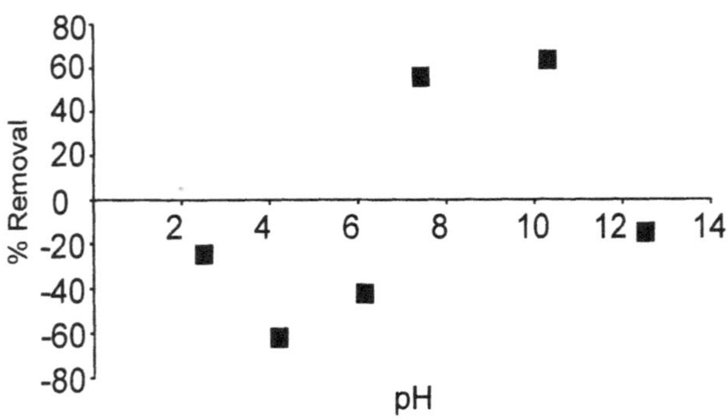

Fig. 3. Effect of pH on the removal of turbidity from grey water coagulated with 120 mg/l alum

organic carbon was removed by adsorption in the first hour. The solution was left for 22 hours with no further removal. The UV light was switched on after 23 hours and a further 20% of the organic carbon was removed. Experiments on a strong grey water (TOC_0 = 248 mg/l) showed that 70 % of TOC was removed within 5 minutes of adding the TiO_2 to the sample irrespective of UV light, indicating an adsorption reaction. A further 10-20 % could then be removed by PCO to form CO_2 using TiO_2 and UV.

The effect of catalyst concentration and initial TOC on adsorption onto suspended TiO_2 can be seen in Figure 5. The removal of TOC increases with increasing TiO_2 concentration with over 20 % of the TOC being adsorbed onto 2 g/l of catalyst compared to 80 % onto 10 g/l. The removal efficiency of the catalyst as an adsorbent is dependent on the TOC loading with much greater removal being achieved with higher loading rates. Adsorption removes only 10 % of the TOC of a weak grey water (TOC_0 = 37 mg/l) compared to 80 % of a strong grey water (TOC_0 = 248 mg/l). A comparison of process performance at different TOC_0 concentrations is shown in Table 2.

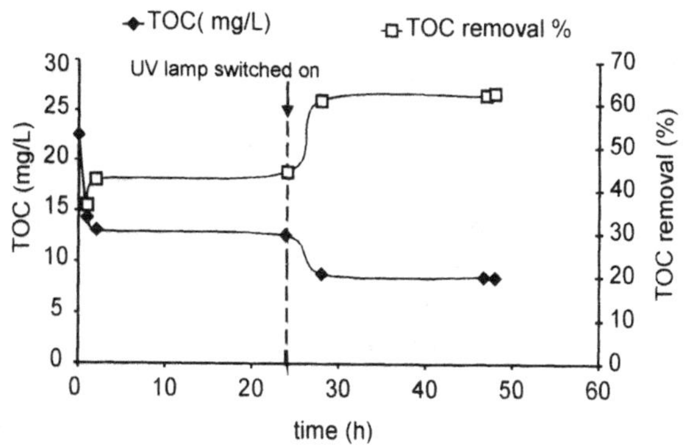

Fig. 4. Removal of TOC from grey water by adsorption onto 2 g/l TiO_2 catalyst followed by photocatalytical oxidation

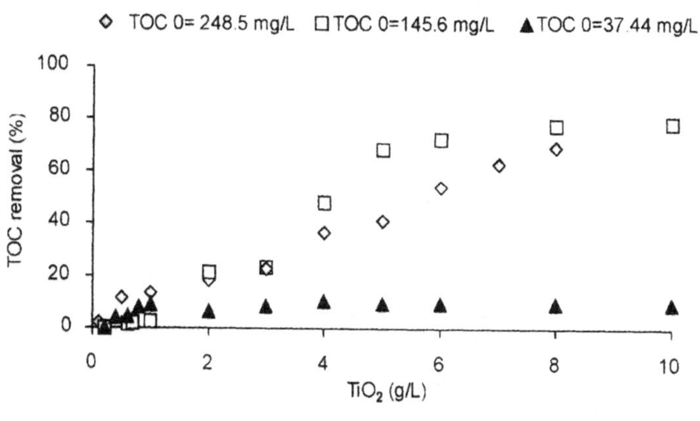

Fig. 5. Adsorption of TOC onto suspended TiO_2 catalyst

Many previous reports of PCO have shown that the rate of oxidation of organic carbon to CO_2 follows Langmuir-Hinshelwood kinetics [1,2]. The Langmuir-Hinshelwood rate form is

$$v = \frac{dC}{dt} = \frac{kKC}{1+KC} \tag{1}$$

Where:
v = the oxidation rate (mg/l·min)
C = concentration of reactant (mg/l)
t = illumination time (min)
k = reaction rate constant (mg/l·min)
K = adsorption coefficient of the reactant (l/mg)

Which can be rearranged (see [1]) to give

$$\ln\frac{C_0}{C} = kKt = K't \tag{2}$$

The apparent first-order rate constant, K', can be found by plotting $\ln(C_0/C)$ as a function of illumination time. The calculated values of the apparent first-order rate constant, K', regression constant and % TOC removal are shown for a range of real grey waters in Table 2.

The values of K' increase with increasing initial concentration. This trend is consistent with previous reported work on complex organic materials [1]. The change in rate and % TOC reduction indicates that the mineralisation of organic carbon to CO_2 is different over the concentration range studied.

Table 2. Simplified Langmuir-Hinshelwood kinetics at different initial TOC concentrations

Initial TOC ($mg.L^{-1}$)	K' (min^{-1})	R^2	% TOC Removal
25	0.0034	0.8	61
69	0.0033	0.9	65
139	0.0046	0.95	91
221	0.0074	0.96	89
660	0.0084	0.89	91

Chemical vs. Biological Processes

Biological processes are currently accepted as the best available treatment for grey water [4]. Table 3 compares the relative performance of two biological treatment processes with that of coagulation and photocatalytical oxidation. Membrane bioreactors (MBRs) are establishing themselves as the best treatment for solids and organic carbon removal as well as disinfection [4], and none of the chemical processes investigated could achieve the overall performance shown by the MBR

Table 3. Comparison of chemical and biological oxidation of grey water

Process	% Solids Removal	% Organic Carbon Removal	Coliform Removal	Retention time (hr)
BAF	87	89	2 log	2
MBR	98	98	6 log	12
Coag/Floc	97	80	-	1
TiO_2 (adsorp.)	-	82	2 log	< 0.1
TiO_2/UV	-	90	6 log	< 0.5

processes. Coagulation with alum and ferric could achieve similar turbidity and organic carbon removal as biological activated filters (BAFs) in a shorter retention time. Adsorption of organic carbon onto TiO_2 catalysts could achieve over 80 % organic carbon removal with a further 10 % being removed by oxidation, and the TiO_2/UV process matched the disinfection performance of the MBR.

Whilst the two processes investigated do not match the performance of the MBRs they are less likely to be affected by changes in loading. Figure 6 shows how the MBR performance decreased with increased organic strength of the grey water being treated. It has been shown however that the TiO_2/UV process performance actually increased with an increase in loading meaning it is more able to deal with shock loading or organics. The retention time required for the removal of turbidity and TOC with chemical processes is significantly shorter than that required for both the BAF and MBR processes making it suitable for schemes where space may be a problem.

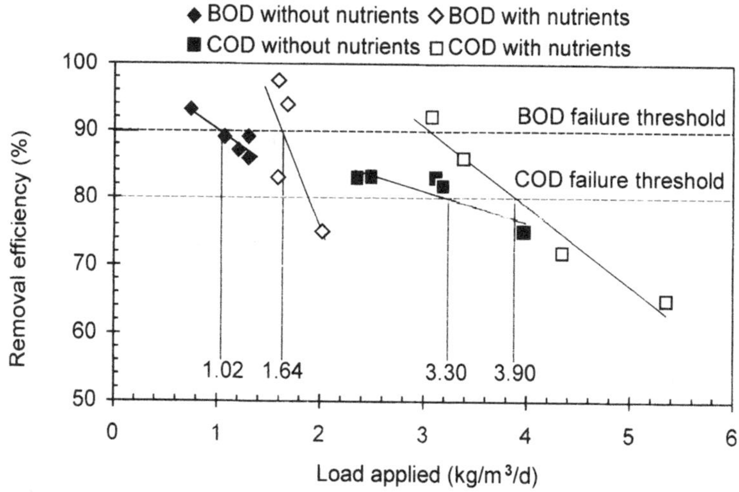

Fig. 6. Effect of organic loading rate on MBR performance

Conclusions

- Coagulation of grey water with aluminum or iron salts removed more than 90 % of grey water turbidity and 85 % of TOC. The removal of turbidity is dependent on pH, with the optimum performance occurring in the pH range where sweep flocculation is the dominant mechanism.
- Oxidation of grey water with TiO_2/UV can remove up to 91 % of the organic carbon. The rate of oxidation increases with increased TOC as does the overall removal.
- Membrane bioreactor processes produce better quality effluent than both chemical processes investigated here, but the chemical processes can deal better with significant increases in organic loading.
- The retention time for chemical processes is much shorter than for both biological processes investigated, making the chemical processes more suitable for small systems with variable organic loading.

References

1. Chen H., Jeng C. H.: Kinetics of photocatalytic oxidation of trace organic compounds over titanium dioxide, Environment International, 24(8) (1998) 871-879
2. Serpone N., Salinaro A., Emeline A., Ryabchuk V.: Turnovers and photocatalysis: A mathematical description. Journal of Photochemistry and Photobiology A: Chemistry, 130 (2000) 83-94
3. Jefferson B., Laine A.T., Brindle K., Judd S., Stephenson T.: Towards a sewerless city. Proc. of the ICE conference: Water environment 98, Maintaining the flow. London, 26 March, (1998) 100-107
4. Jefferson B., Laine A., Parsons S. A., Stephenson T., Judd S.: Technologies for domestic wastewater recycling. Urban Wat. (2000). (in press)

Author Index

FSC
www.fsc.org
MIX
Papier | Fördert
gute Waldnutzung
FSC® C083411

Kolibri 360 GmbH
Ferdinand-Jühlke-Straße 7
99095 Erfurt, Deutschland
produktsicherheit@kolibri360.de